X-ray equipment maintenance and repairs workbook

for

radiographers &
radiological technologists

by
Ian R McClelland
Chief technical support engineer (retired)

Diagnostic Imaging and Laboratory Technology
Essential Health Technologies
Health Technology and Pharmaceuticals
WORLD HEALTH ORGANIZATION
Geneva

WHO Library Cataloguing-in-Publication Data

McClelland, Ian R.
 X-ray equipment maintenance and repairs workbook for radiographers and
 radiological technologists / Ian R. McClelland.

 1.X-rays 2.Radiography 3.Technology, Radiologic 4.Maintenance—methods
 5.Problems and exercises I.Title.

 ISBN 13: 978-9-2415-9163-8   (NLM classification: WN 150)

© **World Health Organization 2004**

All rights reserved. Publications of the World Health Organization can be obtained from
Marketing and Dissemination, World Health Organization, 20 Avenue Appia, 1211 Geneva 27,
Switzerland (tel: +41 22 791 2476; fax: +41 22 791 4857; email: bookorders@who.int).
Requests for permission to reproduce or translate WHO publications—whether for sale or for
noncommercial distribution—should be addressed to Publications, at the above address (fax:
+41 22 791 4806; email: permissions@who.int).

The designations employed and the presentation of the material in this publication do not imply
the expression of any opinion whatsoever on the part of the World Health Organization
concerning the legal status of any country, territory, city or area or of its authorities, or
concerning the delimitation of its frontiers or boundaries. Dotted lines on maps represent
approximate border lines for which there may not yet be full agreement.

The mention of specific companies or of certain manufacturers' products does not imply that
they are endorsed or recommended by the World Health Organization in preference to others of
a similar nature that are not mentioned. Errors and omissions excepted, the names of
proprietary products are distinguished by initial capital letters.

The World Health Organization does not warrant that the information contained in this
publication is complete and correct and shall not be liable for any damages incurred as a result
of its use.

The named authors alone are responsible for the views expressed in this publication.

Contents

Introductory remarks		v
Acknowledgements		vi

Part I. Introduction — 1

Introduction		3
Questionnaire—Student's own department		6
Pre test		8

Part II. Routine maintenance modules — 11

Module 1.0	Routine maintenance overview		13
	Task 1.	Maintenance survey for an X-ray room	17
Module 1.1	X-ray generator maintenance, fixed installation		19
	Task 2.	X-ray control familiarization. Part 1	24
	Task 3.	X-ray control familiarization. Part 2	26
	Task 4.	Test for X-ray tube overload calibration. Part 1	28
	Task 5.	Test for X-ray tube overload calibration. Part 2	31
Module 1.2	X-ray generator maintenance, mobile unit		32
Module 1.3	X-ray generator maintenance, C D mobile		37
Module 1.4	X-ray generator maintenance, portable unit		41
Module 2.0	X-ray tube stand maintenance		44
	Task 6.	X-ray tube-stand maintenance	47
Module 2.1	X-ray tube maintenance		48
Module 2.2	Collimator maintenance		50
	Task 7.	X-ray tube and collimator maintenance	52
Module 3.0	Bucky table & vertical Bucky maintenance		53
Module 3.1	Tomography attachment maintenance		55
Module 4.0	Fluoroscopy table maintenance		57
Module 4.1	Fluoroscopy TV maintenance		60

Part III. Fault diagnosis and repair modules — 63

Module 5.0	Common procedures, for fault diagnosis and repairs		65
	Task 8.	Fuse identification	70
Module 6.0	X-ray generator repairs, fixed installation		71
	Task 9.	No Preparation, Part 1	86
	Task 10.	No Preparation. Part 2	87
	Task 11.	No Exposure	88
	Task 12.	X-ray output linearity	89
Module 6.1	Mobile or portable-generator repairs		90
Module 6.2	C D mobile repairs		94
Module 7.0	X-ray tube stand repairs		99
	Task 13.	Bucky tabletop and tube-stand centre	103

Module 7.1	X-ray tube repairs	104
Module 7.2	Collimator repairs	110
	Task 14. Help! No spare globe for the collimator	116
Module 7.3	High-tension cable repairs	117
Module 8.0	Bucky and Bucky table repairs	121
	Task 15. A film exhibits grid lines	126
Module 8.1	Tomography attachment repairs	127
Module 9.0	Fluoroscopy table repairs	130
Module 9.1	Fluoroscopy TV repairs	135
Module 10.0	Automatic exposure control, operation and problems	140

Part IV. Automatic film processor — 145

Module 11.0	Automatic film processor, routine maintenance	147
Module 11.1	Automatic film processor repairs	151
	Task 16. Films appear too dark	156
	Task 17. Films exhibit symptoms of low fixer	157
Module 11.2	The Film ID printer	158

Part V. Appendices — 161

Appendix A. Sensitometry	163
Appendix B. Recommended tools and test equipment	169
Appendix C. Graphs, check sheets and record sheets	177
Appendix D. Routine maintenance check sheets	186
Appendix E. X-ray equipment operation	205
Appendix F. Teaching techniques	248
Appendix G. Health and safety	253

Part VI. Post test and glossary — 257

Post test	259
Glossary	262

Introductory remarks

This document, which is developed by the International Society of Radiographers and Radiological Technologists (ISRRT) under the umbrella of the WHO Global Steering Group for Education and Training in Diagnostic Imaging, is the second in a series targeting technical aspects, including quality control of diagnostic imaging services. The document is primarily aiming at assisting radiographers and radiological technologists working in small and mid-size hospitals where resources often are limited, to optimize and improve diagnostic imaging, and to ensure the best possible use of resources according to local needs.

The document can be obtained by contacting the following address:

Team of Diagnostic Imaging and Laboratory Technology (DIL)
World Health Organization
20, Avenue Appia
CH-1211 Geneva 27
Switzerland

Fax: +41 22 7914836
e-mail: ingolfsdottirg@who.int

Harald Ostensen, MD
Geneva, July 2004

Acknowledgements

For their considerable input and assistance in producing this workbook, special thanks are due to:

Peter J Lloyd, Peter Hayward, Brett Richards, Sue Salthouse, Peter K Mutua, M Jean Harvey, Leonie Munro, Martin K West, Jiro Takashima, and to Graham English.

PART I
Introduction

Introduction

It is preferred to call this a **workbook** rather than a manual or textbook, because the intent is to, not only give technical information, but to set practical exercises that students can work through, responding to specific questions. Above all, the students should feel that they have actually carried out the tasks themselves and will be more confident to teach others and ensure that these exercises continue to be carried out in their respective areas.

The topic of this workbook is routine maintenance and repairs. The material is designed to assist in the maintenance of equipment, and provide guidelines for locating equipment problems. In many cases this will allow local correction of fault situations. Where external assistance is required, good communication of the diagnosed problem will assist in reducing delay, or multiple service calls.

> **Routine maintenance**
> The overall maintenance programme; put in place to ensure that a comprehensive range of maintenance procedures are systematically carried out.
>
> **Fault diagnosis and repairs**
> The means by which the cause of incorrect equipment operation may be located. This includes adjustment where required, and simple repairs.

A Routine Maintenance Programme should be comprehensive, looking at all aspects of the work involved in ensuring equipment is properly maintained, and capable of producing accurate results. Such a programme can be cost effective, and contribute to minimum failure of equipment. By encouraging local staff to be actively involved in maintenance or minor repairs, delays and expensive service calls may be considerably reduced.

The ultimate responsibility for setting up, running, evaluating and taking remedial action lies with the head of department, although appropriate delegation may be necessary. It is important that someone accepts that responsibility and ensures that the programme happens effectively.

This workbook will be used by radiographers and radiological technologists as well as other medical and technical staff members involved in diagnostic imaging, to:

- achieve a good working knowledge of equipment maintenance routines;
- adopt a logical and practical approach to diagnosing equipment problems;
- on returning to their respective areas after completed training, teach other members of their staff to carry out the routines or techniques that they have learned;
- assist in establishing, or implementing, a suitable routine maintenance programme;
- be encouraged to directly carry out adjustments or minor repairs, or provide suitable assistance to other staff as needed;
- provide accurate reporting of problems to seniors or service engineers;
- assist in establishing criteria for equipment replacement, where it is not cost effective to continue maintenance.

Expected benefits

It is expected that after going through the training and experiences discussed in this workbook, the knowledge and skills will be put into practise. If so:

- Heads of departments will find that the standard of radiography will be maintained at the highest level.
- There will be fewer equipment failures. This reduces costs.
- Where a failure does occur, local hospital staff may be able to repair without an expensive service call.
- When an external service call is required, the service technician can arrive fully informed to deal with the situation, together with appropriate equipment or parts. This will reduce the possibility, and expense, of repeated visits.

- Where work is carried out by an external service organization, the maintenance inspection will ensure this has been carried out fully and effectively.
- Work environments will be improved. Tasks will become easier.
- Repeat films will be kept to a minimum. Staff job satisfaction will increase.
- Patients will receive less radiation and less inconvenience.
- A record and audit trail will exist as proof of high standards.

Achieve some of these, and this workbook has been worthwhile!

What this workbook aims to achieve

- Provide the knowledge and skills required for maintenance of imaging equipment.
- Increase awareness, interest and understanding of maintenance issues.
- Enable radiographers to establish and continue to carry out an effective preventive maintenance programme.
- Provide the knowledge tools to assist in diagnosing equipment problems.
- Provide the knowledge and encouragement to carry out adjustments and minor repairs.
- Raise equipment performance standards.
- Reduce maintenance and service costs.
- Reduce the possibility of equipment malfunction causing injury.
- Improve job satisfaction through correctly functioning equipment.

Summary of the content of this workbook

- Background information.
- A questionnaire seeking information about each student, and their own department.
- A pre test of student's knowledge.
- 13 modules related to routine maintenance of X-ray equipment.
- 14 modules concerned with fault diagnosis and repairs.
- A separate module, concentrating on the film processor.
- A revision of X-ray equipment design and operation.
- 17 tasks the student must perform.
- A list of suitable tools and test equipment, and how to make simple test tools.
- Useful charts and forms.
- Copies of routine maintenance check sheets.
- Advice on teaching methods.
- Health and safety issues.
- A post test of student's knowledge.
- Glossary of terms.
- Reference list.

How to use this workbook

The entire workbook can be used for self-study or self-assessment, ideally as working material during a work shop or a seminar with individual tutors for the students. In either case, however, the book should be approached as indicated below.

The section headed STUDENT'S OWN DEPARTMENT, should be completed by the student *before commencement of the study or course*. This takes the form of a questionnaire which, when completed should give the tutor a background knowledge of the student and their work environment. This background information will allow the tutor to apply the correct emphasis when providing and supervising the training.

The student must complete a PRE TEST prior to starting the course. This is an assessment of the student's relevant knowledge *before the* course. This will be compared to the results of a similar POST TEST completed by the student after completion of the course. These tests are for student information and course evaluation only and are not used in student assessment.

The section on TEACHING TECHNIQUES first gives a broad overview of teaching methods. This is followed by the recommended approach to teaching with this workbook. Both tutor and student should read this section. This is a reprint from the WHO 'Quality assurance workbook' and is included here for convenience.

The section on HEALTH AND SAFETY draws attention to all the health and safety issues appropriate to an X-ray department and how to make the work environment a safe and healthy one. This is an extract from the WHO 'Quality assurance workbook', edited for use with this workbook.

The workbook is divided into modules

- The modules are in three groups.
  - **a.** Routine maintenance of X-ray equipment.
  - **b.** Fault diagnosis and repairs of X-ray equipment.
  - **c.** The automatic film processor, routine maintenance, fault diagnosis and repairs.
- The student should work through one module at a time, studying the technical information and testing methods.

PART I. INTRODUCTION

- **Note.** Although the modules are designed as individual projects, due to the complexity of this workbook, it will be necessary at times to refer to other modules. Where this is required, a note is inserted pointing to the first page of the reference module.
- At the end of each module, tasks have been set. The student must carry out each task and answer the questions asked, and the teacher/tutor where available, will comment or correct these.

The APPENDICES contain information on making **simple test tools**, **report forms**, **record sheets**, **and test result sheets** for use in the student's own department. The APPENDICES also contain information from the WHO 'Quality assurance workbook'. This includes **sensitometry**, **teaching techniques**, and **health and safety**. The GLOSSARY contains a list of terms found in the text, with meanings. The REFERENCES provide a source of further reading.

When used in the scope of a work shop or seminar, The POST TEST must be answered, on completion of the course. The workbook is then handed to the tutor for final assessment.

The student is encouraged to:

- complete all pre reading, discuss the material with colleagues and fill in the questionnaire, 'STUDENT'S OWN DEPARTMENT', before starting the course;
- carry out the PRETEST immediately before starting the course;
- carry out the POST TEST immediately upon completion of the course/self-study.

In daily routine work the workbook and newly gained knowledge and expertise should be used to establish a routine maintenance programme, and train colleagues under the direction of their department manager.

Questionnaire
Student's own department

In order for this course to meet your needs, your tutor must know something about yourself, and the department in which you work. Please answer the following questions in the spaces provided, before you commence the course.

1. Hospital name and address. _____
2. Your name and address. _____
3. Your education qualifications. Please include details of any additional training. _____
   _____
   _____
   _____

4. How many X-ray examination rooms are there? _____
5. How many mobile or portable X-ray generators are there? _____
6. Tell us what X-ray equipment you have. Eg, general-purpose table with Bucky etc. Please include the type, make and model number.

   Room 1 _____
   _____
   _____
   _____

   Room 2 _____
   _____
   _____
   _____

   Room 3 _____
   _____
   _____

   Mobile or portable units _____
   _____
   _____

4. How many darkrooms are there? _____

PART I. INTRODUCTION

5. State the type of film processor in each darkroom. e.g. type, make, model, processing cycle.

   Darkroom 1 _____

   _____

   _____

   Darkroom 2 _____

   _____

   _____

6. How many staff act as radiographers? _____ Qualified radiographers? _____ Others? _____
7. How many darkroom technicians are there? _____
8. Does your hospital have access to an electronics service technician? _____
9. Does your hospital have an electrician? _____ With electronics knowledge? _____
10. Do you already run any form of preventive maintenance programme. Yes/No _____
11. If 'Yes', state here what you do, including details of any external assistance.

    _____

    _____

    _____

13. Is all the X-ray / processor equipment operating at a satisfactory level? _____ If not, describe areas requiring attention. Include equipment waiting parts or further service.

    _____

    _____

    _____

    _____

14. Including all equipment, what is the total number of days 'out of action' for last year? _____
15. What was the average time 'out of action'? _____ And the maximum time? _____
16. List any test equipment or quality control test tools you have.

    _____

    _____

    _____

    _____

17. If you have any issues relating to maintenance, or fault diagnosis / repairs, please state them here.

    _____

    _____

    _____

    _____

Pre test

The student must complete this test before starting the course. The intention is to test your knowledge on the topics covered by this workbook before the course, or before starting studying the work book.

Name and address

_____

_____

_____

Hospital name and address

_____

_____

_____

Instructions

This is a multiple-choice test. In each question you are given three possible answers.

Read each question carefully.

Indicate the answer that you feel is the most accurate by placing an 'X' in front of the letter preceding it.

Example:
A personal radiation monitor (TLD) should be worn
   a) Outside a lead rubber apron.
X b) Under a lead rubber apron.
   c) There is no need to use one when wearing a lead rubber apron.
Answer: b)

All questions must be answered

1. What is meant by the term 'preventive maintenance'?
   a) Maintenance must be carried out by tradesmen only.
   b) Maintenance is not required.
   c) Maintenance to reduce equipment failure.

2. What is meant by the term 'quality assurance'?
   a) The equipment is covered by a maintenance policy.
   b) A repair is guaranteed for three months.
   c) A system that attempts to maintain a high standard of work in all areas.

3. What is meant by the term 'trouble shooting'?
   a) A method to diagnose a problem.
   b) How to deal with under performing staff.
   c) The X-ray generator fails to expose.

4. To check if the light field and the X-ray field of a collimator are correctly aligned:
   a) Look into the collimator mirror.
   b) Place metal markers on the face of a loaded cassette to indicate the light field and make an exposure.
   c) Adjust the collimator knob to the film size as indicated on the collimator.

5. An aluminium disc is often inserted between the collimator and the X-ray tube. The purpose of this disc is to:
   a) Adjust the spacing between the collimator and the X-ray tube.
   b) Prevent light from the tube filament shining through the collimator.
   c) Filter low energy X-ray photons.

6. An 'mAs' control is used in:
   a) An automatic exposure control.
   b) Single knob adjustment of time and mA.
   c) Automatic regulation of mA as kV is adjusted.

7. The term 'space charge' relates to:
   a) High-energy cosmic radiation, causing film fog.
   b) A 'cloud' of electrons around a heated filament.
   c) The rental fee for a private X-ray practice.

8. A 'relay' is:
   a) A competitive sports event.
   b) A replacement of floor covering.
   c) An electromagnetic switch.

9. An 'interlock' is:
   a) A safety device to prevent incorrect operation.
   b) The door locks for entry to the X-ray room.
   c) A device to prevent tampering.

PART I. INTRODUCTION

10. A 'spinning top' is used to:
    a) Check kV accuracy.
    b) Check exposure time accuracy.
    c) Check mA accuracy.

11. A 'stepwedge' is used to:
    a) Calibrate the height of the tomograph fulcrum.
    b) Make comparative measurements of radiation output.
    c) Assist in patient positioning.

12. Before attempting to replace a fuse, you should:
    a) Check the fuse rating.
    b) Inform the chief radiographer.
    c) Ensure the main power isolation switch is turned off.

13. When making an exposure via a Bucky:
    a) The grid should move immediately X-rays are produced.
    b) The grid should commence movement when the exposure button is pressed, and exposure occurs after the grid is in position.
    c) The Bucky should commence operation as soon as the preparation button is pressed.

14. The first aid treatment for a processing chemical splash in the eye is:
    a) Blink continuously for at least 30 seconds.
    b) Wash the eye thoroughly with water.
    c) Wipe the eye with a tissue.

15. Developer temperature should be checked:
    a) Only if the film densities appear different.
    b) Daily.
    c) Once a week.

16. Replenishment rates in automatic processors are checked by:
    a) Referring to the operator's manual.
    b) Measure the relative decrease of level in the replenishment supply tank.
    c) Diverting the replenishment pump output into a graduated flask.

17. Automatic processing temperature should be:
    a) 20°C
    b) 25°C
    c) 35°C

18. A sensitometer:
    a) Is used to measure sensitivity of film to light.
    b) Is used for printing a test strip onto film for film processor monitoring.
    c) Checks the developer concentration in the processor.

19. If the focal length of the grid fitted to a wall Bucky is too short for the required distance:
    a) The film will be dark in the centre, and light at the sides.
    b) The film will be light in the centre, and dark at the sides.
    c) Grid lines will appear.

20. Before using a multimeter to test continuity, you should:
    a) Touch the test leads together to ensure the meter reads 'Zero ohms'.
    b) Fit a new battery to the meter.
    c) Ensure the power isolation switch for the X-ray room is turned off.

21. During preparation for an exposure, which item should *not* occur?
    a) X-ray tube anode rotation.
    b) Adjustment of kV output.
    c) X-ray tube filament heating.

22. The maximum possible anode rotation speed for a low-speed tube operating at 50 Hz is:
    a) 2500 RPM
    b) 3000 RPM
    c) 5000 RPM

23. A high-speed tube has a 'brake' cycle at the end of an exposure. This is to:
    a) Reduce unnecessary noise in the X-ray room.
    b) Increase bearing life.
    c) Prevent damage caused by 'resonant periods' as the anode slows down.

24. The woven metal sheath of the high-tension cable is to:
    a) Shield against the possibility of transmitting electronic interference into other equipment during an exposure, especially with a high-frequency generator.
    b) Increase the resistance to wear, and act as a deterrent to rat damage.
    c) In case there is a fault in the cable insulation, the metal sheath provides a safe electrical conduction to ground.

25. When exposing a large format film, a reduction in film density to one side of the film is noticed. This is probably due to:
    a) The dwell time of the processor is incorrect.
    b) Heel-effect of the X-ray tube.
    c) Poor film screen contact.

26. All exposures suddenly show an increase in film density. The most possible cause is:
    a) An increase in developer temperature.
    b) The generator kV calibration is incorrect.
    c) The processor fixer has become diluted.

27. You have a choice of three X-ray tubes as a replacement. They are identical except for the anode angle. Which tube will have the highest output?
    a) 7 degrees
    b) 12 degrees
    c) 16 degrees

28. Again, considering the same X-ray tube, which will give the best film coverage?
    a) 7 degrees
    b) 12 degrees
    c) 16 degrees

29. A capacitor discharge mobile has:
    a) A non-linear X-ray output.
    b) A similar output to a single-phase generator.
    c) Long exposure times.

30. A spinning top test is made of a single-phase self-rectified (Half wave) portable generator operating on 50 Hz. Twenty dots appear on the film, indicating an exposure time of:
    a) 0.02 seconds
    b) 0.2 seconds
    c) 0.4 seconds

PART II
Routine maintenance modules

MODULE 1.0
Routine maintenance overview

Aim

The aim is to provide an overview of routine maintenance requirements. This includes requirements to commence, or carry out, a routine maintenance programme.

Objectives

On completion of the routine maintenance modules, the student will have developed knowledge and skills to apply a practical maintenance programme for X-ray equipment. This includes keeping proper records of tests, and ensuring all documents and required manuals are available.

Task-1 'Maintenance survey for an X-ray room' should be performed on completion of this module.

Contents

a. What is 'routine maintenance'?
b. Who should carry out routine maintenance?
c. Objections to routine maintenance
d. How often should maintenance be carried out?
e. Typical objectives of routine maintenance
f. Familiarization of equipment
g. Routine maintenance programme
h. Keeping a logbook
i. Routine maintenance modules

a. What is 'routine maintenance'?

Routine maintenance is a procedure to ensure equipment is kept in good condition, and provide a long operating life. Routine maintenance may also discover potential problems, which could cause equipment failure. Potential problems can then be corrected, with a minimum of down time. 'Quality control' procedures, to ensure correct operation and calibration, are also a part of routine maintenance.

Carrying out routine maintenance produces good knowledge of the equipment, and in case of a problem, this knowledge will help to locate the cause. Where there is a more serious problem, accurate reporting for assistance will allow faster and more economic response. **For example, the service engineer can then arrive with suitable parts and test equipment.**

A major part of routine maintenance is just inspection of equipment. This should be done as if seeing the equipment for the first time. At the same time, make note of less understood operation areas, and refer to the operation manual for explanation.

This is also an opportunity to correct any 'legacy' problems. As an example of a legacy problem, a generator might have a notice, 'Do not use fine focus'. This notice may have been there for some time. And, due to staff movements, the reason is not known. As part of maintenance, this long accepted problem should be investigated. It could be an X-ray tube that has a failed fine focus. (Then the tube should be replaced) But more often, this might have been due to some other problem, or even operator error. So, although a note was attached, no action was taken to correct the problem, or to investigate further. Sometimes a part is required, but the service provider has forgotten to come back with this part. In which case routine maintenance inspection ensures that:

- The nature of the problem is properly investigated and documented.
- If needed, a 'follow up' reminder is sent to the service provider.

b. Who should carry out routine maintenance?

This depends on the size of the department, and available staff. In a district hospital, which has just one radiographer, then perhaps that radiographer is 'it'. However, external assistance should be made available if required, for example, assistance provided by an electrician.

When the department has a number of staff, one member should be selected as maintenance co-ordinator. Other staff members may be allocated specific areas or items of equipment to be checked. Where possible, these duties should be rotated. This allows all staff to become familiar with the equipment.

In some hospitals, an electronics technician may be available. But, as the technician does not use the equipment, problems may go undiscovered. For this reason a staff member needs to assist the electronics technician, during maintenance or repairs.

c. Objections to routine maintenance

Existing staff may regard routine maintenance as an unwanted extra duty. This list provides answers for possible objections.

- This is boring.
  Yes, it can be. But even more boring, or frustrating, is using equipment that does not function correctly.
- This is not my responsibility.
  Even when a specific member of staff **does** carry out a comprehensive maintenance programme, your own input will be appreciated. This can be as simple as reporting a problem area, to taking direct action. For example, tighten that loose screw on a Bucky tray handle, before the handle falls off.
- The department is kept very busy. There is no spare time.
  And to make matters worse, you have to use equipment that does not operate correctly. A maintenance programme does not have to take the room out of action. Instead, just one section at a time can be checked. This may take only ten minutes for each section. After the reported problem areas have been fixed, the room will become more efficient.
- The hospital has a paid routine maintenance contract with an X-ray service company. This is carried out every six months.
  You are lucky. However you still need to carry out an inspection to ensure the service is completed as required. A service technician may find a problem, which requires immediate attention. This reduces the time available for the remainder of the maintenance. As a result, some areas are not checked. Your own inspection, together with suitable record keeping, will help ensure contract service is carried out correctly.

d. How often should maintenance be carried out?

- Equipment in heavy use, for example a mobile travelling to different parts of the hospital, should be checked every four months.
- Other equipment, such as a Bucky or Fluoroscopy room, every six months.
- However, in many respects, maintenance in the form of **observation** is a continuous process. If a minor problem occurs, always enter this in the logbook, so it **will** receive attention during the next opportunity.

e. Typical objectives of routine maintenance

- A complete operation and function inspection, list any incorrect operation or area requiring further attention.
- By means of prepared checklists, ensure all required areas are covered. The results are to be retained in a suitable folder. Any problems or areas requiring further attention are entered in the logbook.
- When a problem is located, if this is minor, correct the problem immediately. In case of a larger problem, still attempt to complete the rest of the routine maintenance, while waiting to have the specific problem corrected.
- In case of a specific problem outside local resource to immediately correct, and then request an electrician, or the service department, for assistance. If such a problem is found, be sure to file a report and enter specific details in the logbook.
- Inspection of all electrical plugs, cables, and other electrical connections.
- A full mechanical inspection, adjustment and lubrication as required.
- Tests for calibration of equipment. This may also be part of a quality control programme.
- Cleaning of equipment. Remove pieces of sticky tape, old sticking plaster marks etc.

f. Familiarization of equipment

With much equipment there is often a legacy of misunderstanding. As a result many functions may be ignored or incorrectly interpreted.

Common reasons are:

- The operation manual has been lost. Or is only referred to as a last resort.

- Replacement staff may be incorrectly instructed in the use of the equipment. This legacy tends to be passed on.
- A pre-existing fault is accepted as part of the normal operation of the equipment, and possible correction is ignored.
- The equipment was modified to interface with a non standard system. As a result, some controls are different to those described in the operation manual.
- Where there are any specific limitations or precautions, these should be listed in the logbook. This will assist any replacement staff. This information should also be shown to the service technician

During routine maintenance, you should attempt to be familiar with **all** operation modes of your equipment. This includes.

- Indicator lights. When do they light up? What do they mean?
- Audible signals. When do they occur? What is the reason?
- During preparation, or an exposure, do you hear a contactor or relay operating?
- What is the normal sound produced by the X-ray tube anode, during preparation?
- What sound does the Bucky make during an exposure?
- Meters or indicators can display different readings depending on the generator operation. For example, a meter might first display a percentage of X-ray tube load, then display the mAs obtained after an exposure.

g. Routine maintenance programme

To commence a routine maintenance programme, the following is suggested.

- Where a quality control programme is to be established, ensure that all required procedures for routine maintenance are properly documented.
- Make a list of replacement parts or materials that may be required.
- Suitable tools. A list of suggested tools is provided in appendix 'B' page 169.
- Test tools. Simple test tools that may be constructed are also described in appendix 'B' page 169.
- Be familiar with all operation modes of the equipment to be maintained. (Study the operation manual).
- Allocation of a specified time period to carry out maintenance. This will depend on the patient workload, and will need to be flexible.

- Record keeping.
  i. A separate logbook should be kept for each room of equipment. In the case of mobile or portable equipment, an individual logbook should be kept for each unit. This may be in the form of a loose-leaf binder, and allow insertion of checklists as maintenance is carried out.
  ii. A sample logbook page is provided in appendix 'C' page 177.
  iii. Proper identification of all equipment is required. This includes make, model and serial number, date of installation, and any special features, such as a high-speed starter option or AEC. This information should be entered into the logbook.
  iv. An individual checklist for each item of equipment helps ensure all-important areas are covered. External contractors may also use these lists to assist correct service. Completed checklists, together with service job sheets, should be kept in a master file, or logbook, for each X-ray room. This provides quick reference for future service, or else a previous problem.
  v. Sample checklists are provided in appendix 'D' page 186.
  vi. If an additional repair or service is required, mark this for attention in the logbook. Otherwise, if there is a long delay, the repair or service might be forgotten.
- Equipment manuals.
  Manuals need to be kept in a safe designated area. If manuals are lost or stolen, then every effort should be made to obtain a replacement. **Manual title and publication numbers should be recorded in the logbook.** Depending on the manufacturer, manuals may be presented in many different formats. Some may be in separate folders, while other manufacturers may combine all in the one folder. Typical manuals, which should be available, are listed below.
  i. Operation.
  ii. Specification.
  iii. Parts list.
  iv. Installation, including calibration procedures.
  v. Data sheets for the X-ray tube.
  vi. Service. (*)
  vii. Circuits or connection diagrams.
  viii. Technical explanation of operation. (*)
  ix. (*) Indicates these manuals might be restricted to a service department, or available only on request.
- In some countries, the equipment supplier is required to supply two complete sets of all manuals

originally supplied with the equipment. One set is kept in the X-ray department, and the other retained by the government purchase authority. This provides a backup copy in case of lost or damaged manuals.

- As some information in the manuals can be confidential, manuals need to be kept in a safe place, and restricted to authorized use only.

h. Keeping a logbook

- The logbook may be of any convenient construction.
- The logbook should not confine entries to a single line. Leave space to provide full details.
- If additional columns are required, the logbook may also cover the opposite page.
- The logbook need not be hard bound, but instead be a collection of report sheets kept in an 'insert' folder. This can include the checklists, produced after routine maintenance. Coloured dividers can separate each section.
- A single logbook can contain all the required information for an X-ray room. A separate logbook should be kept for a mobile generator, or a portable system.
- The front page of the logbook should include all details of the equipment, (Make, model, serial No. etc.)
- The logbook should also contain a list of all equipment manuals. This should include any reference numbers, to facilitate re-ordering of lost or damaged manuals.

i. Routine maintenance modules

The modules are for routine maintenance of equipment. Due to the diversity of equipment that may be in use, from very old to the latest technology, not all of the suggestions will apply to your system.

The maintenance modules are designed as individual units, however some cross-reference is required. The reference module name, and title-page number, is indicated in the text as required.

A sample routine maintenance checklist for each module is provided in appendix 'D' page 186.

The equipment covered in the maintenance modules includes the following.

- X-ray generator, fixed installation.
- X-ray generator, mobile unit.
- X-ray generator, capacitor discharge.
- X-ray generator, portable system.
- X-ray tube stand.
- X-ray tube.
- Collimator.
- The Bucky table and vertical Bucky.
- Tomography attachment.
- Fluoroscopy table.
- Fluoroscopy TV systems.
- Automatic film processor.

Data	Requirement	Response	Performedly	Reference No
24-12-02	Routine maintenance of tube stand	Carried out by staff	John Bell	Report no 75
2-2-03	Tube stand has failed bearing	Request attention by X-ray service ltd	X-ray service	Job No X2203
4/2/03	Collimator lamps required	Ordered 3 from Osram supplies	Jean Wells	Order No 45963
10 4-03	No exposure	See service request form. Unit out of action waiting handswitch	John bell	Order No 45964
14/4/03	New handswitch	Fitted new handswitch	John Bell	Report No 76

Fig 1–1. A typical logbook page

TASK 1
Maintenance survey for an X-ray room

You are required to make a basic maintenance survey of a general purpose Bucky room. This may be a room in a different hospital, OR, it may be a room with which you are fully familiar.

The hospital administration has offered to have all defects rectified, but first requires a quick report; a more detailed report can be supplied later.

When making this survey, look for minor defects as well as those affecting performance.

X-ray control:

___

Tube stand:

___

X-ray tube and collimator:

___

Bucky table and Bucky:

___

Wall Bucky:

___

Condition of the room and accessories: Suggestions to make this an efficient environment, and aid patient management.

_____
_____
_____

_____

Tutor's comments

_____	Satisfactory/Unsatisfactory
Signed _____	Date _____
Tutor	

MODULE 1.1
X-ray generator, fixed installation

Aim

The aim is to provide information and procedures for routine maintenance of an X-ray generator, installed as a fixed installation in an X-ray department. Maintenance for the X-ray tube is provided in module 2.1 page 48. Instructions for generator repairs are provided in module 6.0 page 71.

(**Note:** Reference module page numbers refer to the title page.)

Objectives

On completion of this module, the student will be familiar with maintenance procedures for the X-ray generator. These procedures can also be used as a version of quality control, together with the routine maintenance check-sheets provided in the appendix.

Tasks 2, 3, 4, and 5 should be attempted on completion of this module.

Contents

a. Safety precautions
b. Visual inspection of the control panel, power off
c. Operation inspection of the control panel, power on
d. X-ray tube overload protection
e. mA calibration
f. Radiation reproducibility
g. X-ray output linearity

Equipment required

- Basic tool kit.
- Stepwedge.*
- 24/30 cm Cassette.
- Two pieces of lead rubber.
- Aerosol spray lubricant.
- Cleaning solvent.
- Cloth.

* The stepwedge is described in appendix 'B' page 169.

a. Safety precautions

> **Before removing any covers, ensure the generator is switched off, and the room power-isolation switch is also turned off.**

b. Visual inspection of the control desk, power off

- Check all knobs and switches. Where knobs have a pointer attached, check that the pointer aligns correctly at all positions of the indicated scale. **Tip.** Check the pointer at full clockwise and counter clockwise positions of the knob. Look for possible loose knobs, or for push button switches that might tend to stick.
- If controls have had extra labels attached, are these labels still required? If so, are they in good condition?
- Older X-ray controls often have analogue meters instead of digital displays.
   i. With power switched off, the meter needle should be pointing at the 'zero' calibration mark.
   ii. Most meters have a small adjustment screw for zero calibration. If adjusting, first tap gently in case the meter tends to 'stick'.
   iii. **Caution. Contact the service department before adjusting.** In some cases, the meter may be deliberately adjusted 'off zero', as an incorrect method of calibration.

c. Operational inspection of the control desk, power on

- Check all indicator lamps. If necessary, operate different selection techniques to ensure all indicators operate correctly. In particular, pay attention to the following. (Depending on make or model, some of these indicators may not be available).
   i. Small focus / broad focus selection indication. On some controls, the mA selection switches

control selection of the focal spot. Other controls can have a separate focal spot selection switch.

ii. X-ray tube number, or position. This should be linked to technique selection. Some controls may also indicate the actual fine and broad focus size.

iii. X-ray tube overload protection. Select high pre-exposure factors, and check operation of the overload light. (On some controls with a microprocessor, the system might not allow selection of excessive output.)

iv. Automatic Exposure Control (AEC), or Photo-timer. When this option is fitted, check that all chamber and station selection indicators operate correctly.

v. **Note.** On older systems, it may be possible to select an AEC chamber or station combination that is not available. In that case ensure a notice is fitted, to warn against incorrect operation.

vi. Illumination of kV, mA, and time selection. Where a digital readout of selected values is provided, select a number of different values to ensure there are no display errors, or missing segments of the display.

- Older controls may have manual adjustment of power line voltage, with a meter to indicate correct compensation. Check the range of adjustment. It should be possible to reset the voltage by 10%, above, or below, the required voltage.

d. X-ray tube overload protection

Note. Reference to X-ray tube rating charts is required for this section.

Tip. You may select specific values from the chart, and record them separately on a check sheet. This will save time in the future.

- Maximum radiographic kV.
  i. Select a short exposure time, and a low mA station. Increase kV setting till the exposure-prevention, or inhibit, light operates.
  ii. The maximum available kV should not exceed the specified kV for the particular X-ray tube.
  iii. In some cases, the available kV limit may be 10% less than the possible maximum. For example, a 150 kVp tube may be limited to 140 kVp. This is a safety precaution, as 150 kV is the maximum limit only when the tube is in excellent condition.

- Minimum radiographic kV. This will often be set at 40 kV, depending on individual country regulations. Variations will exist where an interlock at the collimator is provided for different filters.
  i. Select a low mA station and a short exposure time. Adjust kV towards the minimum available value. An exposure inhibit should occur if kV is too low.
  ii. Repeat this test for systems that have a removable filter in the collimator. In this case, with the filter removed, an exposure inhibit should occur as kV is increased. (Depending on the system, this may be above 60 kV.)
  iii. **Note.** Although the collimator will have the required minimum filtration for full operation, an additional filter, typically 0.5 mm, may be inserted. This is an option, and does not require an interlock.
  iv. On older generators, especially those with 'stud', or switch selection, and pre-reading kV meters, it may be possible to set kV below the safety requirement. Where this can occur, provide a warning notice, and contact the service provider in case an upgrade is available.

- Minimum kV for filament over-heat protection. Refer to the rating charts, to see if a particular combination of high mA and low kV should be avoided. This is to avoid overheating the filament during preparation.
  i. Select the maximum available mA station and a short exposure time. Reduce kV towards the minimum kV available. Either the kV will not be permitted to extend below the minimum specified value, or else should cause an exposure inhibit to operate.
  ii. As an example, the minimum kV with 500 mA selected may be 55 kV, while if 400 mA is selected, the minimum kV might extend down to 45 kV.
  iii. Repeat for both focal spots.
  iv. **Note.** This protection may not be available on older X-ray controls. If a combination of high mA and low kV is possible, provide a warning notice. In some cases, an upgrade may be available from your service provider. In other cases, a re-allocation of available mA stations may be available.

- Anode maximum heat load. This is the maximum instantaneous heat input to the anode.
  Note. The X-ray tube rating charts assume a cold anode. For this reason, some X-ray controls de-rate the maximum output. This allows for anode heat produced by previous exposures.

For example, on over-table operation, output may be limited to 95% of maximum, while with a fluoroscopy table, this limit be reduced to around 70~80% of maximum output.

i. Select the appropriate anode-rating chart for the X-ray tube in use.
   The anode speed is normally controlled by the power frequency.
   Take care to select between 50 or 60 hz for low-speed operation, or between 150 or 180 hz for high-speed operation.
ii. In addition to anode speed, select either single or three phase operation, depending on the type of generator.
iii. If you have a high frequency generator, select the three-phase chart. This will still apply if the generator is supplied by single-phase mains power.
iv. **Note.** The rating charts provide a family of curves. It is not required to use the same mA or kV for testing. For example, 0.1 sec', 125 kV & 360 mA is the same as 90 kV & 500 mA.
v. On the rating charts, select suitable time periods. (For example. 0.02, 0.1, 0.3, 1.0, 5.0 seconds). At these time selections, determine a suitable mA station, and the maximum kV that can be used with that mA selection. Adjust the kV towards this maximum value. The exposure inhibit should occur before this value is reached. Repeat this test for each of the preselected time settings.
vi. Repeat this test for each mA station; together with both fine and broad focus spot selection.
vii. Some X-ray controls may have provision for both high **and** low speed operation. In these cases, the maximum load available for low speed operation, should not exceed 85~90% of the value indicated in the low speed chart.

- **Note.** Many microprocessor-controlled systems have the rating charts pre-installed in computer memory. On selection of the manufacturers X-ray tube, a code for that tube is entered into the computer. If the manufacturer of the X-ray control does not supply the X-ray tube, a good match of a rating chart may not be possible. In this situation, contact the service department for advice.

e. mA calibration

(For this test, the X-ray control is required to display the actual mA, or mAs, resulting from an exposure. The control may have either an mAs meter, or a quick acting mA meter).

Microprocessor controlled systems have an internal switch, which is set to 'calibration mode'. This should only be adjusted on direct advice from the service department.

When the X-ray control has an mAs meter:

i. mAs meters may be of two types. Type one is 'ballistic'. With this version, watch for the maximum reading on exposure, before the needle returns to zero.
ii. The other version is a true integrating mAs meter. This type will hold the reading for a period of time, often while the preparation button is kept pressed at the end of exposure. With this type of meter, ignore the peak needle deflection, and only record the steady reading.
iii. mAs meters may be dual function. In some controls, the meter will first indicate the % of anode load, and on preparation change over to the mAs function. Another type first indicates the preselected mAs, and on exposure indicates the actual mAs. Actual mAs remains displayed until preparation is released.
iv. When choosing an exposure time, avoid uneven times like 0.01, 0.03, etc. This avoids timer problems that can exist on older units. Select an exposure time of 0.1 second for easy calculation.
v. Test mAs output using two kV positions. Values suggested are 60 kV and 90 kV. Repeat this test for all mA stations and focal spots.
vi. The test mAs output should be within 10% for older systems, and in modern equipment within 5%. Variations of mAs between adjacent mA stations should be less than 5%, including older designs.
vii. When preparation is complete, allow another half to one second before exposing. This is to eliminate possible errors due to incorrect pre-heating of the filament.
viii. To check for a possible filament pre-heating problem, select 60 kV, and the largest mA station. Make an exposure immediately preparation is completed, and record the mAs output. Now make another exposure, but this time wait for about one second after preparation is completed, then make an exposure.
ix. If the difference between the two tests is more than 5%, contact the service department for advice. The generator should have the filament pre-heating adjusted, or else a small increase in preparation time.

When the X-ray control has a mA meter only:

i. Select a low kV, between 60 and 70 kV.
ii. Make an assessment of tube loading with the selected mA station by selecting an exposure time of two seconds.
iii. Assuming a two second time would permit an exposure; now select a time of 0.8–1.0 second. This time allows the mA meter to reach a steady reading, during the exposure.
iv. When preparation is complete, allow another half to one second before exposing. This is to eliminate possible errors due to incorrect pre-heating of the filament.
v. On exposing, watch the mA meter needle arrive at the expected value. Record the steady reading. (Ignore any bounce or overshoot.)
vi. MA should be within 10% of the required value.
vii. Repeat this test on both focal spots. Test only the mA stations that are well within the anode load safety limit, at the exposure times of 0.8–1.0 second.
viii. Between test exposures, allow at least three to five minutes for anode cooling.
ix. To check for a possible pre-heating problem, select 60 kV, and the largest mA station that was previously tested. Make an exposure immediately preparation is completed, and record the mA output. If the change in mA is more than 5%, contact the service department for advice. The generator should have the pre-heating adjusted, or else a small increase in preparation time.

f. Radiation reproducibility tests, using a step-wedge

- This test should be carried out after the film processor has received its general maintenance.
- Adjust the FFD to 100 cm.
- Place the stepwedge on a 24/30 cm cassette.
- Several exposures can be made on the one piece of film. Place two pieces of lead rubber on top of the cassette, positioned against either side of the stepwedge. As the stepwedge is repositioned, the lead rubber prevents unwanted radiation entering the cassette.
- Select a suitable mAs and kV combination, and make a total of four exposures.
  i. Allow about 0.5–1.0 second delay after preparation is completed, before making each exposure. This is to ensure the filament has reached a stable temperature.
  ii. After each exposure, reposition the stepwedge and lead rubber on the cassette.
  iii. Develop the film. As the exposure settings are the same for all exposures, the film should show very little variation.
  iv. If necessary, change kV or mAs so the film displays a good range of densities, then repeat this test.
- Make another series of four exposures, using the same settings as before.
  i. This time, do not delay the exposure, but expose immediately preparation is completed.
  ii. This is a test for filament pre-heating, or temperature stability.
- Compare all eight exposures. If available, use a densitometer. As the same output settings were used, the exposures should show very little variation.
  i. If the second group is lighter, or darker, than the first group, the filament pre-heating or preparation time should be adjusted. Contact the service department for advice.
  ii. In case there is a general variation of densities in either group, this may be due to power mains voltage fluctuations. If suspect, repeat this test at a later time when power is more stable.
  iii. Variable output can be caused by a poor connection to the X-ray tube filament. This is due to a problem with the cathode high-tension cable, where the cable-end plugs into the X-ray tube housing. See **module 7.3 page 117**.
- Repeat the test for each focal spot.
- Record the settings used in the maintenance logbook for future use. Include which cassette used. Retain the test films for comparison with future tests.

g. X-ray output linearity test, using a step-wedge

This is an important check on overall performance. By using a stepwedge, a comparison test may be made, not only between the mA stations of the unit under maintenance, but also with other units in the department.

- This test should be carried out after the film processor has received its general maintenance.
- **Note.** This test will indicate variations in kV output as well as mAs.
- For this test, select an mAs value that can be repeated over a number of mA stations by changing time factor only. (To avoid possible errors due to kV rise and fall time, avoid exposure times below 0.02 seconds.)

- Set 80 kV, and a FFD of 100 cm.
- Position the stepwedge on a 24/30 cm cassette.
- Several exposures can be made on the one piece of film. Place two pieces of lead rubber on top of the cassette, positioned against either side of the stepwedge. As the stepwedge is repositioned, the lead rubber prevents unwanted radiation entering the cassette.
- Using the selected value of kV and mAs make a series of exposures. Change the mA station after each exposure, and adjust the time to obtain the same mAs.
    i. Allow about 0.5–1.0 second delay after preparation is completed, before exposing. This is to ensure the filament has reached a stable temperature.
    ii. If the film is too light, increase the kV, and repeat the test.
    iii. If the film is too dark, add extra aluminium under the step wedge. Or, place a sheet of paper between one side of the film, and the intensifying screen in the cassette.
- It may not be possible to obtain the same mAs value for all mA stations. In this case, select a different mAs value, but include one of the mA stations previously tested. Repeat the test with the new selection of mA values.

  This is illustrated in table 1–a, where 100 ma is used for both 20 mAs and 30 mAs comparisons.

Table 1–a. Selection of mAs values for test

mA	Time	mAs	mA	Time	MAs
500	0.04	20	100	0.3	30
400	0.05	20	150	0.2	30
200	0.1	20	300	0.1	30
100	0.2	20			

- If one of the mA stations shows a significant change in density, make another test with that mA station, this time change kV to obtain the required film density.
    i. Providing the required kV change is not more than 2~3%, the station is within tolerance.
    ii. If no more than 3~5% it is still within tolerance. However, make a note in the maintenance record, and have the calibration checked next time the service department pays a visit.
    iii. If greater than 5%, then that station is out of tolerance. This may be due to mA or kV calibration. If significant, then place that mA station 'out of operation' and contact the service department for advice.
    iv. An estimation of mA calibration error can be made by a comparison exposure, changing time only. This needs an initial time setting of 0.1 second or greater. For example, if the suspect mA station of 200 mA showed a low output, and on changing the exposure time to 0.11 second still showed a slightly low output, then the mA station is more than 10% out of calibration.
    v. Besides a possible change of mA or kV calibration, the timer may not be accurate. A single-phase generator can have an error of plus or minus 0.01 seconds. This is a large error at short time settings. A 'Spinning top' test can indicate single-phase generator exposure times. See appendix 'B' page 169.
- Record all calibration settings used with the stepwedge in the logbook. Include the kV, mAs, FFD, and the cassette used. This will allow a quick set-up when this test is repeated. Save the films for comparison with future tests.

TASK 2

X-ray control familiarization Part I

You have just been transferred to an X-ray department in another hospital, and have been requested to commence a routine maintenance programme. You are not familiar with this particular X-ray control. The control is situated in a standard Bucky table room, with an over-table X-ray tube only.

1. Locate the manufacturer model and serial number, for recording in the logbook. (In some cases, it may be necessary to look behind the control desk or X-ray control cabinet.) _____

2. Check carefully the range and type of controls. Some are unfamiliar. Suggest a way that the function of these controls may be verified. _____

3. Check and list the full range of mA stations available. Which mA stations are available for the broad focus, and for the fine focus? _____

4. Is individual selection of fine and broad focus available? If so, which mA stations may be used on *either* fine or broad focus? _____

5. It is possible that initial inspection indicates the control operates on selection of mAs and kV only. ('Two knob' technique.) Is there a control switch to enable operation by individual selection of mA, time, and kV? ('Three knob' technique.) _____

6. Older basic X-ray controls often have a meter and knob to adjust line voltage. If your unit has such a system, does the meter have a calibration mark? Is it possible to adjust line voltage so the meter indicates excessive voltage, as well as low line voltage? _____

PART II. ROUTINE MAINTENANCE MODULES

7. kV selection and method of generation depends greatly on the type, model and age of X-ray control being investigated. From your general inspection, which method of kV generation is applicable to this control?

   Single phase, self rectified. _____

   Single phase, full wave rectified. _____

   Three phase, six or twelve pulse. _____

   High frequency inverter system. _____

   Capacitor discharge. _____

8. Discuss possible methods to identify the other versions of kV generation.

   _____
   _____
   _____
   _____
   _____
   _____
   _____
   _____

9. What is the maximum and minimum possible kV to select with this X-ray control? (Make this test after selecting low mA and a short exposure time.) _____
   _____

10. Is it possible to obtain a simultaneous selection of *maximum* kV, mA, and time? In this case an overload or exposure inhibit signal should be indicated. What form does this take? _____
    _____
    _____

11. On some controls, selection of a high mA station and low kV may generate an overload or inhibit signal, even for very short exposure times. This inhibit signal disappears on *increasing* kV, or reducing mA. Does this apply to this X-ray control? _____

    Discuss the reason for such a protection, and to which part of the X-ray tube this is applicable. _____
    _____
    _____
    _____
    _____

Tutor's comments

Satisfactory/Unsatisfactory

Signed _____ Date _____
   Tutor

TASK 3

X-ray control familiarization Part 2

You have identified the functions of the various controls on the X-ray generator control desk. In addition, you have made a test to ensure the overload protection system is working, and correct line voltage can be obtained. It is now time to make test exposures, and carefully observe the system in operation.

1. Ensure the X-ray tube collimator is closed, and the tube is angled away from the control desk.

2. Select 100 mA, 60 kV and 0.1 s time. (Or 10 mAs if individual selection is not available). Ensure there are no warning lights or signals displayed. Select non-Bucky operation.

3. Press the preparation switch. Note; if a single button controls both preparation and exposure, at this point, only press it half way.

4. Carefully observe the control panel. Did any meters change their reading, or indicator lamps immediately signal a different operation mode? _____

_____

_____

_____

5. Shortly after pressing the preparation switch, the control should indicate 'Ready for exposure' How is this indicated? Approximately how long is the delay time before 'ready' is indicated? _____

_____

_____

6. Release the preparation switch, and again press to go into preparation. This time listen carefully for sounds of a relay or contactor. It is possible several may operate. Hint. If the X-ray control has a door that may be opened, this will allow better observation. *Take care not to touch or open any of the internal sections.* _____

_____

_____

_____

7. Once again carry out preparation. This time listen carefully at the X-ray tube. It should be possible to hear the anode speed up, and when preparation is released, to slowly slow down.

   You may need an assistant to carry out preparation for you, as you will need to be close to the X-ray tube.

   Is the acceleration of the anode during preparation clearly audible? _____

   _____

   _____

   Does the anode gradually slow down when preparation is released? _____

   _____

   _____

PART II. ROUTINE MAINTENANCE MODULES

8. If the X-ray tube is operated at high-speed, you should hear a fast drop in anode speed when preparation is released. This is the brake cycle. There are two types of brake cycle. The DC brake cycle quickly slows the anode to a complete stop. A dynamic brake cycle will quickly reduce the anode speed to about 3000 RPM, after which the anode gradually slows down to a full stop.

   Assuming high-speed operation.
   On release of preparation, does the anode come quickly to a full stop? _____

   _____

   _____

   Or, does the anode quickly brake to a slow speed, then very slowly coast to a full stop? _____

   _____

   _____

9. Once again, go into preparation mode. This time, when 'ready' appears, press the exposure button. Keep pressing this button after the exposure ends. At the same time carefully observe the control panel.

   Does a radiation 'On' indicator light up on the control panel? _____

   _____

   _____

   Does an audible signal occur during the exposure? _____

   _____

   _____

   Is there an indication of the mA or mAs generated during the exposure? _____

   _____

   _____

   If the control has an mAs meter, does the reading of the mAs meter remain until the exposure, or preparation, switch is released? _____

   _____

Tutor's comments

_____

Satisfactory/Unsatisfactory

_____

Signed _____   Date _____
           Tutor

TASK 4

Test for X-ray tube overload calibration
Part I

During the routine maintenance check, you decide to ensure the X-ray tube operating parameters are within safe limits. You also want to check if this X-ray tube allows optimum use of the generator output power.

Please note; this test assumes the generator has independent selections of mA and time. Some controls may allow an mAs mode as well as individual selection of mA and time. In that case switch off the mAs mode.

If the control only provides mAs selection, this test is still valid, providing the control indicates which mA position is actually in use.

1. Locate the make, model and serial number, also the focal spot sizes, of the X-ray tube.
   a. Note, in some cases the label with this information may be on the side of the X-ray tube throat. You may need a mirror and torch to read the label.
   b. Enter this information into the routine maintenance logbook, and check sheet.

   _____

   _____

2. Depending on the mains power-supply frequency, what is the theoretical maximum anode speed for low speed operation?

   a. Input power-supply frequency. _____

   b. Anode speed, low speed operation. _____

3. Does this generator have a high-speed starter? If so;
   a. What is the frequency generated by the high-speed starter? (Two common frequencies are 150 hz and 180 hz)
   Hint. Refer to the specifications for the starter, in either the operation or installation manual for the starter, or the generator. The frequency generated by the starter need not be related to the power frequency.

   _____

   b. Depending on the starter frequency, what is the possible maximum anode speed?

   _____

   c. Is high speed anode rotation:
      i. Always high-speed?

   _____

      ii. Automatic selection between high or low speed depending on the X-ray tube load?

   _____

      iii. High or low-speed is individually selected by the operator?

   _____

4. Is the generator single or three-phase operation? (A high frequency generator is considered three-phase.)

   _____

PART II. ROUTINE MAINTENANCE MODULES

5. From the information obtained in the preceding parts, select the appropriate anode load charts from the X-ray tube specification or operation manual. **Note.** To avoid mistakes, tick the appropriate charts, and place a cross against the unwanted charts.

6. Select a suitable series of exposure times. Times of 0.01, 0.03, 0.1, 0.3, 1.0, and 3.0 seconds are suggested. Using these times, determine from the charts the maximum kV/mA product that is allowed.
   - kV **max** is the maximum kV indicated on the charts for an individual mA/time selection.
   - kV **test** refers to the maximum kV the control would allow for an exposure, at the same selections of mA/time. This should be less than kV max.

   a. Broad focus
      i. Time _____ mA _____ kV-max _____ kV-test _____
      ii. Time _____ mA _____ kV-max _____ kV-test _____
      iii. Time _____ mA _____ kV-max _____ kV-test _____
      iv. Time _____ mA _____ kV-max _____ kV-test _____
      v. Time _____ mA _____ kV-max _____ kV-test _____
      vi. Time _____ mA _____ kV-max _____ kV-test _____

   b. Fine focus.
      i. Time _____ mA _____ kV-max _____ kV-test _____
      ii. Time _____ mA _____ kV-max _____ kV-test _____
      iii. Time _____ mA _____ kV-max _____ kV-test _____
      iv. Time _____ mA _____ kV-max _____ kV-test _____
      v. Time _____ mA _____ kV-max _____ kV-test _____
      vi. Time _____ mA _____ kV-max _____ kV-test _____

7. Using the values for time, mA, and kV max, set the control to the predetermined mA and time. Advance the kV setting until the control indicates an inhibit signal. Enter that value for kV-test.
   a. If the control has optional high and low speed operation, the above test should first be made with high-speed selected. After which make a second test for low speed, using data from the appropriate load charts.
   b. Some later model controls may have a selection for 'Load full' or 'Maximum load'. Ensure this selection is made for the above test.

8. Compare kV-max and kV-test. kV-test should be less than kV-max. Are there any points where kV-test is just under or slightly over kV-max?

   _____
   _____

   Express an opinion if this could be a reason for concern. For example, are normal exposures close to these limits?

   _____
   _____
   _____
   _____

9. Compare the generators rated maximum output, at 0.1 second, with that of the X-ray tube. (Also at 0.1 second). Does the present tube make optimum use of the generator power?

_____

_____

_____

10. Based on (9), if the X-ray tube was replaced, would you prefer any change in the X-ray tube specifications? Discuss the reasons why.

_____

_____

_____

_____

_____

Tutor's comments

_____

Satisfactory/Unsatisfactory

Signed _____   Date _____
           Tutor

TASK 5

Test for X-ray tube overload calibration Part 2

You have verified the anode load parameters are operating within the X-ray tube specifications. There remain two areas to test.

1. Maximum kV protection.
    a. Select the large focal spot, combined with the lowest mA station and a short time.
    b. From the X-ray tube specifications, what is the maximum kV that may be used? _____
    c. Increase the kV selection at the generator. Is an exposure inhibit generated before the maximum kV is reached? _____

2. Minimum kV protection.
    a. From the X-ray tube rating charts, examine the fine and broad focus characteristics, and look for a possible mA limitation related to kV. As an example, a 60 kV curve may be shown part dotted, or have a cut off line. This indicates the maximum mA available for that kV value.
    b. Take care to use the charts related to either single or three-phase operation, depending on the generator mode of operation.

        i. From the chart data, are any generator mA stations affected by this restriction? _____

        _____

        ii. What is the minimum kV for these mA stations? _____

        _____

    c. Select the affected mA station and set a short time. Try reducing kV below the indicated minimum level for that station. Is an exposure inhibit generated before that level is reached? _____

    _____

    d. If it is possible to adjust kV below the minimum kV requirement, describe how accidental operation could be reduced, or what action should be taken. _____

    _____

    _____

    _____

_____

Tutor's comments

Satisfactory/Unsatisfactory

Signed _____   Date _____
        Tutor

MODULE 1.2

X-ray generator, mobile unit

Aim

The aim is to provide routine maintenance procedures for a mobile X-ray generator. Within the mobile generator capabilities, a similar check is made as used for a fixed installation. Maintenance includes mechanical operation of the mobile, together with the X-ray tube and collimator. Instructions for repairs to the mobile are provided in module 6.1 page 90.

Capacitor discharge mobile procedures are provided in module 1.3 page 37.

(**Note:** Reference module page numbers refer to the title page.)

Objectives

On completion of this module, the student will be familiar with maintenance procedures for a mobile X-ray generator. These procedures can be used as a version of quality control, together with the routine maintenance check-sheets provided in the appendix.

Contents

a. General precautions
b. Visual inspection of the control panel, power off
c. Mechanical and electrical inspection, power off
d. Operation inspection of the control panel, power on
e. Mechanical and electrical inspection, power on
f. X-ray tube and collimator
g. Milliampere calibration
h. Radiation reproducibility
i. Radiation linearity

Equipment required

- Basic tool kit.
- X-ray alignment template. *
- Stepwedge. *
- 24/30 cm Cassette.
- Two pieces of lead rubber.
- Aerosol spray lubricant.
- Cleaning solvent.
- Cloth.

* The template and stepwedge is described in appendix 'B' page 169.

a. General precautions

- **Before removing any covers, or testing any wires or connections, ensure the system is switched off, and unplugged from the power point.**
- **Mobile high-frequency generators may be battery operated. The batteries in these are connected in series, and may have a total voltage of up to 240 V DC. Refer to the operating or installation manuals for the position of the battery isolation switch, and ensure this is switched off before removing any covers.**
- **If the power plug has loose connections, have an electrician check the plug. The plug may be assembled incorrectly.**

b. Visual inspection of the control panel, power off

- Check all knobs and switches. Where knobs have a pointer attached, check that the pointer aligns correctly at all positions of the indicated scale.
  Tip. Check the pointer at full clockwise and counter clockwise positions of the knob. Look for possible loose knobs, or for push button switches that may tend to stick.
- Where controls have had extra labels attached, are these labels still relevant? If so, are they in good condition?
- Older X-ray mobiles often have analogue meters instead of digital displays.

i. With power switched off, the meter needle should be on the 'zero' calibration mark.
ii. If not, first tap gently in case the meter tends to 'stick'. If the needle is not sitting on the zero mark, most meters have a small adjustment screw in the middle for zero calibration.
iii. Caution. Contact the service department before adjusting. In some cases, the meter may be deliberately adjusted 'off zero', as an incorrect method of calibration.

- With the aid of a suitable solvent, clean off the residue left behind from sticking plaster, and pieces of sticky tape.

c. Mechanical and electrical inspection, power off

- Look for any loose panels or sections. Pay particular attention to the mounting of the collimator. With a screwdriver, check for possible loose screws, particularly with the tube support arm and the vertical bearing tracks.
- With the X-ray tube set to minimum height, check the vertical suspension wire rope for possible broken strands. CAUTION, do not test with bare fingers, and instead test by rubbing the cables up and down using a piece of cloth.
- Check the action of the tube-stand bearings. Are there any visible gaps between the bearings and the track surface? Also are there any 'clunking' noises or 'jerking' when moved, which can indicate damaged bearings.
- Spray the tube stand tracks and bearings with a light aerosol lubricant. Wipe down afterward, so only a very small film is left on the tube stand tracks.
- Check for possible loose lock handles, and ensure manually operated locks have an adequate range of adjustment.
- Ensure the mobile brakes operate in a positive fashion when the hand is released from the handle, and that they are fully released while the mobile is travelling.
- Pay particular attention to the cabling from the X-ray tube and tube stand. All movements of the system should not cause any stress or pulling of the cables.
- Inspect the HT cables for any sign of damage to the safety earth shield at the X-ray tube cable ends. Ensure the cable ends are firmly inserted into the X-ray tube, and the securing ring nut is not loose.
- Where there is evidence of twisting or pulling on the HT cables, particularly at the X-ray tube end, investigate means of providing additional support. If necessary, discuss with the service department.
- Examine carefully all plugs and sockets attached to cable ends. The outer insulation of cables should not be pulled out from the cable clamp.
- Check the condition of the power cable. If necessary, remove the plug cover, and ensure terminations are tight, and no connections are stretched or have broken strands. Should the cable exhibit excessive twisting, or have cracks in the outer sheath, ask an electrician for to replace the cable or plug.
- Older mobiles with battery operated power assistance should have the battery electrolyte level checked.
  i. Ensure first that the power cable is unplugged from the power point.
  ii. To gain access to the battery, refer to the operation or service manual.
  iii. Top up with either distilled water or else fresh rainwater.
  iv. Later systems use sealed or 'low maintenance' batteries. This includes high-frequency mobile generators.

d. Operation inspection of the control panel, power on

- Check all indicator lamps etc. If necessary, operate different selection techniques to ensure all required status indicators operate correctly.
- Where a digital readout of radiograph settings is provided, select a number of different values to ensure there are no display errors or missing segments.
- Older controls may have manual adjustment of power line voltage, with a meter indicating correct compensation. Check the range of adjustment. It should be possible to reset the voltage by 10%, above or below, the required voltage.
- With battery-operated equipment, check the status of the battery charge indicator. If low, place the unit on charge, and check that after a reasonable time the system indicates fully charged. This time should not be greater than an overnight period.

e. Mechanical and electrical inspection, power on

- Test operation of the electromagnetic locks. There should be no hesitation in operation, nor should the lock 'stick on'. In some cases the surface of the lock may require cleaning, to obtain a better 'grip'

- With power assisted mobiles, check for correct operation in all forward and reverse modes. Where there is an anti-crash bumper, manually operate the bumper. This should stop motor drive. (Do NOT test by standing in front while the unit is moving forward)

f. X-ray tube and collimator

- Inspect the X-ray tube housing for possible oil leaks.
- When in preparation, listen for excessive X-ray tube bearing noise.
- Check the operation of the collimator lamp timer. With mechanical timers, listen for possible sticking of the clockwork.
- Check the alignment of the collimator lamp and X-ray beam. This should be checked through 180 degrees rotation of the collimator.
- The collimator has a scale associated with the adjustment knob to indicate the field size. The knob can slip on the shaft, or not be correctly positioned after replacing a collimator globe.
    i. Place a 24/30 cm cassette on the tabletop. Adjust the FFD to 100 cm
    ii. With the collimator light switched on, check the knob pointer indicates the correct position on the scale.
    iii. If necessary, reposition the knob on the shaft.
    iv. Repeat this test for other cassettes in use.
    v. If the scale is worn or not legible, use a marker pen to indicate positions for common cassettes in use. Order a new scale from the service department.
    vi. While waiting for a new scale, ensure you have a spare collimator globe in stock.
- To test the rotation accuracy of the light beam;
    i. Rotate the collimator 90 degrees in either direction.
    ii. With the light on, open the collimator so an average size field is projected on the tabletop. For example, a 24/30 cm cassette size.
    iii. Place markers to indicate the light beam position.
    iv. Now rotate the collimator 180 degrees in the opposite direction. The light field should be within 1% or better, compared to the previous position. If not, see module 7.2 page 110.
- To test the alignment of the X-ray to the light beam;
    i. Place the X-ray alignment template on a 24/30 cm cassette.
    ii. Adjust the FFD to 100 cm.
    iii. Adjust the light beam to the template markers.
    iv. Make a low kV and mAs exposure.
    v. Develop the film.
    vi. Measure the distance where the X-ray does not coincide with the markers. Any error should be inside the compliance requirements for the country. See module 7.2 page 110.
    vii. Repeat this test, with the collimator rotated 90 degrees clockwise, and 90 degrees counter clockwise.
    viii. If there is an alignment problem, see module 7.2 page 110.

g. mA calibration

- Some microprocessor-controlled mobiles allow mA to be checked and also calibrated via the front panel. These systems require either an internal switch to be operated, or else a code entry at the panel. To test or adjust calibration, the procedure in the manufacturer's service manual **must** be referred to.
- This test only applies to older mobiles that allow individual selection of mA and time. A panel mounted mA meter is also required.
    i. Use 70~80 kV and an exposure time of around 1.0 second. (The actual exposure time used should allow the mA meter to just reach a steady reading)
    ii. Select the mA station to be tested, and obtain preparation. Wait about 1.0 second after 'ready' is obtained, then expose. Record the reading obtained from the mA meter.
    iii. Milliampere should be within 10% of the required value.
- Repeat the above for all mA stations, and both focal spots. Caution, as the X-ray tube for the mobile will have a small heat capacity, allow cooling time between exposures.
- Filament pre-heat test.
    i. Select the highest mA station, and 60 kV. Go into preparation, and immediately 'ready' is obtained, make an exposure.
    ii. Make another exposure. This time wait about one second after 'ready' is obtained, before exposing.
    iii. If the difference between the two tests is more than 5%, contact the service department for advice. The generator should have the filament pre-heating adjusted, or else a small increase in preparation time.

h. Radiation reproducibility

- This test should be carried out after the film processor has received its routine maintenance.

- Position the stepwedge on a 24/30 cm cassette.
- Several exposures can be made on the one piece of film. Place two pieces of lead rubber on top of the cassette, positioned against either side of the stepwedge. As the stepwedge is repositioned, the lead rubber prevents unwanted radiation entering the cassette.
- Adjust the FFD to 100 cm
- Select a suitable mAs and kV combination, and make a total of four exposures.
  i. Allow about 0.5~1.0 second delay after preparation is completed, before making each exposure. This is to ensure the filament has reached a stable temperature.
  ii. After each exposure, reposition the stepwedge and lead rubber on the cassette
  iii. Develop the film. As the exposure settings are the same for all exposures, the film should show very little variation.
  iv. If necessary, change kV or mAs so the film displays a good range of densities, then repeat this test.
- To test filament-heating stability, make another series of four exposures, using the same settings as before. This time expose immediately 'ready' is obtained.
- Develop the film and compare to the first series of exposures. If any significant difference is obtained, either filament pre-heating or preparation time may need adjustment. Contact the service department for advice.

i. X-ray output linearity test, using a step-wedge

Note. *This test is only applicable to mobiles with independent selection of mA and time settings.*

This is an important check on overall performance. By using a stepwedge, a comparison test may be made, not only between the mA stations of the unit under maintenance, but also with other units in the department.

Note. This test will indicate variations in kV output, as well as mAs.

- This test should be carried out after the film processor has received its routine maintenance.
- Position the stepwedge on a 24/30 cm cassette.
- Several exposures can be made on the one piece of film. Place two pieces of lead rubber on top of the cassette, positioned against either side of the stepwedge. As the stepwedge is repositioned, the lead rubber prevents unwanted radiation entering the cassette.
- For this test, select an mAs value that can be repeated over a number of mA stations by changing time factor only.
- Set 80 kV, and a FFD of 100 cm.
- Using the selected value of kV and FFD make a series of exposures, changing the mA station, and adjusting time setting to obtain the same mAs.
  i. Allow about 0.5~1.0 second delay after preparation is completed, before exposing. This is to ensure the filament has reached a stable temperature.
  ii. If the film is too light, select a different mAs value, or kV, and repeat the test.
  iii. If the film is too dark, add extra aluminium under the step wedge. Or, place a sheet of paper between one side of the film, and the intensifying screen in the cassette.
- It may not be possible to obtain the same mAs value for all mA stations. In this case, select a different mAs value, but include one of the mA stations previously tested. Repeat the test with the new selection of mA values.
- This is illustrated in table 1–b, where 10 ma is used for both the 20 mAs and 30 mAs comparisons.

Table 1–b. Selection of mAs values for test

mA	Time	mAs	mA	Time	MAs
50	0.4	20	10	3.0	30
40	0.5	20	15	2.0	30
20	1.0	20	30	1.0	30
10	2.0	20			

- If one of the mA stations shows a significant change in density, make another test with that mA station only, but this time change kV.
  i. Providing the required kV change is not more than 2~3%, the station is within tolerance.
  ii. If no more than 3~5% it is still within tolerance. However, make a note in the maintenance record, and have the calibration checked next time the service department pays a visit.
  iii. If greater than 5%, then that station is out of tolerance. This may be due to mA or kV calibration. If significant, then place that mA station 'out of operation'. Contact the service department for advice.

iv. Besides a possible change of mA or kV calibration, the timer may not be accurate. A single-phase generator can have an error of plus or minus 0.01 seconds. This is a large error at short time settings. A 'Spinning Top' can check single-phase generator exposure times. See appendix 'B' page 169.

Record all calibration settings used with the step-wedge in the logbook. Include the kV, mAs, FFD, and the cassette used. This will allow a quick set-up when this test is repeated. Save the films for future comparison tests.

MODULE 1.3

X-ray generator, capacitor discharge mobile

Aim

The aim is to provide routine maintenance procedures for a capacitor discharge (CD) mobile generator. Within the CD mobile capabilities, a similar check is made as for a conventional mobile. There are however, some important differences. These relate to the non-linear output due to the kV/mAs relationship, plus other modes of operation. Maintenance includes mechanical operation of the CD mobile, together with the X-ray tube and collimator. Instructions for repairs to the CD mobile are provided in module 6.2 page 94.
(**Note:** Reference module page numbers refer to the title page.)

Objectives

On completion of this module, the student will be familiar with maintenance procedures for a portable X-ray generator. These procedures can be used as a version of quality control, together with the routine maintenance check-sheets provided in the appendix.

Contents

a. General precautions
b. Visual inspection of the control panel, power off
c. Mechanical and electrical inspection, power off
d. Operational inspection of the control panel, power on
e. Mechanical and electrical inspection, power on
f. X-ray tube and collimator
g. mAs calibration
h. Radiation reproducibility

Equipment required

- Basic tool kit.
- X-ray alignment template.*
- Stepwedge.*
- 24/30 cm Cassette.
- Two pieces of lead rubber.
- Aerosol spray lubricant.
- Cleaning solvent.
- Cloth.

* The template and stepwedge is described in appendix 'B' page 169.

a. General precautions

- *Before removing any covers, or testing any wires or connections, ensure the system is switched off, and unplugged from the power point.*
- *If the power plug has loose connections, have an electrician check the plug. The plug may be assembled incorrectly.*
- *Do NOT make any adjustments to the HT cables without first discharging the capacitor. See module 6.2 page 94.*

b. Visual inspection of the control panel, power off

- Check all knobs and switches. Where knobs have a pointer attached, check that the pointer aligns correctly at all positions of the indicated scale.
  Tip. Check the pointer at full clockwise and counter clockwise positions of the knob. Look for possible loose knobs, or for push button switches that may tend to stick.
- Where some mobiles have had extra labels attached, are these labels still relevant? If so, are they in good condition?

- Older CD mobiles have an analogue kV meter instead of a digital display. With power off, the meter needle should be on the 'zero' calibration mark, **providing the capacitor is fully discharged**. If not, first tap gently in case the meter tends to 'stick'. **Do not attempt to adjust the meter zero position without reference to the service manual, and operation of the internal capacitor discharge device.**
- Other CD mobiles may have a line voltage meter. This meter should read zero on power off. If necessary, the meter zero position may be adjusted by the centre screw. Tap the meter gently first, to ensure the meter is not sticking.
- With the aid of a suitable solvent, clean off the residue left behind from sticking plaster, and pieces of sticky tape.

c. Mechanical and electrical inspection, power off

- Inspect for any loose panels or sections. Pay particular attention to the mounting of the collimator. With a screwdriver, check for possible loose screws, particularly with the tube support arm and the vertical bearing tracks.
- With the X-ray tube set to minimum height, check the vertical suspension wire rope for possible broken strands. **CAUTION**, do not test with bare fingers. Test instead by rubbing the cables up and down with a piece of cloth.
- Check the action of the tube-stand bearings. Are there any visible gaps between the bearings and the track surface?
- Are there any 'clunking' noises or 'jerking' movements, when the X-ray tube is positioned? This can indicate damaged bearings.
- Spray the tube stand tracks and bearings with a light aerosol lubricant. Wipe down afterward, so only a thin oil film is left on the tube stand tracks.
- Check for possible loose lock handles, and ensure manually operated locks have an adequate range of adjustment.
- Ensure the mobile brakes operate in a positive fashion when the hand is released from the handle, and that they are fully released while the mobile is travelling.
- Pay particular attention to the cabling from the X-ray tube and tube stand. All movements of the system should not cause any stress or stretching of the cables.
- Inspect the HT cables for any sign of damage to the safety earth shield at the X-ray tube cable ends. Ensure the cable ends are firmly inserted into the X-ray tube, and the securing ring nut is not loose.
- **Note. Never remove the HT cable ends unless the capacitor is fully discharged, and the capacitor safety switches are operated. See module 6.2 page 94.**
- Where there is evidence of twisting or pulling of the HT cables, particularly at the X-ray tube end, investigate means of providing additional support. If necessary, contact the service department for advice.
- Examine carefully all plugs and sockets attached to cable ends. The cable outer insulation should not be pulled out from the cable clamp.
- Check the condition of the power cable. If necessary, remove the plug cover, and ensure terminations are tight, and no connections are stretched or have broken strands. Should the cable exhibit excessive twisting, or have cracks in the insulation, replacement is required. **An electrician should carry out any repairs to the power cable or plug.**
- With motorized mobiles check the battery electrolyte level.
  i. *Ensure the power cable is unplugged from the power point.*
  ii. To gain access to the battery, refer to the operation or service manual.
  iii. Top up with distilled water or else fresh rainwater.
  iv. If the mobile is fitted with a 'low maintenance' battery, contact the service department for advice.

d. Operational inspection of the control panel, power on

- Check all indicator lamps operate.
- For CD mobiles equipped with an analogue or digital kV meter, test the kV adjustment for correct operation.
  i. Set the required kV to 60 kV and press the charge button.
  ii. The charge light should illuminate. Once the required kV is reached, then the 'ready' lamp should light up.
  iii. Check that the kV displayed on the meter closely agrees with that indicated at the kV control knob.
  iv. Observe the kV meter for a few minutes. The kV should slowly drop back by about 2~3 kV, then return briefly to the charge mode. (This is called 'topping up')

v. Increase the set kV to 90 kV. The charge light should illuminate, until the kV meter reaches 90 kV.
vi. Now reset the required kV back to 60 kV. The X-ray ON light should illuminate. At the same time the indicated kV should quickly drop down to the required value.
vii. **Note.** A low mA X-ray exposure is produced when the kV is reset. Radiation is blocked in this mode by a lead shutter in the collimator.
viii. In case the kV resets very slowly, similar to discharge prior to topping-up, this can indicate a problem. See **module 6.2 page 94**.

- Some CD mobiles have an adjustment for power line voltage. A meter indicates when the voltage is correct. Check the range of adjustment, to ensure compensation may be set approximately 10% above or below the required voltage.
- **Note.** The power line voltage adjustment can directly affect the kV charge on the capacitor.
- With motorized mobiles, check the status of the battery charge indicator. If low, place the unit on charge. Charging time should not be greater than an overnight period.

e. Mechanical and electrical inspection, power on

- Test operation of the electromagnetic locks. There should be no hesitation in operation, nor should the lock 'stick on'. In some cases the surface of the lock may require cleaning to obtain a better 'grip'
- With motorized mobiles, check for correct operation in all forward and reverse modes. Where there is an anti crash bumper, manually operate the bumper. This should stop motor drive. (Do NOT test by standing in front while the unit is moving forward)

f. X-ray tube and collimator

- Inspect the X-ray tube housing for possible oil leaks.
- During preparation, listen for excessive X-ray tube bearing noise.
- Check the operation of the collimator lamp timer. With mechanical systems, listen for possible sticking of the clockwork.
- The collimator normally has a scale associated with the adjustment knob to indicate the field size. The knob can slip on the shaft, or not be correctly positioned after replacing a collimator globe.
  i. Place a 24/30 cm cassette on the tabletop. Adjust the FFD to 100 cm
  ii. With the collimator light switched on, check the knob pointer indicates the correct position on the scale.
  iii. If necessary, reposition the knob on the shaft.
  iv. Repeat this test for other cassettes in use.
  v. If the scale is worn or not legible, use a marker pen to indicate positions for common cassettes in use. Order a new scale from the service department.
  vi. While waiting for a new scale, ensure you have a spare collimator globe in stock
- Check the alignment of the collimator lamp and X-ray beam. This should be checked through 180 degrees rotation of the collimator.
- To test the rotation accuracy of the light beam;
  i. Rotate the collimator 90 degrees in either direction.
  ii. With the light on, open the collimator so an average size field is projected on the tabletop. For example, a 24/30 cm cassette size.
  iii. Place markers to indicate the light beam position.
  iv. Now rotate the collimator 180 degrees in the opposite direction. The light field should be within 1% or better, compared to the previous position. If not, refer to the collimator service notes.
- To test the alignment of the X-ray to the light beam;
  i. Place the X-ray alignment template on a 24/30 cm cassette.
  ii. Adjust the FFD to 100 cm.
  iii. Collimate the light beam to the outer 20 by 26 cm rectangle.
  iv. Make a low kV and mAs exposure.
  v. Develop the film.
  vi. Measure the distance where the X-ray does not coincide with the markers. Any error should be inside the compliance requirements for the country. See **module 7.2 page 110**.
  vii. Repeat this test, with the collimator rotated 90 degrees clockwise, and 90 Degrees counter clockwise.
  viii. If there is an alignment problem, see **module 7.2 page 110**.
- Check operation of the 'dark current' shutter. This is an additional lead shutter fitted close to the focal spot. Its purpose is to block all radiation except when making a radiographic exposure. (In some mobiles, the shutter is retracted immediately prior to the exposure, in other mobiles the shutter is retracted during preparation.)
  i. Place the X-ray alignment template on a 24/30 cm cassette.

ii. Adjust the FFD to 100 cm
iii. Adjust the light beam to the central markers.
iv. Set 90 kV on the control panel, and press the charge button.
v. When the mobile indicates 'ready', or 'charging completed', do not expose or enter preparation. Instead, reset the kV down to 60 kV.
vi. A low mA X-ray exposure is produced when the kV is reset. Radiation is blocked in this mode by the dark-current shutter in the collimator.
vii. Wait till the control again indicates 'ready', and then process the film. The film should be clear, and not indicate any patterns from the alignment test phantom.

g. mAs calibration

- **Note.** This only applies to mobiles fitted with an mAs control, and a kV meter.
- A CD mobile has a direct relationship of mAs and kV. During an exposure, the kV will drop by one kV per mAs. (This is for a one microfarad mobile)
- Ensure the collimator is fully closed.
- Select 90 kV and 20 mAs
- Once charging is completed, make an exposure, observing the kV meter. There should be a drop from 90 kV to 70 kV.
- (In some cases a smaller drop of kV may occur. For example, from 90 to 72 kV. This is due to capacitor manufacturing tolerance.)
- Select several other combinations of kV and mAs and repeat the above test.

h. Radiation reproducibility

- This test should be carried out after the film processor has received its general maintenance. The test films can be used for comparison with future tests.
- Adjust the FFD to 100 cm.
- Place the stepwedge on a 24/30 cm cassette.
- Several exposures can be made on the one piece of film. Place two pieces of lead rubber on top of the cassette, positioned against either side of the stepwedge. As the stepwedge is repositioned, the lead rubber prevents unwanted radiation entering the cassette.
- Select a suitable mAs and kV combination, and make a total of four exposures.
  i. After each exposure, reposition the stepwedge and lead rubber on the cassette.
  ii. Develop the film. As the exposure settings are the same for all exposures, the film should show very little variation.
  iii. If necessary, change kV or mAs so the film displays a good range of densities, then repeat this test.
- **Note.** If a step wedge is not available, a water phantom may be used. In which case a series of four films are required.
- Record all calibration settings used with the stepwedge in the logbook. Include the kV, mAs, FFD, and the cassette used. This will allow a quick set-up when this test is repeated. Save the films for future comparison tests.

MODULE 1.4

X-ray generator, portable unit

Aim

The aim is to provide routine maintenance procedures for a portable X-ray generator. Within the portable generator capabilities, a similar check is made as for a mobile generator. The portable generator may be either self-rectified, or else full wave rectified. The stationary anode X-ray tube and HT transformer are contained in a single housing. Maintenance includes mechanical operation of the portable, together with the X-ray tube and collimator. Instructions for repairs to the portable generator are provided in module 6.1 page 90.
(**Note:** Reference module page numbers refer to the title page.)

Objectives

On completion of this module, the student will be familiar with maintenance procedures for a portable X-ray generator. These procedures can be used as a version of quality control, together with the routine maintenance check-sheets provided in the appendix.

Contents

a. General precautions
b. Visual inspection of the control panel, power off
c. Mechanical and electrical inspection, power off
d. Operation inspection of the control panel, power on
e. X-ray tube and collimator
f. mA calibration
g. Radiation reproducibility

Equipment required

- Basic tool kit.
- X-ray alignment template.*
- Stepwedge.*
- 24/30 cm Cassette.
- Two pieces of lead rubber.
- Aerosol spray lubricant.
- Cleaning solvent.
- Cloth.

* The template and stepwedge is described in appendix 'B' page 169.

a. General precautions

- **Before removing any covers, or testing any wires or connections, ensure the system is switched off, and unplugged from the power point.**
- **If the power plug has loose connections, have an electrician check the plug. The plug may be assembled incorrectly.**

b. Visual inspection of the control panel, power off

- Check all knobs and switches. Where knobs have a pointer attached, check that the pointer aligns correctly at all positions of the indicated scale.
  Tip. Check the pointer at full clockwise and counter clockwise positions of the knob. Look for possible loose knobs, or for push button switches that may tend to stick.
- Where extra labels have been attached, are these labels still relevant? If so, are they in good condition?
- The generator will often have an analogue line-voltage and mA meter. These meters should read zero on power off. If necessary, the meter zero position may be adjusted by the centre screw. Tap the meter gently first, to ensure the meter is not sticking.
- **Caution;** check the meter zero position when the control is mounted, or placed, in its usual position.
- With the aid of a suitable solvent, clean off the residue left behind from sticking plaster, and pieces of sticky tape.

c. Mechanical and electrical inspection, power off

- Look for any loose panels or sections. Pay particular attention to the mounting of the collimator. With a screwdriver, check for possible loose screws, particularly with the tube support arm and the vertical bearing tracks.
- Check the operation of the height adjustment system. Does it operate smoothly without binding or sticking?
- Check the action of the height adjustment bearings. Are there any visible gaps between the bearings and the track surface?
- Are there any 'clunking' noises or 'jerking' movements, when the X-ray tube is positioned? This can indicate damaged bearings.
- Check for possible loose lock handles, and ensure the locks have an adequate range of adjustment.
- Spray the height adjustment tracks and bearings with a light aerosol lubricant. Wipe down afterward, so only a thin oil film is left on the height adjustment tracks.
- Examine carefully all plugs and sockets attached to cable ends. The outer insulation of cables should not be pulled out from the cable clamp.
- Check the condition of the power cable. If necessary, remove the plug cover, and ensure terminations are tight, and no connections are stretched or have broken strands. *Should the cable exhibit excessive twisting, or have cracks in the insulation, ask an electrician to replace the cable or plug.*

d. Operation inspection of the control panel, power on

- Check all indicator lamps operate in each mode of operation.
- Check the range of adjustment of line voltage, to ensure this may be set approximately 10% above or below the optimum position.
  Note. Adjustment of the line voltage can directly affect the radiographic kV.

e. X-ray tube head and collimator

- Look for possible oil leaks.
- Check operation of the collimator lamp timer. (If fitted)
- With clockwork lamp timers, check for possible sticking of the clockwork.
- To test the alignment of the X-ray to the light beam;
  i. Place the X-ray alignment template on a 24/30 cm cassette.
  ii. Adjust the FFD to 100 cm.
  iii. Collimate the light beam to the outer 20 by 26 cm rectangle.
  iv. Make a low kV and mAs exposure.
  v. Develop the film.
  vi. Measure the distance where the X-ray does not coincide with the markers. Any error should be inside the compliance requirements for the country. See **module 7.2 page 110**.
  vii. Some units may be fitted with a rotating collimator. Repeat this test, with the collimator rotated 90 degrees clockwise, and 90 degrees counter clockwise.
  viii. If there is an alignment problem, see **module 7.2 page 110**.

f. mA calibration

- **Note.** Depending on system design, mA selection may be linked to the kV knob. Other units may have an independent selection of mA.
- Ensure the collimator is fully closed.
- Select 60 kV and the maximum associated mA station. Select an exposure time of 1.0 second.
- Commence preparation, and expose when preparation is complete. Observe the mA meter, and record the indicated value.
- Repeat this test for all other mA and kV combinations. Record the results.
- If mA on any position has an error of more than 10%, recalibration is required. On some systems, this may be accessed with a screwdriver through an access hole. Please refer to the operation or service manual before adjusting.
- If in doubt, contact the service department for advice.

g. Radiation reproducibility

- This test should be carried out after the film processor has received its general maintenance.
- Place a stepwedge on a 24/30 cm cassette.
- Several exposures can be made on the one piece of film. Place two pieces of lead rubber on top of the cassette, positioned against either side of the stepwedge. As the stepwedge is repositioned, the lead rubber prevents unwanted radiation entering the cassette.
- Adjust the FFD to 100 cm.
- Select a suitable mAs and kV combination, and make a total of four exposures.
  i. After each exposure, reposition the stepwedge and lead rubber on the cassette.

ii. Develop the film. As the exposure settings are the same for all exposures, the film should show very little variation.
iii. If necessary, change kV or mAs so the film displays a good range of densities, then repeat this test.

- **Note.** If a step wedge is not available, a water phantom may be used. In which case a series of four films are required.

Record all calibration settings used with the step-wedge in the logbook. Include the kV, mAs, FFD, and the cassette used. This will allow a quick set-up when this test is repeated. Save the films for future comparison tests.

MODULE 2.0
X-ray tube-stand

Aim

The aim is to provide routine maintenance procedures for the X-ray tube-stand, or suspension. The instructions provided are for the floor ceiling tube stand. Most of these procedures can also be applied to a ceiling mounted tube suspension. Repair procedures are provided in module 7.0 page 99.

(**Note:** Reference module page numbers refer to the title page.)

Objectives

On completion of this module, the student will be familiar with maintenance procedures for the X-ray tube-stand. These procedures can be used as a version of quality control, together with the routine maintenance check-sheets provided in the appendix.

Task 6, 'X-ray tube-stand maintenance', should be attempted on completion of this module.

Contents

a. General precautions
b. Mechanical and electrical inspection
c. Tube-stand lateral centre
d. Tube stand command arm, or panel

Equipment required

- Basic tool kit.
- X-ray alignment template. *
- 24/30 cm cassette.
- Aerosol spray lubricant.
- Cleaning solvent.
- Cloth, for cleaning.

* The template is described in appendix 'B' page 169.

a. General precautions

- **Electrical safety.**
  i. In most installations the tube-stand power will come from the generator, but in some installations, switching off the generator does not remove power from the tube stand.
  ii. **Before removing any covers, ensure the generator is switched off, and the room power isolation switch is also turned off.**
  iii. This also applies if testing wiring connections, or electrical components.
- If removing an X-ray tube, or collimator.
  i. **See module 7.1 page 104, and module 7.2 page 110.**
  ii. Ask an electrician or electronics technician for assistance.
  iii. Do **not** rely on the vertical lock system.
  iv. Attach a rope so that the system cannot move upwards, once the weight of the collimator or X-ray tube is removed.
  v. The X-ray tube is heavy. Removal or replacement requires two people.
  vi. Make a diagram of electrical connections. Attach labels to wires or high-tension cables. This is to ensure correct connection when an X-ray tube or collimator is replaced.
  vii. Place all screws or other small parts in a box, so they are not lost.
- Do not place a ladder against a tube stand. The tube stand may suddenly move.
- An adjustment to any tube-stand bearing requires skill, and good mechanical knowledge. When a problem is identified, request a mechanic, or the service department, to make the required adjustments.

b. Mechanical and electrical inspection

- Inspect for any loose panels or sections. Pay particular attention to the collimator and the control panel.
- Check the tube-stand suspension, tracks and bearings.

PART II. ROUTINE MAINTENANCE MODULES

i. With the X-ray tube set to minimum height, check the vertical suspension wire rope for broken strands. CAUTION, do not test with bare fingers. Test by rubbing the cables up and down with a piece of rag.
ii. With the vertical lock released, the X-ray tube should balance in the vertical direction. It should need the same effort to move either up or down.
iii. Check the action of the tube-stand bearings. Are there any visible gaps between the bearings and the track surface?
iv. Are there any 'clunking' noises or 'jerking' movements, when the X-ray tube is positioned? This can indicate damaged bearings.
v. Check the vertical guide rails. Look for loose mounting screws.
vi. Spray the tube stand vertical guides and bearings with a light aerosol lubricant. Wipe down afterward, so only a thin oil film is left on the vertical guide rails.
vii. Clean any accumulated dirt on and inside the floor track. Spray the track and tube-stand floor-bearings with a light aerosol lubricant. Wipe off any excess.
viii. Look for loose mounting screws along the floor rail.
ix. Observe the position of the bearings on the ceiling rail. These should be fully engaged along the full length of the rail. Check the rail is properly fastened in place, and does not move.

- Check for loose mechanical lock handles, and ensure manually operated locks have an adequate range of adjustment.
- Test operation of the electromagnetic locks. There should be no hesitation in operation, nor should the lock 'stick on'. In some cases the surface of the lock will require cleaning, to obtain a better 'grip' when operated.
- HT cables should be supported at the X-ray tube, to minimize twisting or pulling at the cable end.
- Pay attention to the cabling from the X-ray tube and tube stand. Any movements of the system should not cause any stress or stretching of the cables. Nylon or plastic cable-ties should be used to secure loose cables, not sticking plaster.
- Examine carefully all places where cables pass into different sections of the tube-stand.
  i. Where the cables pass though holes in a metal cover, there should be protective inserts to avoid damaging the cable insulation.
  ii. Look for possible damage to the outer insulation of electrical cables.
  iii. Check electrical cables where they enter plugs and sockets, the outer insulation may be pulled back, exposing individual conductors.

c. Tube-stand lateral centre

Lateral centring over the Bucky table should be checked in both directions. In some cases this may be accurate only when approached from one direction.

i. Tape a thin piece of wire, or a paper clip, to the centre of a 24/30 cm cassette. Place the cassette in the Bucky.
ii. Position the X-ray tube to the lateral centre position.
iii. Adjust the table top also to the lateral centre position.
iv. Bring the collimator face to rest on the tabletop, and ensure it is flat against the tabletop. Then rise to 100 cm S.I.D.
v. As the collimator moves away from the tabletop, check that the light beam remains central to the tabletop. If not, adjust the tube angle a small amount so the light beam remains in position.
vi. If the tube-stand centre position appears incorrect, this may need adjustment. Before adjusting, continue with the rest of these checks.
vii. Place the X-ray alignment template on the centre of the tabletop.
viii. Adjust the light beam to the template markers.
ix. Select a low kV and mAs, expose and develop the film.
x. The radiation field should be centred to the template markers. If not, the collimator requires adjustment. This should be corrected before any adjustment to the tube-stand centre. See **module 7.2 page 110**.
xi. The position of the template marker is compared to the wire marker on the cassette. This checks the tabletop centre accuracy. If not correct, see **module 8.0 page 121**.
xii. If the tube stand centre is not correct, see **module 7.0 page 99**.

- After checking lateral centring to the table Bucky, check the centring to the wall Bucky.
  i. Keep the X-ray tube rotation in the trunnion rings at the same setting for the Table Bucky.
  ii. Bring the collimator close to, or up against the wall Bucky. The light field should be centred to the Bucky-centre mark.
  iii. Move the tube stand away from the Bucky to the distance normally used. The light beam should remain centred.

iv. In case of a small error, a compromise adjustment of the tube rotation in the trunnion rings may be made. Otherwise, see **module 7.0 page 99**, and **module 8.0 page 121**.

d. Tube stand command arm, or panel

- The X-ray tube trunnion-ring rotation-lock should operate firmly, and prevent unwanted rotation.
- Hand grips should not be loose.
- The indicator for tube angle should rotate smoothly, and not hesitate before changing its position.
- Check all indicator lamps and switches for correct operation.
- Check alignment of the Bucky centre light. (When fitted.)
- Ensure all labels are legible.
- With a tape measure, check for correct indication of the focal spot to Bucky distance. (FFD.)
- Rotate the tube head and check the angulation indicator operates smoothly and does not stick.
- Clean and remove remains of sticking plaster, adhesive tape, etc.

PART II. ROUTINE MAINTENANCE MODULES

TASK 6
X-ray tube-stand maintenance

The X-ray tube stand in room 1 is due for its maintenance check.
Carry out this check, using the maintenance checklist provided in appendix 'D' of this workbook as a guide.

What is the tube stand model No? _____ Serial No? _____

Was it necessary to make any adjustments? (Provide details) _____
_____
_____
_____

Are there any areas still requiring attention? _____
_____
_____
_____

Tutor's comments

                                                            Satisfactory/Unsatisfactory
_____   _____
Signed   Date
            Tutor

MODULE 2.1

X-ray tube

Aim

The aim is to provide routine maintenance procedures for the X-ray tube. This includes techniques to improve the high-voltage performance or reliability of the X-ray tube. Fault diagnostic procedures are provided in module 7.1 page 104.
(**Note:** Reference module page numbers refer to the title page.)

Objectives

On completion of this module, the student will be familiar with maintenance procedures for the X-ray tube. These procedures can be used as a version of quality control, together with the routine maintenance check-sheets provided in the appendix.

Contents

a. General precautions
b. X-ray tube inspection
c. X-ray tube 'seasoning'

Equipment required

- Basic tool kit.
- Cable ties.

a. General precautions

- **Before disconnecting any wires, or removing a cover, always ensure power is turned off and unplugged from the power point. If the equipment is part of a fixed installation, besides switching the generator power off, ensure the isolation power switch for the room is also switched off.**
- During the seasoning technique, make sure the collimator is closed. Aim the X-ray tube away from the X-ray control.
- If a problem occurs during seasoning, **stop**. Depending on the symptoms, see **module 7.1 page 104**, or **module 7.3 page 117**.

b. X-ray tube inspection

- Check rotation of the X-ray tube in the trunnion rings. The locking device should hold the housing firmly in place, but allow free rotation on release.
- Ensure no attachments, such as a command arm control panel, or collimator, have become loose.
- Examine electrical cables to the X-ray tube. Ensure they are securely clamped into position, and not subject to being pulled. Where cables pass into the housing, they should be protected from sharp edges.
- Inspect the HT cables for any sign of damage to the safety earth shield, at the X-ray tube cable ends.
- Ensure the HT cable ends are firmly inserted into the X-ray tube, and the securing ring nut is not loose.
  Note. In some systems there is a locking screw on the side of the ring nut. Undo this screw first, and then check the ring nut is fully tightened, then refasten the locking screw. This check is most important if the X-ray tube or cables have recently been replaced.
- Where there is evidence of twisting or pulling on the HT cables, particularly at the X-ray tube receptacle, investigate means of providing additional support. If necessary, contact the service department.
- Examine the X-ray tube housing for any oil leaks.
- At the generator, go into preparation, then release preparation without exposing. Listen to the anode rotation for excessive noise.

- If a high-speed tube, check that the anode brake cycle operates normally.
    i. Some systems use direct current (DC brake) to bring the anode to rest.
    ii. Other use alternating current, to bring the speed down to 3000 rpm, after which the anode coasts to rest.

c. X-ray tube seasoning

- This is also called 'ageing', and is a process to reduce residual gas in the X-ray tube. Seasoning improves the stability of the tube, when operated at high kV.
- Seasoning should always be performed if a new X-ray tube is installed, or has not been used for more than one month. The same applies where the tube has not been used over 80~90 kV for some time, and then it is desired to use 110 kV or higher.
- If using an X-ray tube of 125 kV capacity at 110 kV or higher, seasoning should be performed each day prior to use. If the tube is rated at 150 kV, the same applies if operating above 125 kV.
- During seasoning, an X-ray tube may at first appear unstable. After two to three exposures, the tube should now be stable. If not, see **module 7.1 page 104**.
- Many manufacturers specify a seasoning procedure. This can be found in the X-ray tube operation or installation manual. Operation or installation manuals for mobile generators may include a seasoning procedure.
- Table 2–a is suggested for a 150 kVp X-ray tube with a fixed installation.
- Table 2–b is typical for a mobile generator, with a 125 kVp tube.

Table 2–a. Seasoning technique for a 150 kVp X-ray tube

Step	kV	mAs	Times	Pause time (Seconds)
1 #	60	20	2	40
2 #	70	20	2	40
3 #	80	20	2	40
4	90	20	2	40
5	100	20	2	40
6	110	20	2	40
7	115	20	2	40
8	120	20	2	60
9	125	20	2	60
10	130	10	4	60
11	135*	10	4	60
12	140**	10	2	60

Note
i. Recommended output is 200 mA and 0.1 sec for 20 mAs. For 10 mAs use 100 mA and 0.1 sec.
ii. # Steps 1, 2, and 3 are only required if the tube is just installed, or has not been in use for more than one month.
iii. * If, for example, you never use above 125 kV, then ignore steps 11 and 12.
iv. ** Although an X-ray tube is rated at 150 kV, this is the absolute maximum rating. This can reduce as the tube becomes worn, and especially as metal evaporation collects on the glass. Operation above 140 kV may result in premature failure.

Table 2–b. Seasoning technique for a 125 kVp tube

Step	kV	mAs	Times	Pause time (Seconds)
1 #	60	10	2	40
2 #	70	10	2	40
3 #	80	10	2	40
4	90	10	2	40
5	100	10	2	40
6	110	10	2	50
7	115	10	2	50
8	120*	10	2	60
9	125*	10	2	60

Note
i. The majority of mobiles provide selection of mAs only. If mA and time selection is available, then aim for exposure times less than 0.3 sec.
ii. # Steps 1, 2, and 3 are only required if the tube is just installed, or has not been in use for more than one month.
iii. * If, for example, you never use above 110 kV, then ignore steps 8 and 9.

MODULE 2.2
X-ray collimator

Aim

The aim is to provide routine maintenance procedures for the X-ray collimator. These maintenance suggestions are for a standard collimator used on a standard tube-stand. Adjustment procedures are provided in module 7.2 page 110.
(**Note:** Reference module page numbers refer to the title page.)

Objectives

On completion of this module, the student will be familiar with maintenance and performance checks for the X-ray collimator. These procedures can be used as a version of quality control, together with the routine maintenance check-sheets provided in the appendix.

Task 7,'X-ray tube and collimator maintenance', should be attempted on completion of this module.

Contents

a. General precautions
b. General maintenance
c. Alignment tests

Equipment required

- Basic tool kit.
- X-ray alignment template.*
- 24/30 cm cassette.
- Cloth, for cleaning.
- Detergent.

* The template is described in appendix 'B' page 169.

a. General precautions

Whenever changing a collimator lamp, always ensure power is turned off and equipment unplugged from the power point. If the equipment is part of a fixed installation, besides switching the generator power off, ensure the isolation power switch for the room is also switched off.

b. General maintenance

- Check electrical cable to the collimator. Ensure the cable entry is protected against sharp edges of the collimator housing. Rotation of the collimator should not stretch or pull the cable.
- Check the operation of the collimator blades. These should stay in position when adjusted, and not slip if the X-ray tube is repositioned. If adjustment of the clutch or brake is required, see **module 7.2 page 110**. (Do **not** rely on adhesive tape to hold the knob in position.)
- Check the operation of the collimator lamp timer. With clockwork systems, look for possible sticking of the mechanism.
- Evaluate the intensity of the light beam from the collimator. If too dim, see **module 7.2 page 110**.
- Check the type of globe fitted, and make sure you have a spare globe in stock. **Note.** Some globes may appear similar but have the filament in a different position. See **module 7.2 page 110**.
- The following precautions should be observed if changing the globe.
  i. **Ensure power to the generator and/or tube stand is turned off.**
  ii. If a globe has just failed, wait for it to cool down.
  iii. When unpacking and inserting a new globe, do not handle it directly. Instead use a tissue or a piece of cloth so your fingers do not touch the globe. This is very important when handling Quartz-Halogen globes.

c. Alignment tests

- Check the crosshair alignment of the front transparent cover.
  i. With the collimator blades almost closed, the crosshair should be in the centre of the light field. Check at both horizontal and vertical settings.
  ii. If adjustment is required, on most collimators, the cover may be moved after loosening the four retaining screws. (In some cases, the cover may at first stick in place.)
- Check the Bucky centre light, if fitted. If out of alignment, see **module 7.2 page 110**.
- The collimator has a scale combined with the adjustment knob to indicate the field size. The knob can slip on the shaft, or not be correctly positioned after replacing a collimator globe.
  i. Place a 24/30 cm film on the tabletop, and position the X-ray tube 100 cm above the tabletop.
  ii. With the collimator light switched on, adjust the light field to the film size.
  iii. Check that the knob pointer indicates the correct position on the scale.
  iv. If necessary, reposition the knob on the shaft.
  v. Repeat this test for other films in use.
  vi. If the scale is worn or not legible, contact the service department and obtain a new scale. In the meantime, use a marker pen to indicate positions for common cassettes in use.
  vii. Attention to the scale is important. The lamp might fail, and the spare globe has already been used.
- To test the alignment of the X-ray to the light beam;
  i. Place the X-ray alignment template on a 24/30 cm cassette.
  ii. Collimate the light beam to the outer 20 by 26 cm rectangle.
  iii. Make a low KV and mAs exposure.
  iv. Develop the film.
  v. In many cases the collimator is enabled to rotate. Repeat the above test, with the collimator rotated 90 degrees clockwise, and then 90 degrees counter clockwise.
  vi. If there is an alignment problem, see **module 7.2 page 110**.
- Does the alignment meet the required compliance? Two versions are provided as an example only. The actual compliance requirement will depend on individual country regulations.
  i. *The X-ray field edges should not deviate by more than 2% of the distance between the plane of the light field and the focal spot.*
  $[a1] + [a2] \leq 0.02 \times S.$
  $[b1] + [b2] \leq 0.02 \times S.$
  Where S is the distance from the focal spot, a1 and a2 are the two sides on one axis, and b1 and b2 are the two sides of the other axis.
  For example, at a FFD of 100 cm, if the two vertical edges of the light field were displaced by 1.0 cm, this would be at the limit of acceptance. *If only one edge was displaced, then 2.0 cm is at the limit of acceptance.*
  ii. Another version has a different requirement.
  The total misalignment of any edge of the light field with the respective edge of the irradiated field must not exceed 1% of the distance between the plane of the light field and the focal spot.
  For example, at a FFD of 100 cm, the maximum displacement of **any** edge should be less than 1.0 cm.

TASK 7

X-ray tube and collimator maintenance

The X-ray tube and collimator in room 1 is due for its maintenance check.
Carry out this check, using the maintenance checklists provided in appendix 'D' of this workbook as a guide.

What is the X-ray tube model No? _____ Serial No? _____

Focal spot sizes are; Broad focus. _____ Fine focus. _____

Collimator model No? _____ Serial No? _____

Examine carefully the HT cables as they enter the X-ray tube. Note, you will need to undo the retaining ring nut, and slide back the cable support clamp.

Is the safety earth shield in good condition? _____
_____

Examine the collimator globe. How would you describe this globe if requesting a replacement? (**Hint**, look in the parts manual) _____
_____

Carry out tests for X-ray beam and light beam alignment. Is this correct for all rotation positions of the collimator? _____
_____

If alignment is outside acceptable limits, explain how to readjust the collimator. _____
_____
_____
_____

Were any adjustments required? Provide details. _____
_____
_____

Do any areas still require attention? _____
_____

Tutor's comments

_____

Satisfactory/Unsatisfactory

Signed _____   Date _____
       Tutor

PART II. ROUTINE MAINTENANCE MODULES

MODULE 3.0
Bucky table and vertical Bucky

Aim

Aim is to provide routine maintenance procedures for the Bucky table and vertical Bucky. Adjustment and repair procedures are provided in module 8.0 page 121.
(**Note:** Reference module page numbers refer to the title page.)

Objectives

On completion of this module, the student will be familiar with maintenance and performance checks for the Bucky table and vertical Bucky. These procedures can be used as a version of quality control, together with the routine maintenance check-sheets provided in the appendix.

Contents

a. General precautions.
b. Bucky table.
c. Potter Bucky.
d. Vertical Potter Bucky.

Equipment required

- Basic tool kit.
- Torch.
- Aerosol spray lubricant.
- Cloth, for cleaning.
- Detergent.

a. General precautions

Please take the following precautions.

- **Before removing a cover, always switch the generator power off, and ensure the isolation power switch for the room is also switched off.**
- When removing the cover from a vertical Bucky, make sure the Bucky cannot move upwards when the cover is removed. For example, attach a rope to hold it in position, or remove the cover with the Bucky set to maximum height.
- Keep all screws, or other small parts in a container, to avoid loss.

b. Bucky table

- Examine the physical condition of the table. Clean the remains of adhesive tape etc from the table body.
- **Hint.** Car polish, designed to 'rejuvenate' faded and oxidised paint, can often improve the appearance of an older table. Silicon furniture polish can assist in removing scuffmarks and fingerprints etc.
- Check for loose screws on the tabletop profile rails. The rails can become loose due to using a compression device.
- Examine the condition of the compression device. Check for correct operation. Remove the band from the mechanism, and have it laundered.
- Check the operation of switches and indicator lamps.
- With an elevating Bucky table, use a tape measure to check the table height at the centre stop position.
- Check the operation of the magnetic locks.
  Note. Some movements may have two or more magnetic locks. Carefully observe these locks and ensure all locks are actually in operation. If adjustment is required, see **module 8.0 page 121**.
- Check the operation of the tabletop lateral centre-stop. Where this is mechanical, the spring tension may need adjustment. In case of operation by the magnetic locks, the stop position is normally

controlled by a microswitch. Adjustment of this microswitch can control the width and position of the centre-stop operating position. See **module 8.0 page 121**.
- Move the tabletop in all positions. If there is scraping or binding in some positions, check the position of the locks. Look also for a faulty bearing.
- Spray the bearing tracks and bearings with a light aerosol lubricant, then clean the residue away from the tracks, so only a thin film is left.

c. Potter Bucky

- Move the Bucky to both ends of the table. Check that the Bucky carriage operates smoothly, and that the Bucky lock operates correctly in all positions across the table.
- Electrical cables to the Bucky should be firmly attached at the Bucky, and no twisting or puling occurs on the cable, at any position of the Bucky.
- Where there is a folding support arm for the connecting cable, look for possible binding or excessive 'droop'. This can indicate loose mounting screws.
- Spray the Bucky track with a light aerosol lubricant, and then wipe the residue from the track.
- Remove the Bucky tray. With a torch, examine the Bucky interior for lost film markers.
- Look for loose screws holding the tray handle. Take care not to over-tighten, as this might damage the thread.
- Test the action of the Bucky tray cassette clamps. If they do not hold the cassette firmly, see **module 8.0 page 121**.
- Spray the moving sections on the underside of the Bucky tray with a light aerosol lubricant, move the cassette clamps in and out, and then clean off all residue. Take care that no residue appears on top of the tray.
- Test the grid oscillation.
  i. At the generator, select the lowest mA station, 50 kV, and exposure time of 1~2 seconds.
  ii. Ensure the collimator is closed, and the tube is positioned away from the Bucky. Then make an exposure with the Bucky selected.
  iii. During the exposure, check for smooth operation of the grid.
  iv. Watch for any shaking, or vibration of the Bucky, as the grid reverses its movement. Or else, just as the grid first starts to move.
  v. Should shaking or vibration occur, this can cause reduced sharpness of the radiograph. If this occurs, contact the service department for advice.

d. Vertical Potter Bucky

- The vertical Bucky should be checked in the same manner as the table Bucky, but with the following provision for retrieving lost film markers. These markers can fall into the motor section at the bottom of the Bucky, and may cause a problem. To inspect, it is necessary to remove the front cover.
  i. With power off, ensure the vertical lock firmly holds the Bucky in place. If not, keep the Bucky in position by tying with a rope, or by adding extra weights.
  ii. Carefully examine the method of attaching the front cover. In most cases this is a series of screws around the front cover. Other systems may attach by screws on the top and bottom sides.
  iii. If separating profile rails from the front cover, make a small mark so they can be returned to the same position, including left and right, on re-assembly.
  iv. After removal of the front cover, look carefully for any film markers. A torch will help to locate them.
  v. Re-assemble the front cover, taking care not to over-tighten any screws.
- Check operation of the vertical lock.
- Vertical movement. Check and lubricate the vertical track. Wipe off any excess.
- Check the rotation or tilt lock. (Only on Bucky with this option)

PART II. ROUTINE MAINTENANCE MODULES

MODULE 3.1
Tomography attachment

Aim

Aim is to provide routine maintenance procedures for a tomographic attachment. This may be fitted to a standard tube stand, or integrated with a Bucky table. Adjustment and repair procedures are provided in module 8.1 page 127.
(**Note:** Reference module page numbers refer to the title page.)

Objectives

On completion of this module, the student will be familiar with maintenance and performance checks for the tomographic attachment. These procedures can be used as a version of quality control, together with the routine maintenance check-sheets provided in the appendix.

Contents

a. General precautions.
b. Mechanical and electrical inspection
c. Operation test.
d. Performance test.

Equipment required

- Basic tool kit.
- Tomography resolution tool.*
- Or, tomography test tool.*

* The tomography resolution tool is described in appendix 'B' page 169.
* A tomography test tool is described in the WHO 'Quality assurance workbook.'

a. General precautions

Please take the following precautions.

- **Before removing a cover, always switch the generator power off, and ensure the isolation power switch for the room is also switched off.**
- Keep all screws, or other small parts in a container, to avoid loss.

b. Mechanical and electrical inspection

- Inspect all sections of the attachment for loose or missing screws. This can be a common problem with some coupling arms.
- Inspect connecting cables. The outer insulation of electrical cables should not be pulled out from the plug or socket cable clamps. Plugs or sockets should be in good condition and fit firmly into position.
- Check the rotation lock on the tube stand. With the rotation lock 'off', the X-ray tube should rotate easily.
- The Bucky lock should disengage, and the Bucky move freely along the guide rails.
- When assembling the unit, pay attention to bearings or pivot points, which the coupling arm passes through. The arm should move up and down freely.
- When attaching the coupling arm to the tube stand, ensure the clamp holds the arm firmly.
- With the coupling arm attached, switch off the tube stand longitudinal-lock. Disengage the tomographic motor, and push the tube stand by hand to each end of the normal tomographic travel. Check that the tube stand and Bucky moves smoothly, and there is no sudden jerking to the movement. (This test depends on individual system design, and may not be possible with some units)
- Check the fulcrum height adjustment. This should operate smoothly, and have a clear indication of setting height. This may be a scale and pointer, or a digital readout in later systems.

c. Operation test

- This test may be performed by using the 'Reset' switch to drive the unit to either direction. If this facility is not available, or is only available at a slow speed, then it is necessary to produce a radiographic exposure. In this case, select minium kV and mA, an exposure time sufficient to allow full operation, and ensure the collimator is closed.
- Operate the tube stand travel in all speeds and modes of operation. Check for the following possibilities.
    i. Lack of positive drive during the start of travel. For example, the drive system is slipping.
    ii. Poor stopping position or overshoot at the end of travel, especially when operated at maximum speed and angle.
    iii. Shaking or uneven travel through the active area of movement. Some initial shaking may occur at the start of travel, but should stop before reaching the exposure position.

d. Performance test

- This is a test for layer height, and providing the tomography resolution test is used, an evaluation of image sharpness.
    i. Select a low kV and mA position.
    ii. Select a medium angle. Set exposure time to suit the speed and angle.
    iii. Adjust the fulcrum height to the centre height position of the test piece.
    iv. Perform a tomographic exposure.
    v. Examine the film. There should be even blurring of the test objects above and below the test piece centre, which should be sharply defined.
    vi. If the top and bottom paper clips are not equally blurred, then adjust the fulcrum height a small amount, and repeat this test.
    vii. If the central paper clips are blurred in any direction, this can indicate shaking or uneven movement of the Bucky, or tube stand. See **module 8.1 page 127**.
    viii. Repeat this test using the maximum angle and speed, then again using minimum speed.
    ix. Retain the films as a record. Include on the films the exposure setting and mode of operation.
- In case the fulcrum height is incorrect, or the image sharpness is poor, record the test procedures used. Include the tomograph speed and angle. Contact the service department for advice.

MODULE 4.0
Fluoroscopy table

Aim

Aim is to provide routine maintenance procedures for the fluoroscopy table. These procedures are intended for an under-table tube fluoroscopy table. The under-table Bucky maintenance procedures are described in **module 3.0 page 53**.

Fluoroscopy TV maintenance is provided in **module 4.1 page 60**. Adjustment and repair procedures are provided in **module 9.0 page 130**.
(**Note:** Reference module page numbers refer to the title page.)

Objectives

On completion of this module, the student will be familiar with maintenance and performance checks for the fluoroscopy table. These procedures can be used as a version of quality control, together with the routine maintenance check-sheets provided in the appendix.

Contents

a. General precautions.
b. Mechanical and electrical inspection.
c. Operation test, table body.
d. Operation test, serial-changer (Spot filmer).
e. X-ray beam alignment.
f. Cleaning.

Equipment required

- Basic tool kit.
- Torch.
- Spirit level.
- Cloth, for cleaning.
- Detergent.

a. General precautions

- **Before removing any cover, always switch the generator power off, and ensure the isolation power switch for the room is also switched off.**
- If removing a cover, or dismantling any section, place the screws in a container to avoid loss.

b. Mechanical and electrical inspection

- Make a general inspection of the table body and serial-changer. Tighten any loose screws, or panels and fittings.
- Check the rails holding the tabletop in place. Tighten any loose screws.
- Examine all suspension system cables and chains for any sign of wear, or uneven tension in the case of dual systems.
- Check electrical cables, particularly at the rear of the table. Pay special attention to cables that may be twisted or tangled. It may be necessary to remove existing cable ties, reposition the cables, and then install fresh cable ties. (Use plastic or nylon cable ties only)
- Where electrical cables enter the table, check that the cable clamps properly secure them. The protective outer insulation of the cable should not be pulled back, exposing inner conductors.
- The serial-changer should be able to move smoothly from the 'park' position, and lock firmly into place.
- Check the operation of all switches and operation lights on the serial-changer.
- Additional control switches are sometimes placed on the side of the table body. Patient trolleys can damage these switches.
- Older systems with a demountable image intensifier (II).
  i. The II should balance vertically when removed from the serial-changer.
  ii. The II should be held firmly in position when clamped to the serial-changer. Look for loose clamps.
- Is the serial-changer lead-rubber radiation shield in good condition? Can it be easily positioned?

- With the aid of a torch, make a careful examination for lost film markers inside the serial-changer.
- **The footrest requires careful attention. If there is any tendency for the locking mechanism to slip, this will require urgent correction.**

c. Operation test, table body

- Check operation of all locks. The serial-changer vertical lock should operate quickly when operated, and not 'stick on' when released.
- Check all switches and indicator lamps for correct operation. This especially includes those on the table body, which may be damaged by patient trolleys.
- Operate the tabletop in all positions. Listen for any unusual squeaks or bearing rattles.
- On tabletops fitted with a lateral centre stop, the stop position should be checked. Use the procedures described in **module 2.0 page 44**.
- Rotate the table to the vertical and Trendelenburg positions. Listen for unusual bearing or motor noise while the table is rotating. Does the table stop quickly after the rotation control is released, or does it tend to continue or 'coast' for a short while? This could indicate a failure of the motor brake system. Contact the service department for advice.
- The table should stop in the horizontal position, on returning from the vertical position. Use a spirit level to check the horizontal position.
- With the table tilted at maximum position in either direction, check the electrical cables are not pulled tight, or restrict the movement of the serial-changer up and down the table.
- Some tables have power assistance for movements, such as longitudinal movement of the serial-changer; this should be smooth and free from sudden jerks. See **module 9.0 page 130**.
- Some tables have safety anti-crash bars or flaps. Pushing against these safety devices should prevent the table from rotating.
- With the vertical or compression lock 'on', operate the tabletop longitudinal movement. On some tables, movement should not operate, and on others, the vertical lock should release on movement of the tabletop. If not, contact the service department. There may be a safety upgrade available.
- Place a standard 24/30 cm cassette in the serial-changer. With the vertical lock 'off', check for vertical balance when moved up and down.
- Rotate the table into the vertical position. With the serial-changer longitudinal lock 'off', check the serial-changer and tower assembly for balance, while moving it along the table body.

d. Operation test, serial-changer (Spot filmer)

- Compression cone movement should operate smoothly, and lock into position.
  i. If the movement is stiff or hesitant, try cleaning the slide tracks, then spray with an aerosol lubricant.
  ii. Operate several times, and wipe off any excess spray.
- Some serial-changers have manually operated 'close to film' shutters. These are coupled to the film format selection. If stiff or hesitant, clean and lubricate the tracks in the same manner as for the compression cone.
- Manually operated cassette movement.
  i. Load a 24/30 cm cassette in the tray.
  ii. Select a low kV and mAs setting at the X-ray control.
  iii. At the serial-changer, select the four-spot mode.
  iv. Set the collimator control to 'Automatic'.
  v. Advance the cassette, and expose all four divisions. Watch the cassette as it moves into the 3rd spot division. If the cassette 'slams' into position, a pneumatic 'damper' may need replacement. Contact the service department for advice.
  vi. Process the film, and check that the four spots are evenly distributed around the film, are the same size, and without overlap.
  vii. If incorrect, see **module 9.0 page 130**.
  viii. Provide a similar test for other split formats.
- Motorized cassette movement.
  i. Place a 24/30 cm cassette in the cassette carriage.
  ii. Press the 'load' button. The cassette should be withdrawn and stop smoothly. A 'bang' at the end of travel can indicate an adjustment problem, caused by the cassette carriage overshooting, and hitting the end stop. Contact the service department for advice.
  iii. Press the 'eject' button. The cassette should move to the 'eject/load' position, and stop smoothly. There should be no 'bang' at the end of travel to indicate travel 'overshoot'. Enough of the cassette should extend to allow easy removal.
  iv. Reload the 24/30 cm cassette.
  v. Select a low kV and mAs setting at the X-ray control.

vi. At the serial-changer, select the four-spot mode.
vii. Set the collimator control to 'Automatic'.
viii. Advance the cassette, and expose all four divisions. Watch the cassette as it moves into the 3rd spot division, for possible failure of the movement damper. (Only on some older systems)
ix. Process the film, and check that the four spots are evenly distributed around the film, are the same size, and without overlap. If incorrect, see **module 9.0 page 130**.
x. Repeat the above test for other split formats and film sizes.

e. X-ray beam alignment

- Install a 24/30 cm cassette, and select the four-spot mode.
- At the X-ray control, select a low level of fluoroscopy kV and mA. This should be just sufficient to see the position of the X-ray beam shown on the TV monitor. (Or the fluorescent screen)
- With fluoroscopy 'on', manually collimate the beam to the maximum four spot size.
  i. The edge of the beam should be sharply defined by the four-spot cone or by the 'close to film' shutters.
  ii. In case one side is less sharp, and shows movement with only a small adjustment of the collimator, then beam alignment is incorrect.
- Observe the beam alignment with the serial-changer at both minimum and maximum height positions above the tabletop.
- Repeat the above test with the table tilted to vertical position.
- **Note.** In most cases, beam alignment will shift a little when the table is moved from horizontal to vertical. The amount of misalignment is usually due to flexing of the table framework. In some cases this may be due to incorrectly adjusted bearings. Contact the service department for advice.
- If beam alignment is incorrect, see **module 9.0 page 130**.

f. Cleaning

- Due to the types of examinations, barium spills can leave deposits under the tabletop, or on the protective cover under the tabletop. Inspect with the tabletop moved to its maximum position in each direction.
- **Hint.** Shine a torch beam between the tabletop and the protective cover.
- Another cause of artefacts is contrast media from an IVP examination. This is sometimes found underneath the serial-changer, as well as the tabletop. The contrast media is not easy to see when it has dried out.
- The contrast residues may be cleaned with a mixture of household detergent and warm water. Use sufficient to just dampen the cleaning cloth.
- Fluoroscopy tables are subject to marks from patient trolleys etc.
  i. These marks and scratches can be reduced with the aid of car polish.
  ii. The type to use is one with a mild abrasive, advertized to 'Restore faded or chalky paint'.
  iii. Silicone furniture polish can help remove scuff-marks, fingerprints etc.

MODULE 4.1
Fluoroscopy TV

Aim

Aim is to provide routine maintenance procedures for the fluoroscopy TV system. These include checks for image sharpness, and the automatic brightness control. The basic TV imaging system consists of an image intensifier (II), TV camera, and monitor. Systems with greater complexity, such as DSA and electronic radiography, are not included. Adjustment and repair procedures are provided in module 9.1 page 135.
(**Note:** Reference module page numbers refer to the title page.)

Objectives

On completion of this module, the student will be familiar with maintenance and performance checks for the fluoroscopy TV system. These procedures can be used as a version of quality control, together with the routine maintenance check-sheets provided in the appendix.

Contents

a. General precautions.
b. Mechanical and electrical inspection, image intensifier.
c. Mechanical and electrical inspection, TV system.
d. Image sharpness.
e. Automatic brightness/kV control.

Equipment required

- Basic tool kit.
- Resolution test piece.*
- Plastic container for water phantom.

* A 'V' pattern test piece is described in appendix 'B' page 169.

a. General precautions

- **Before removing any cover, always switch the generator power off, and ensure the isolation power switch for the room is also switched off.**
- Do not remove the cover of the TV monitor. Dangerous voltages can exist for a considerable time after the monitor is switched off.
- Do not attempt any adjustments to the TV camera, unless under instruction by the service department.
- If removing a cover, or dismantling any section, place the screws in a container to avoid loss.

b. Mechanical and electrical inspection, image intensifier (II)

- On older systems with a ceiling suspension.
  i. If the unit is dismounted from the serial-changer, the vertical balance should be neutral.
  ii. When attached to the serial-changer, the clamp or latch system should hold the II securely, with minimum movement.
  iii. Up and down movement should be free, without binding or unusual noises.
  iv. The ceiling suspension unit should travel freely. (Some systems may hesitate before moving, this is normal.)
- Electrical cables to the II and TV camera should be securely attached. The cables should not be pulled, or stretched, during table movements.
- Push buttons and selection switches should have their function clearly marked. **Note.** Some knobs or control settings may be for an option only, and not be installed. This should be noted in the logbook.

c. Mechanical and electrical inspection, TV system

- Electrical cables should be securely attached to the monitor trolley. There should be no possibility of pulling against cable connections. This also applies in case of wall mounted plugs and sockets.
- Examine the video cable connection, both at the monitor and TV camera. The cable should be firmly

attached to the plug, including the outer earth shield. Moving the cable at the plug should not cause any 'flicker' or change in the TV image.
- Check the video-input 75 ohm termination-switch. In the case of a single video connection only, this should be 'ON'. If two or more monitors, then the switch on the end monitor should be 'ON', while the middle monitor, with two video connections, should be switched to the 'OFF' position. (In some cases, a termination plug is fitted to the unused video 'out' connection.)
- Ensure the monitor is securely fastened to the monitor trolley. This especially applies to monitor trolleys with a tilting platform.

d. Image sharpness

- An evaluation of image sharpness or focus is best carried out with a 'Line pair' gauge. The industry standard is one made from 0.1 mm lead.
- An alternate test piece for testing focus may be constructed from long sewing needles arranged in a 'V' pattern. See appendix 'B' page 169.
- There are several methods used to evaluate performance of an imaging system. The basic method described here is to indicate if resolution has drifted below an acceptable level.
    i. Tape the gauge onto the centre of the input face of the image intensifier. If access is difficult, then tape the gauge to the under surface of the serial-changer.
    ii. To avoid interaction with grid lines, attach the gauge so it is rotated approximately 25~45 degrees. (On some CCD TV cameras, this also avoids interaction between pixels.)
    iii. Lift the serial-changer to maximum height above the tabletop.
    iv. If the system has automatic kV control, this should be turned off. Set manual fluoroscopy kV to 50~55 kV.
    v. With fluoroscopy 'on', adjust kV or mA to obtain a normal brightness and contrast image on the monitor.
    vi. Carefully observe the line-pair patterns. The limiting definition is the line pair group that is reasonably visible, while the next group is completely blurred out. (This can sometimes be a good test of individual eyesight.)
    vii. If using the 'V' pattern test piece, measure the distance to the apex before blurring occurs.
    viii. Record the line pair resolution, or 'V' pattern distance obtained, and compare with any earlier tests.
    ix. Repeat this test for other field sizes if a multi-field II is installed.
    x. As a guide, with a 9" image intensifier, resolution should not be less than 1.0 line pairs/mm. (Typical resolutions for current systems are 1.4 line pairs/mm minimum for a standard CCD camera, while some higher performance systems may achieve resolutions of more than 2.0 line pairs/mm.
    xi. If image sharpness, or image quality, is not good, see module 9.1 page 135.

e. Automatic brightness/kV control

- Automatic brightness adjusts the TV image as different sections of anatomy are examined.
- Automatic brightness normally controls the fluoroscopic output, either by kV or mA, or else a combination of both kV and mA. Older methods operate by direct compensation in the TV camera only. Current systems often use a combination of both methods.
- Systems with automatic control of fluoroscopy kV or mA.
    i. Place a plastic bucket or container with about 3.0 cm (1.25") of water on the tabletop.
    ii. Bring the serial-changer down close to the water phantom.
    iii. With fluoroscopy 'on', the image on the monitor should be a normal brightness level.
    iv. If automatic control is by kV only, or mA only, this may not provide complete compensation. If the image has excess brightness, it may be necessary to reset the manual adjustment. For example.
        —If automatic control is by kV only, and kV has reached its minimum value of 50 kV, but the mA indicates 3.0 mA, then reduce the fluoroscopic mA setting.
        —If automatic control is by mA only, and the manual kV was set to 120 kV, then of course, reduce kV.
    v. Increase the height of water in the container to about 18.0 cm (7") of water.
    vi. With fluoroscopy 'on' kV or mA should automatically adjust to maintain the correct brightness.
    vii. Close the collimator. Then with fluoroscopy 'on', kV or mA should now reach its maximum value.
    viii. **Note.** For the above tests, the brightness should stabilize without oscillating up and down in brightness level. (This means the 'settling time' is not stable.)

- Systems with automatic control of brightness by the camera only.
    i. Place a plastic bucket or container with about 5.0 cm (2.0") of water on the tabletop.
    ii. Set Fluoroscopy kV to about 60 kV, and the mA control to about 1.0 mA.
    iii. Bring the serial-changer down close to the water phantom.
    iv. With fluoroscopy 'on', the image on the monitor should be a normal brightness level.
    v. If overbright, and it is necessary to reduce kV still further, the camera brightness control is not effective.
    vi. If brightness appears normal, increase kV in 10 kV steps, till the image becomes over bright, or 'flaring' occurs. A good system should be able to compensate up to about 100 kV. (This is based on a target-controlled vidicon TV camera. CCD types may have reduced control.

PART III
Fault diagnosis and repair modules

MODULE 5.0

Common procedures for fault diagnosis and repairs

Aim

The aim is to provide information for common procedures involved in diagnosing or repairing a problem. Most of these procedures are applicable to all of the fault diagnosis and repair modules.

Objectives

On completion of this module the student will be able to:

- Adopt a systematic approach to fault diagnosis, as an introduction to the modules for individual equipment.
- Be aware of methods for locating broken wires, and faults in plugs and sockets.
- Test and replace a fuse with a suitable replacement.

Task 8, 'Fuse identification', should be attempted on completion of this module.

Contents

a. Problem diagnosis, or trouble-shooting
  i. Method.
  ii. Typical problems.
  iii. Observation of a problem.
  iv. Equipment manuals.
  v. Request for assistance.
b. Safety first
c. Locating bad connections
  i. Typical areas.
  ii. Locating a broken wire or bad connection.
  iii. A plug or socket may have a fault.
  iv. Power plug connections.
d. Is a fuse open-circuit?
  i. Indications and tests.
  ii. Fuse locations.
  iii. Why has a fuse failed?
  iv. Removing and replacing a fuse.
  v. Common fuse types.
  vi. Circuit breakers.
  vii. Selecting a replacement fuse.
  viii. Special fuses for high frequency generators.

a. Problem diagnosis, or trouble-shooting

Method

Trouble-shooting is a procedure of observation, then making suitable tests to either eliminate, or confirm, a suspect section of equipment. As the area under examination is reduced by further tests, it becomes easier to locate the actual problem.

Note. With any trouble-shooting technique, it is not necessary to approach a problem from any specific direction or set of rules. Rather, you should first observe, consider a possibility and then devise a test to check that assumption.

Consider which items are quick or easy to check. This can save time if first carried out. When a problem occurs, record how the equipment was used just before the problem occurred. This allows a similar procedure to be used as a test, in case the symptoms of the recorded problem are not easily reproduced.

During the process of locating the cause of a problem, record the tests or checks made, and the results. This will provide a valuable record if it becomes necessary to ask advice from the service department.

Typical problems
- Operator error.
- Equipment incorrectly calibrated.
- Faulty connecting plugs, sockets, or cables.
- A safety interlock is preventing equipment operation.
- Electrical or electronic failure.
- High-tension cable or X-ray tube failure.
- Mechanical problems.
- Alignment adjustments.

Observation of a problem

This is important, especially at the time a problem occurs. Possibilities that may be observed are:

- Is there an operator error?
- A burning smell? Where does it come from?
- Is there an increase in temperature? For example:
  —The X-ray tube housing has become very hot to touch.

—A lock coil, in the area where a burning smell is observed, is very hot to touch.
- Unusual sound. What sort of sound? Where from?
- Absence of sound. For example:
  —No anode rotation noise from the X-ray tube.
  —Ventilation fans are quiet
- Wrong mechanical operation. Look for obstructions, or loose sections. For example:
  —An indicator knob on a control panel has slipped into a wrong position.
  —A film is jammed in the processor.
- Visual observation. For example:
  —Appearance of the film immediately as it leaves the processor.
  —Smoke rising from equipment, or a HT cable end.

Equipment manuals

Should be referred to whenever there is a problem. The operation manuals often include a section on fault or problem symptoms, as do the installation or service manuals. The spare parts illustrations can help find the physical positions of parts, such as locating a fuse in equipment.

If during maintenance or other events manuals appear to be missing, replacements should be obtained as soon as possible. Quite often, service engineers attending your equipment will also require these manuals. If not available, this could lead to delays in correcting a problem.

Request for assistance

When requesting advice from the service department, the following information may be required.

- Hospital name, address, fax, and phone number.
- Who to contact at the hospital when discussing the problem. Include the department and phone number.
- Department and room number for equipment location.
- Make and model number of equipment.
- A description of the problem. Include any symptoms
- What tests have been made, and the results.
- Are you asking for advice, or is this a direct request for a service call?
- If a request for a service call, is an order number available?
- Any conclusions that were made regarding the cause of the problem, or what will be needed to correct the problem.

If the request is made via e-mail or by fax, then this information should always be included. If the request is made via a phone call, having the information available will save time.

A record should be kept of service department address details. This should include the address, e-mail, phone, and fax numbers. Where a service engineer is assisting with advice, include the engineers name and contact details.

A sample service request form is provided in appendix 'C' page 177.

b. Safety first

> **Before investigating a possible fuse or wiring problem, always ensure power is turned off and unplugged from the power point. If the equipment is part of a fixed installation, besides switching the generator power off, ensure the isolation power switch for the room is also switched off.**
>
> With battery operated mobiles, ensure the battery isolation switch is in the off position. If this cannot be located, contact the service department for advice before proceeding.

c. Locating bad connections

Typical areas
- The generator handswitch cable.
- The connecting cable to the collimator.
- The fluoroscopy footswitch for a fluoroscopy table.
- Plugs or sockets used with mobile or portable X-ray equipment.
- Power plugs.
- Plugs or sockets for tomographic attachments.
- Cables, which are pulled or twisted.

Locating a broken wire or bad connection
- Remove the cover from the plug or socket, and check if wires have broken away from the contact pins.
- Wires can break inside a cable. This will occur where there is a lot of twisting or stretching. For example, a handswitch cable.
- **Note.** Some wires may seem intact, but can be broken a short distance from a connection point. This is difficult to see, as the wire is covered by the plastic insulation.
- A simple test is to give a gentle to firm tug to a suspect wire. If a particular wire appears to stretch, compared to other wires, it is probably broken

PART III. FAULT DIAGNOSIS AND REPAIR MODULES

inside the insulation. However, this test is not reliable, as a break can occur in another position.
- A multimeter should be used to give an accurate test of a suspect connection or broken wire.
  i. Select the low ohms position of the multimeter.
  ii. Connect the two probes together. The scale should read less than 1.0 ohm. If using an analogue meter, adjust the zero-ohms knob on the meter, so the meter needle is on the zero position.
  iii. Record the minimum value that was indicated when the probes were touched together.
  iv. Connect the meter probes at the two ends of the wire being tested. The meter should indicate less than 2.0 ohms above the value previously recorded with the probes connected together.
  v. A broken wire in a cable may have a partial connection, and cause an intermittent fault. Give the cable under test a tug or twist while watching the meter. If the meter flickers or indicates a changing value while the cable is moved, the wire is faulty.

Check for a faulty handswitch, or handswitch cable.

> *Caution.*
> *Ensure the generator is switched off, and the room power isolation switch also turned off. If this is a mobile or portable generator, ensure it is switched off, and the power cable unplugged from the power point.*

- This test should be performed with the assistance of an electrician or electronics technician.
- Make a diagram, so that when the handswitch is disconnected or unplugged, it can be correctly reconnected.
- A multimeter is required, set to the low ohms position. Check that the meter indicates zero ohms with the meter probes touching together.
- Open the hand switch assembly, and identify the terminals of the suspect switch. With the meter probes touching the switch terminals, operate the switch. The meter should indicate zero ohms. Repeat this test for other switches in the handswitch.
- Identify the connecting wires between the switch terminals and the other end of the handswitch cable. With the meter connected to each end of the cable wires, the meter should indicate less than 2.0 ohms. When making this last test, if at first you have a good result, test again by tugging on the connecting lead, and also move the cable in different positions.
- If there is a broken wire in the handswitch cable, this often occurs close to the handswitch, and sometimes where the cable enters or is attached to the control desk.
- To look for a broken wire, cut open the cable outer insulation to expose the internal wires. Commence where the cable enters the handswitch, and continue about 15 cm down the cable. Test individual conductors by giving a firm pull. If a broken wire is found, shorten the cable past the bad section, and reconnect to the handswitch. *The entire cable should be replaced as soon as possible, in case of other partially broken wires.*
- The above procedures should be repeated for all the switches in a handswitch assembly.

A plug or socket may have a fault
This can occur due to:

- A pin or contact has moved out of position.
  i. This is due to bad assembly during manufacture. However, the plug or socket may have considerable use before this fault occurs.
  ii. Remove the back of the plug or socket assembly.
  iii. Locate the wire attached to the pin or contact.
  iv. Try pushing it firmly back into position. If it sets in place and will not move with a gentle to firm tug on the wire, all is well.
  v. If the pin or contact is not firmly attached, then remove it and look for bent 'hooks' on it. Straighten these out, so that when reinserted they will keep the pin or contact in position.
- The contacts of a socket may have become enlarged, and not provide a reliable or good contact.
  i. The contact can be adjusted by pushing the sides of the contact closer together. A useful tool is a large sewing needle.
  ii. Care must be taken not to damage the contact.
  iii. This problem may occur where plugs and sockets have had a lot of use. For example, with a portable X-ray generator.

Power plug connections
The wires connected to the power plug terminals can become loose. This can cause arcing, and may cause a fuse to become open circuit.

The wires must be held securely inside the plug. If the cable is pulled, this should not pull the wires from the plug terminals. If the plug does not have a suitable method to prevent this from happening, then replace the plug with an improved type.

> If wires have pulled loose from a power plug, this may be due to a previous incorrect assembly. Have the connections checked by an electrician or electronics technician. This is to ensure the active and neutral wires are connected to the correct positions on the power plug, and, most important, correct connection of the earth or ground wire.
>
> While wire colours now conform to international standards, older equipment may use non-standard colours. If this is found, have the power cord replaced by an electrician.

d. Is a fuse open-circuit?

Indications and tests
- Some glass fuses can visually show if they are open circuit by metal deposited on the glass. If the failure is a small break in the wire, this is not easy to see.
- Fuses may have fine wire, again not easy to see, so it may appear open circuit, but still be in good condition.
- A continuity test with a multimeter is the only reliable way of verifying if a fuse is good or open circuit. The meter is set to the low ohms range. If the meter shows no indication when the probes are attached to the fuse, the fuse is open circuit.
- Do not try to test a fuse with a meter while it is still connected in the equipment, this can give a false result.

Fuse locations
The physical location of fuses will vary greatly depending on the manufacturer and model of equipment. Electrical regulations in many countries require a fuse to be protected from access without using a tool. Where fingers may be able to unscrew the cap of a fuse holder, a protective cover must first be removed.

Possible fuse locations are:

- X-ray generator, fixed installation. There will not be any external access fuses. Most fuses will be located in the control cabinet, and a panel will need to be removed to gain access. In some cases, there may be additional fuses under a cover at the HT transformer.
- **Note.** In some installations miniature circuit breakers may be fitted instead of fuses. These have a reset switch, or button, mounted on top of the device.
- Mobile or portable X-ray generators can have external access fuses mounted on a rear panel. Otherwise internal access to the equipment is required.
- The Bucky table may have an external access fuse panel on one side, or else on the side of a small control box. Otherwise, a panel may have to be removed to gain access.
- The vertical Bucky and stand may have a small power supply box for magnetic locks. This may have external access to a fuse, or require removal of the box cover.
- The fluoroscopy table may have external access power fuses located at the table foot. There will be additional fuses inside the control cabinet, or inside electronic control boxes placed on the table body.
- A floor-ceiling tube stand can have a small power supply box mounted at the top of the column. This usually has external access to the fuses.
- A ceiling suspension tube support may have external access to the fuses. In most cases a panel cover will need to be removed.
- The operation or installation manuals often show the location of fuses. The parts manual will also indicate fuse positions, however the diagrams can be very complex. Otherwise, contact the service department for advice.

Why has a fuse failed?
- There is a fault. The fuse has blown to protect the equipment. There will often be a heavy metal deposit on the glass. On replacing the fuse, the new fuse will also fail.
- There was a temporary fault caused by a power surge.
- There is no fault. The fuse has become 'fatigued' or 'tired'. This sometimes happens. If a glass fuse, the wire may show a small broken section.
- An incorrect fuse was previously fitted. For example, a 5A fuse fitted where a label indicates a 7.5A fuse. This may have been a temporary replacement, however first contact the service department for advice.
- When a blown fuse is found, contact the service department for advice. Give full details of the fault symptoms, fuse number or identification, and position in the equipment.

Removing and replacing a fuse
Never attempt to remove or replace a fuse unless all power is switched off, and where applicable, unplugged from the power point. In an X-ray room, also turn off the main power isolation switch.

There are a large variety of fuse holders. Some types are listed below.

- The fuse is held between two spring clips.
  i. Remove by pushing or lifting first at one end. If trying to prise out by lifting in the middle, some glass fuses may break.
  ii. When re-inserting a fuse, the spring clips should have a positive grip. If not, remove the fuse, bend the clips inwards a little, and then reinsert the fuse.
- The fuse is in a container, totally enclosing the fuse.
  i. The front appears as a small square. There is no screwdriver slot. In this case there should be a small gap under the square front. Pushing a small screwdriver into this gap will release the catch, allowing an internal spring to push the fuse out.
  ii. There is a round cap, similar to a large version of a toothpaste tube cap. Grip with the fingers, and turn anticlockwise to undo.
  iii. Two versions may have a screwdriver slot. One version has a bayonet catch. To remove, first push inwards with a screwdriver, then rotate anticlockwise a quarter turn. This should unlatch the top. In the second version, this is a simple threaded type. Undo with a screwdriver, turning anticlockwise.

Common fuse types

Fuses come in a great variety of styles, shapes and sizes. Most common are the larger 3AG sizes, 6 × 32 mm, and the smaller M205 size, 5 × 20 mm. The fuse may have a glass or ceramic body.

Of the above fuses, there are two main types.

- General-purpose fuse.
  i. The element is normally a straight piece of wire, or a thin metal strip.
  ii. European fuses may have 'MT' marked on one end.
- Slow blow or delay fuse.
  i. These fuses are designed to allow a short high surge current. This allows for the momentary peak current demand from motors, transformers, or power supplies, when first switched on.
  ii. The element appears as a piece of wire, attached to a spring.
  iii. Another version has a wire, similar to a general-purpose fuse. However, the wire has two small metal beads spaced along the fuse wire.
  iv. European fuses may have 'T' marked on one end.

Circuit breakers

Circuit breakers may be fitted instead of fuses. In most cases they are designed to handle large currents, such as the main power supply to an X-ray generator. In some cases, small circuit breakers are also fitted instead of standard 5A or 10A fuses.

If a circuit breaker has tripped, the switch at the front of the circuit breaker may be half way between the 'off' and 'on' positions. To reset, push the switch to the off position, then back to the on position.

Caution: Circuit breakers, like fuses, should only be reset after the supply power is switched off.

Selecting a replacement fuse

Providing the fuse has the same rating and characteristic, a ceramic fuse may be replaced with a glass fuse.

If a fuse of the correct current rating is not available, then a smaller rated fuse might sometimes be substituted as a temporary replacement, providing it is close to the original value. Eg, a 2.5A fuse might substitute for a 3A fuse. In case a delay fuse is not available, a standard fuse of the same rating may be tried as a temporary replacement. **Do not** substitute a delay fuse for a standard fuse.

Special fuses for high-frequency generators

These are very large fuses, sometimes rated up to 200 amps, for use in the inverter. The fuse may be a long, large diameter cartridge fuse, seated in heavy-duty brass clips; or else a large ceramic type, retained by a nut and bolt. While inverter fuses can fail due to fatigue, or a temporary problem due to X-ray tube instability, there could be a semiconductor failure in the inverter.

- **Only an electrician or electronics technician should attempt to test or replace an inverter fuse.**
- Before replacing the fuse, a test for a possible short circuit in the inverter should be made.
- **Caution:** High residual voltages can be present in the power supply capacitors.
- **Always consult the service department for advice before attempting to test or replace an inverter fuse.** The service department may require specific tests to be made before replacement, otherwise further damage could occur.

TASK 8

Fuse identification

You have been supplied with a mixed selection of fuses.

- What is the current and voltage rating of each fuse?
- Are any of these fuses a 'delay' or 'slow blow' version?

Fuse No	Voltage	Current	Delay fuse?
1			
2			
3			
4			
5			
6			

One or two of these fuses may be faulty. Test the fuses, using a multimeter. Use the instructions provided in module 8.0

Were any of these fuses faulty? Indicate which fuse No. _____

Was there any difficulty in using the multimeter? _____

Tutor's comments

Satisfactory/Unsatisfactory

Signed _____ Date _____
Tutor

PART III. FAULT DIAGNOSIS AND REPAIR MODULES

MODULE 6.0

X-ray generator: Fixed installation

Aim

The aim is to provide information and procedures for diagnosing, or repairing, a problem with an X-ray generator. The generator is installed as a fixed installation in an X-ray department. Flow charts are provided to indicate logical steps, when diagnosing a problem.

Objectives

On completion of this module, the student will be able to carry out a set of procedures to locate a problem area in the generator. This will allow minor repairs to be carried out. An electrician or electronics technician may also provide assistance. Otherwise an accurate description of the problem can be provided to the service department, if requiring advice or direct assistance.

This module is divided into two sections. The first section looks at problems that occur during preparation for an exposure. The second section looks at problems affecting an exposure, after preparation for an exposure has been completed.

Task 9. 'No Preparation, Part 1' and task 10. 'No Preparation, Part 2', should be attempted after section one of this module.
Task 11. 'No Exposure', and task 12. 'X-ray output linearity', should be attempted after section two of this module.

Contents

Section 1: No preparation
    Part 1. Nothing happens.
    Part 2. Is there a bad connection or fuse?
    Part 3. Warning signals due to a fault condition.
    Part 4. X-ray tube tests.
    Part 5. Other tests.

Section 2: No radiographic exposure
    Part 6. Operation tests.
    Part 7. No mA or kV.
    Part 8. High-tension problems.

(**Note:** Reference module page numbers refer to the title page.)

Section 1: No preparation

This section examines a number of situations where the generator prepares for an exposure. Before the control system will permit preparation to begin, the generator makes a number of safety tests. Once the generator enters into the preparation mode, more tests are made. When all tests are satisfied, the control system signals ready for an exposure. There are a total of five parts, each with an associated flow chart.
Study each flow chart first, and then read the text for additional information.

Part 1. Nothing happens

Refer first to the flow chart, Fig 6–1 page 75, 'Unable to obtain preparation, part one'.

On attempting to have the control prepare for an exposure, nothing happens. There is no sound of X-ray tube anode rotation. At the end of the normal preparation time, there is no sign the generator is ready for an exposure.

Are there any warning lights on the control panel?

The warning light could indicate a fault in the generator, or else an operator error. In some systems there may be two warning lights, one for incorrect setting of the X-ray control, and the other for a fault in the generator.

In the case of a warning light indicating wrong exposure factors, this could be due to:

- The kV set for the exposure is above the maximum X-ray tube rating. Reduce the kV setting.
- The kV is below the minimum kV specified by radiation regulations. Increase the kV setting.
- The mA selection is too high for the required mAs. Select a lower mA position, and a longer exposure time.
- The combination of a high value of mA, and a low value of kV. This could overheat the X-ray tube filament. Either increase kV, or else select a lower value of mA and increase the exposure time.
- With some generators, selecting a very short exposure time combined with a low mA will prevent operation. For example, if the limit is 0.5 mAs, then 500 mA at 0.001 second is accepted. However, if 250 mA is selected, then the minimum exposure time becomes 0.002 second.
- The X-ray control microprocessor has calculated the amount of heat in the X-ray tube anode from previous exposures. A further exposure would cause the anode heat capacity to be exceeded. Wait a few minutes for the anode to cool down.
- **Note.** Most X-ray controls, which calculate total anode heat, display the result on the control panel. This may take the form of a number, or a bar graph. In other cases a small symbol indicates excessive anode heat.

Does the control panel display a message, instead of a warning light?

Modern microprocessor-controlled generators can have special display modes to indicate problems or operator error. Compared to a single indicator light, these systems give increased information. For example:

- A direct message 'Reduce mAs' appears on the control panel. This direct type of message might be displayed with microprocessor-controlled generators, together with a plasma or liquid crystal display panel.
- The kV display does not show the required kV. Instead a code E2 appears in the kV display. This is another way a microprocessor-controlled generator may indicate an error.
- The operation manual has a list of the codes and what they mean. In this case E2 could mean 'Set kV is too high'. **Codes are used to also indicate fault conditions.** A code F3 could mean the generated mA exceeded the calibration limit during the previous exposure. **If a fault code appears, consult the service department for advice.**
- *The actual code numbers, where they appear on the control panel, and their meanings, depend on the make and model of equipment. Only some X-ray controls have these features.*

Does resetting the exposure factors turn off the warning light?

- If the warning light does not turn off, **go to part three**.
- If the warning light is off, then test again for preparation. If preparation can now be obtained, all is well. Otherwise continue with 'Other basic checks'.

Other basic checks

- Is a suitable technique selected? For example, if the fluoroscopy table is selected instead of the table Bucky, handswitch operation may not be permitted.
- Is a suitable focal spot selected? If the exposure factors were suitable for the large focal spot, selecting the small focal spot with the same exposure factors could prevent operation.
- Check all other settings. If unsure of their correct use or position, please refer to the operation manual, or contact the service department for information.
- Was an Automatic Exposure Control (AEC) in use for the last exposure? The AEC may have generated an inhibit signal. Disable the AEC and try again.
- Does the X-ray room have a door-open safety switch or 'interlock'? The operation of this switch may be faulty.
    i. When the door is closed, can you hear the switch operate? In this case, it is probably ok.
    ii. If no indication is heard, check the switch operation with a multimeter.
    iii. **Ensure all power is turned off, including the room power isolation switch.**
    iv. Set the multimeter to the low ohms position. Connect the probes together. The meter should indicate close to zero ohms.
    v. With the probes connected to the switch contacts, test the switch operation by opening and closing the door.

Can preparation now be obtained?

If the answer is yes, the problem is solved. ***Otherwise continue with part two.***

Part 2. Is there a bad connection or fuse?

Refer first to the flow chart, Fig 6–2 page 76, 'Unable to obtain preparation, part two'. **In this situation, the tests indicated in part one have been completed.** No warning lights or error codes are displayed on the control panel.

PART III. FAULT DIAGNOSIS AND REPAIR MODULES

Is there a fault with the handswitch?

- Listen at the X-ray tube while an assistant presses the preparation button. Is there any sound of anode rotation, or other noise, from the X-ray tube when the button is pressed? If there is, then the handswitch is ok. Proceed to part four.
- On pressing the preparation button, can you instead hear a relay operate in the generator control cabinet? If so, the hand switch is not at fault. Proceed to part four.

Check for a faulty preparation switch, or handswitch cable

- *First ensure the generator is switched off, and the room power isolation switch also switched off.*
- This test should be performed with the assistance of an electrician or electronics technician.
- To test the handswitch or cable, refer to **module 5.0 page 65**.
- If there was no problem found with the handswitch, or handswitch cable, proceed to part 5.

Part 3. Warning signals due to a fault condition

Refer first to the flow chart, Fig 6–3 page 77, 'Unable to obtain preparation, part three'.

In this situation, a warning light or code may appear immediately the generator is switched on, or else during the last exposure. The checks indicated in part one have been performed. Operator error or incorrect exposure settings are not the cause of the problem.

A warning light or code appears when switched on

This indicates a serious fault condition. If a code or message is displayed, refer to the operation manual for its meaning. If the code indicates an operator or setting error, then recheck the tests carried out in part one. Otherwise contact the service department. Provide details of the fault code, and any other symptoms that were observed.

- A fuse may have failed, or a circuit breaker tripped.
  i. *Before opening any panels to check a fuse, always ensure the equipment is switched off, and the room power isolation switch also turned off.*
  ii. Procedures for testing or replacing a fuse, are described in **module 5.0, page 65**.
  iii. If a fuse is open circuit, or a circuit breaker has tripped, consult the service department for advice regarding the possible cause.

- Is the mains power supply voltage too low? Or is one phase missing? This could happen after a storm or power failure.
  i. If the generator has a manual adjustment for the mains voltage, check and make sure it is set correctly.
  ii. Other generator designs have automatic compensation for changes in supply voltages. Depending on the design, the mains voltage may have changed outside the adjustment range of the automatic compensation. Ask an electrician to check or measure the supply voltage.
  iii. A three-phase generator may still switch on with only two phases of the mains supply available. Some designs have a fault detector in case this happens. Ask an electrician if there is a problem with the hospital power.

A warning light or code appears during preparation

If a code or message is displayed, refer to the operation manual for its meaning, or contact the service department. Provide details of the fault code, and any other symptoms that were observed.

- Does the warning light appear **immediately** on pressing the preparation switch? This may indicate a serious problem with the X-ray tube filament control section of the generator. Contact the service department before proceeding.
- In case the warning light occurs during or just at the end of preparation, this may indicate a filament connection problem, or failure to pass the anode rotation safety test. **Proceed to part four.** Otherwise contact the service department for advice.
- Did the warning light operate **during** the last exposure? This can indicate an over current (mA) situation. Over-current may be caused by instability in the X-ray tube, or else a fault in the HT cable or cable end.
  i. Try resetting the fault detection by switching the X-ray control OFF then ON again. Then try to obtain preparation.
  ii. If preparation can now be obtained, go to part 8.
  iii. If preparation still cannot be obtained, **proceed to part four**, or contact the service department for advice.

Part 4. X-ray tube tests

Refer first to the flow chart, Fig 6–4 page 78, 'Unable to obtain preparation, part four'. There are no warning lights or message codes. On attempting preparation, some sound was observed at the X-ray tube. This indicates the preparation handswitch is ok. During preparation the X-ray control applies a test for anode rotation, and minimum filament current in the selected focal spot. There may be a problem with anode rotation, or else a poor connection to the X-ray tube filament.

Does anode rotation appear normal?
This should be the normal sound of the anode accelerating to the required preparation speed. Instead it may be just a buzzing or humming noise. The possibilities are:

- One of the three conductors for the X-ray tube stator is broken or has a bad connection.
    i. Check the stator cable where it enters the X-ray tube housing.
    ii. Does the cable show signs of being pulled or stretched? A wire may have been pulled away from a terminal.
    iii. ***Before checking any connections, always ensure the equipment is switched off, and the room power isolation switch also turned off.***
    iv. Some x-ray tube housings have a plug and socket for the stator cable connection. Has this become loose, or is there a bad connection to the plug or socket terminals?
- The X-ray tube itself is faulty, either with seized bearings, or else with broken glass. If the glass is broken, the anode may appear to rotate slowly, and very quickly come to a stop when the preparation switch is released.
- Is there oil leaking from the X-ray tube housing? An arc may have occurred during the last exposure, causing internal damage.
- Is there an open circuit fuse, or tripped circuit breaker, supplying power to the X-ray tube starter? This can be a common problem with some high-speed starters.
    i. ***Before opening any panels to check a fuse, always ensure the equipment is switched off, and the room power isolation switch also turned off.***
    ii. If unsure where such a fuse or circuit breaker may be located, contact the service department for advice.
    iii. If the fuse is faulty, take care to replace with the correct size and type. **See module 5.0 page 65.**
    iv. Examine the stator cable where it enters the X-ray tube housing. If there is a possibility the cable has been pulled, remove the cover plate and check the connections for a possible short circuit.
    v. If on restoring power, or attempting preparation again, the fuse blows, then **stop**, and contact the service dept.

Is the focal spot filament connected?
During preparation the X-ray control tests for a minimum filament current. This test is only to ensure the filament is intact, and is at a minimum temperature.

Note. This test does not ensure the correct mA will be generated. A poor HT cathode-cable connection can cause a drop in the required filament current, but still be above the minimum value for the filament current test. **See part seven, 'No radiographic exposure'.**

- The selected focal spot may be open circuit, or have a bad connection. Try obtaining preparation with the other focal spot.
- There may be a bad connection to that focal spot due to a bad HT cathode-cable cable-end pin connection. There are three pins on the cable end. If there is a bad connection at the centre pin, this will affect both focal spots.
    i. Before proceeding, ensure all power is switched off.
    ii. To check, undo the cathode cable-end retaining ring nut. Partly withdraw the cable end about 2~4mm, then reinsert the cable end. Replace and firmly tighten the cable-end ring nut.
    iii. Test for preparation. If moving the cable end has cleared the preparation problem, the cable end will require further attention. This involves removal of the cable end, cleaning the cable-end pins, and re-sealing. For this procedure, **see module 7.3 page 117**.
    iv. **Caution.** The cable end may have silicon grease as insulation instead of anti-corona silicon pads. Removal of the cable end by more than 2~4mm can reduce the effectiveness of the insulating grease. As a precaution, do not exceed 100 kV until the cable end is re-sealed with fresh silicon grease, **see module 7.3 page 117**.

PART III. FAULT DIAGNOSIS AND REPAIR MODULES

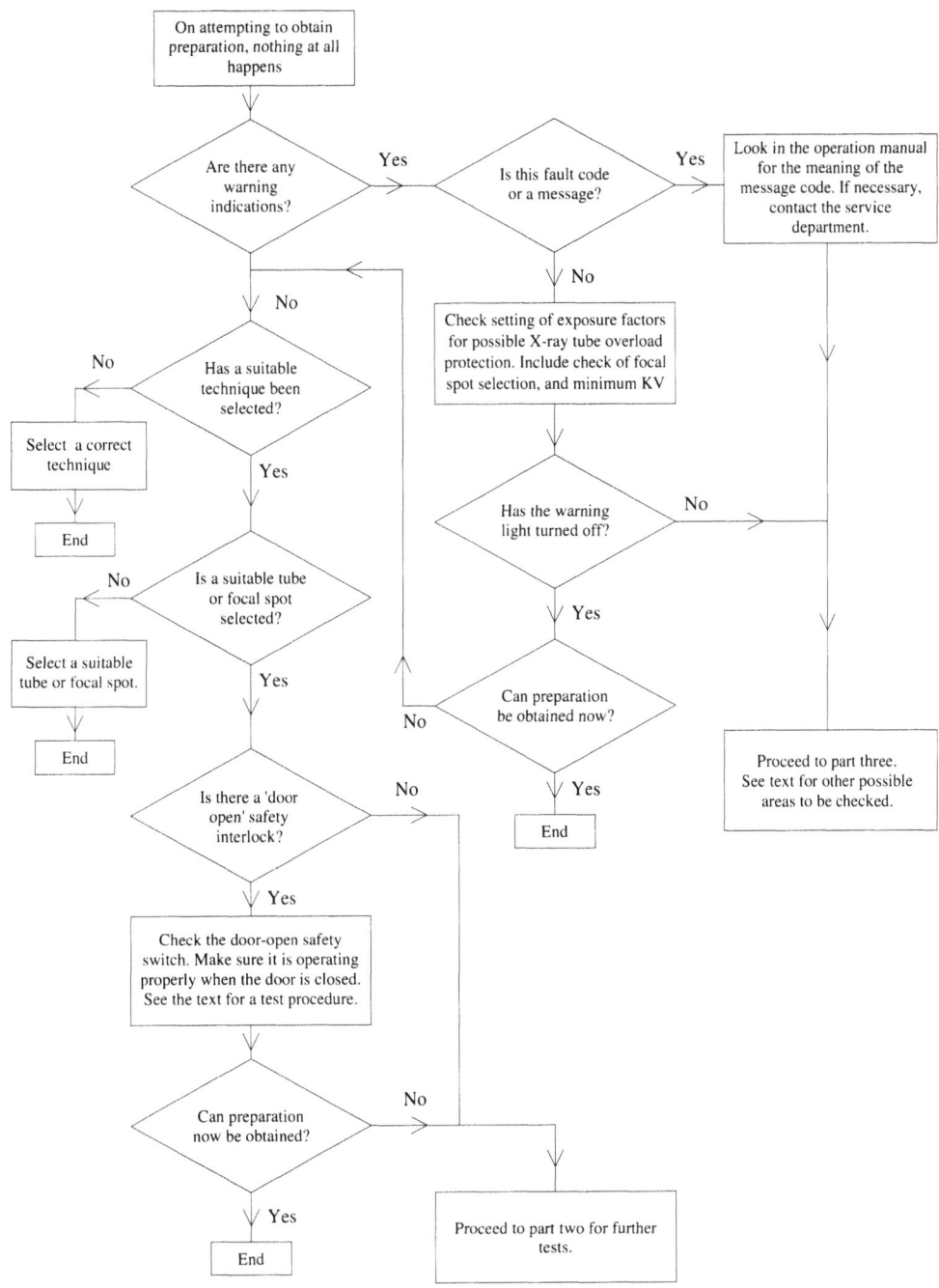

Fig 6-1. Unable to obtain preparation, part one

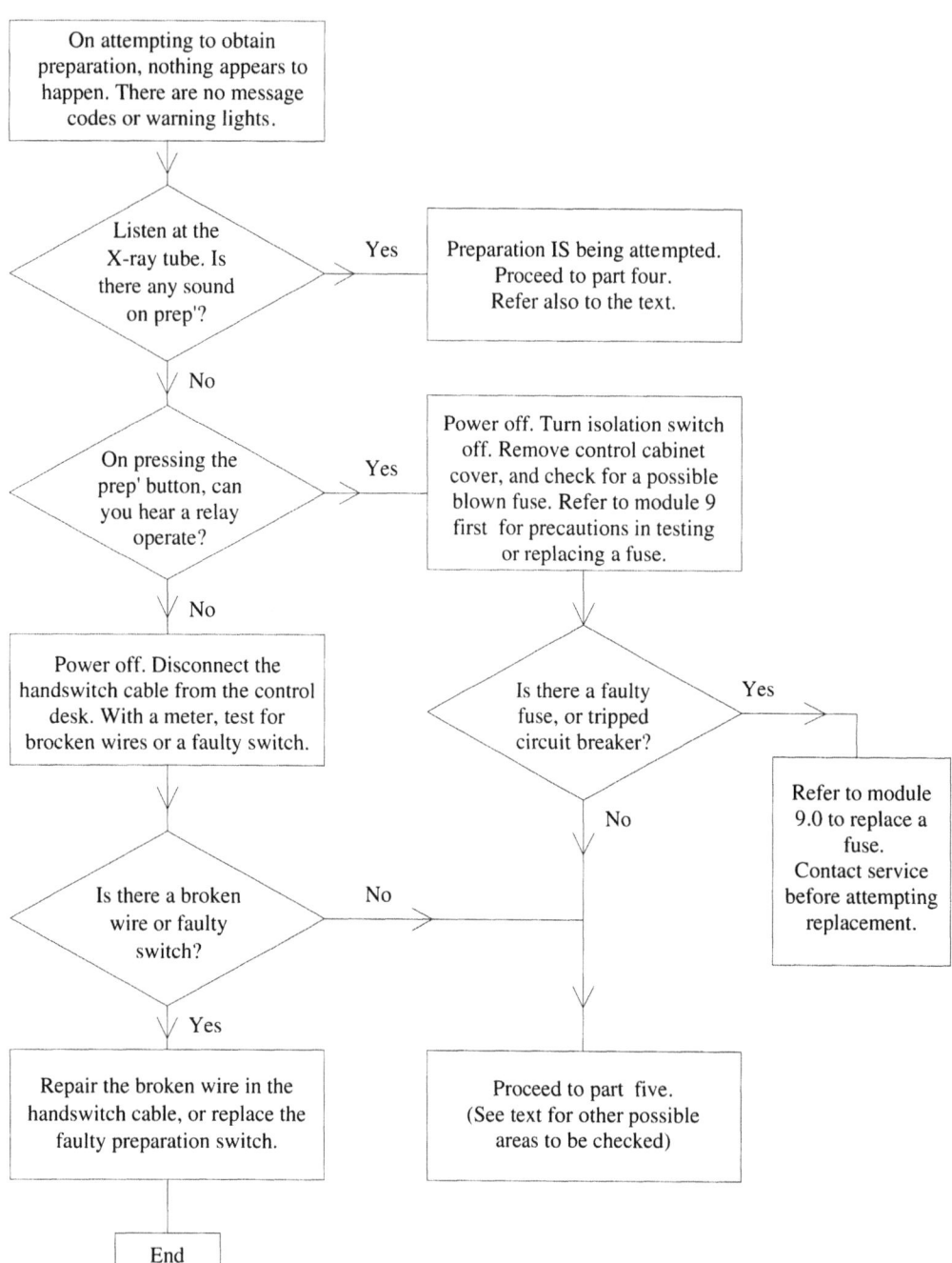

Fig 6-2. Unable to obtain preparation, part two

PART III. FAULT DIAGNOSIS AND REPAIR MODULES

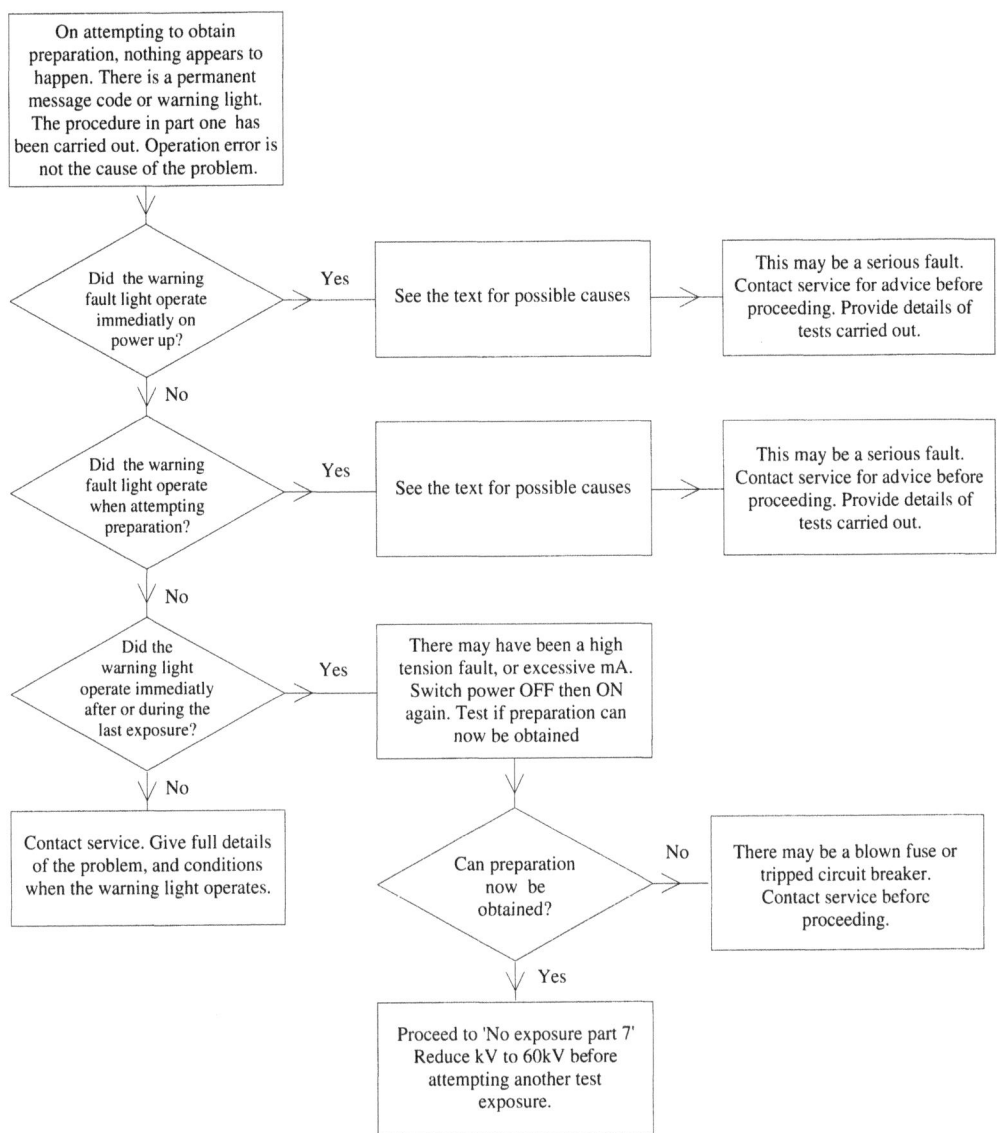

Fig 6–3. Unable to obtain preparation, part three

X-RAY EQUIPMENT MAINTENANCE AND REPAIRS WORKBOOK

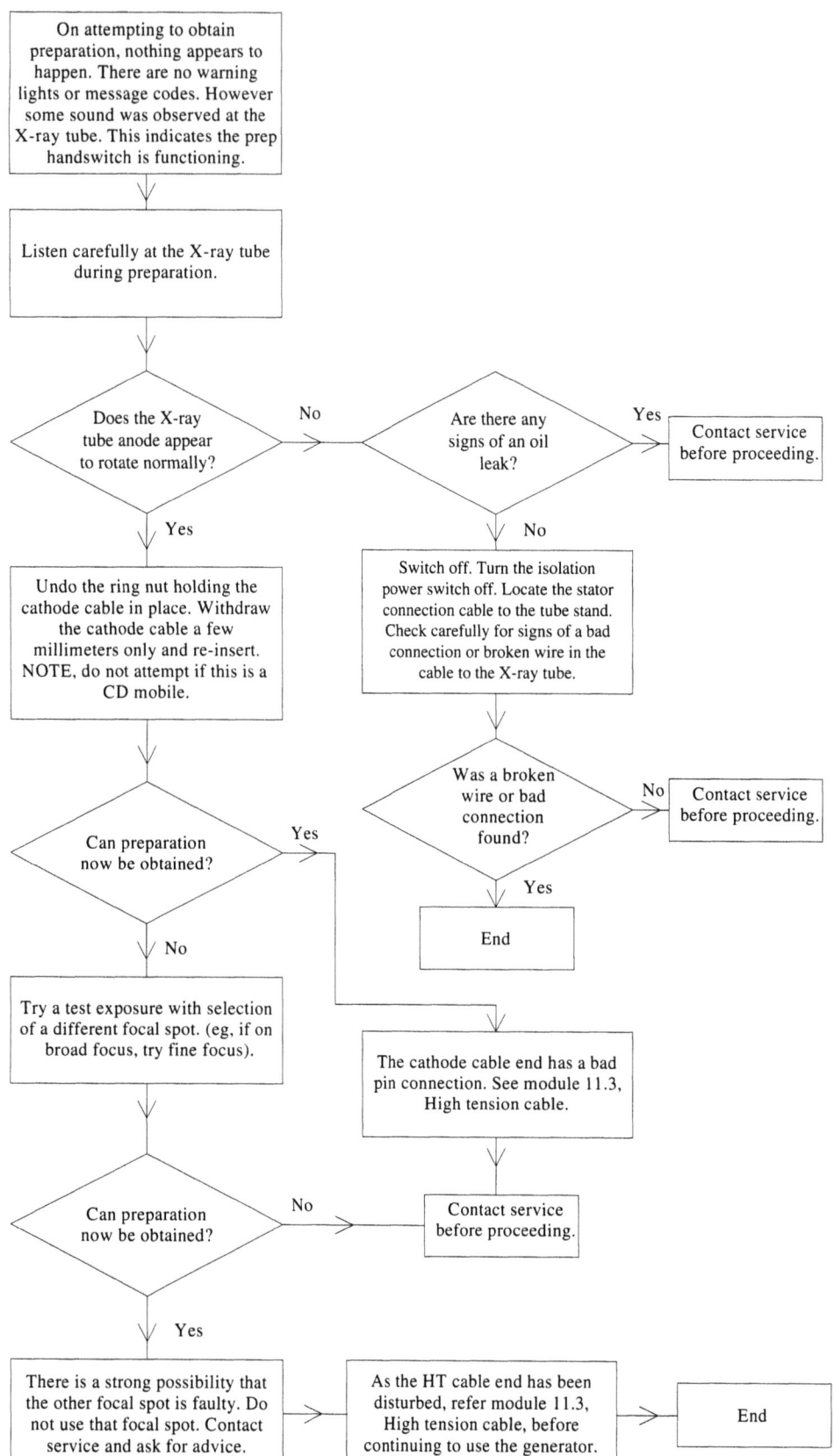

Fig 6–4. Unable to obtain preparation, part four

PART III. FAULT DIAGNOSIS AND REPAIR MODULES

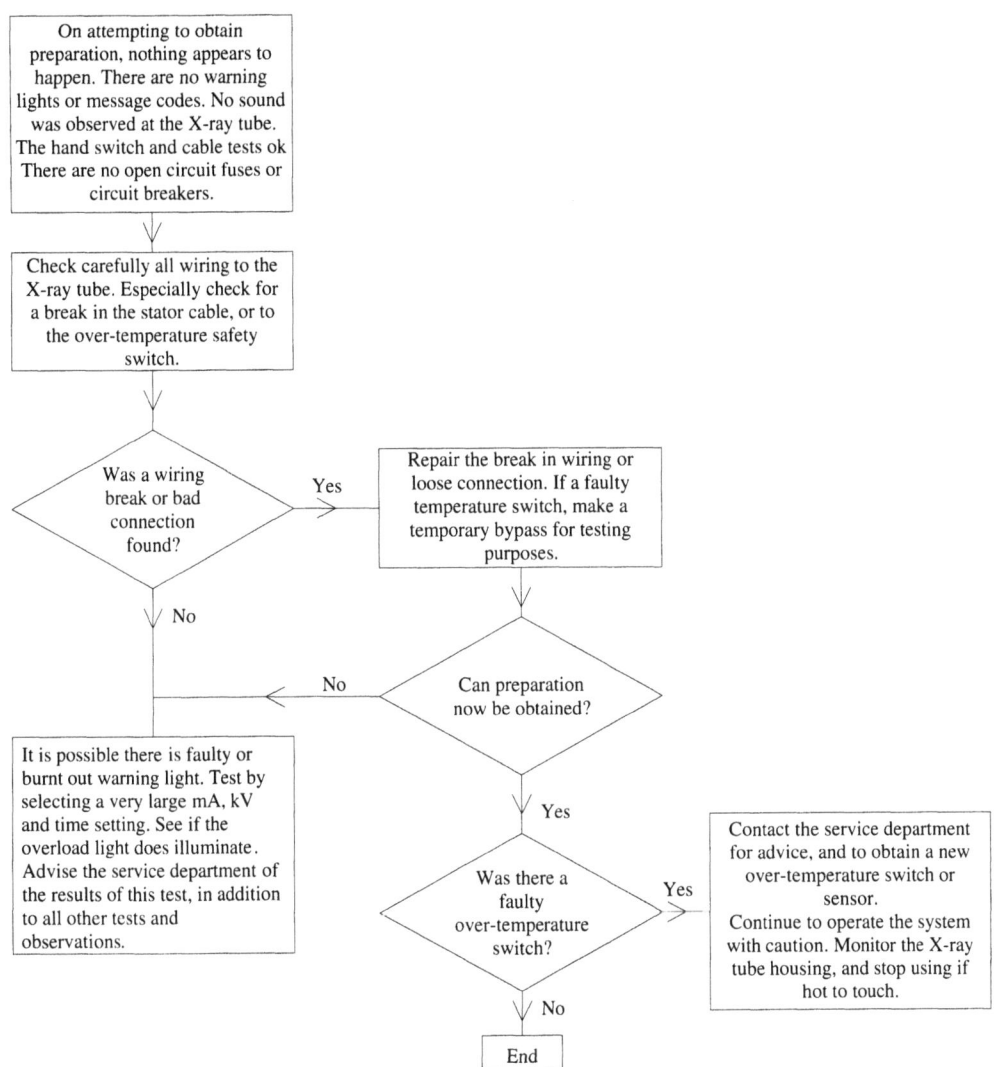

Fig 6–5. Unable to obtain preparation, part five

Part 5. Other tests

Refer first to the flow chart, Fig 6–5 page 79, 'Unable to obtain preparation, part five'.

There remain two possibilities to be checked.

- The wiring to the X-ray tube stator or the housing over-temperature switch is broken or has a bad connection. This could occur where it enters the X-ray tube, or where the cable has received a lot of flexing or twisting. See **module 7.1 page 104**.
- There may be an internal fault or problem in the generator, but the warning light is also faulty or burnt out. Test the light by setting very high exposure factors, which would normally cause a tube overload condition. If no warning light illuminates, then the globe or control circuit is faulty.
- Include the results of all tests, when requesting help from the service department.

Section 2: No radiograph exposure

This section assumes you are able to obtain preparation, the control has indicated preparation is completed, and is ready for an exposure. On attempting to obtain an exposure, nothing happens. Or else, the control appears to expose, but the film is blank or very light. Each part has an associated flow chart. Refer to these flow charts before reading the text.

Part 6. Operation tests

Refer first to the flow chart, Fig 6–6 page 83, 'No exposure, part six'.

The control has indicated it is ready for an exposure. On attempting to obtain an exposure, nothing appears to happen. There is no sound from the Bucky. The exposure light does not operate.

Operation tests

- Check the technique selection to ensure a valid operation. Look for selection buttons or switches that are sticking, and may not have properly operated or released.
- Is an Automatic Exposure Control (AEC) in operation?
    i. The warning light or signal with the AEC reset button may have failed. Press reset and try exposing again.
    ii. Have you selected the correct AEC mode for the particular Bucky or technique?
    iii. Disable AEC operation, select suitable manual exposure factors, and attempt another exposure.
    iv. If you can now expose, continue using the system with AEC switched off. For AEC tests, see **module 10.0 page 140**.
- Select the direct, or non-Bucky, handswitch technique. Try again to expose. If successful, consider the following.
    i. **First ensure all power is turned off, including the room power isolation switch.**
    ii. There is a possible open fuse or circuit breaker suppling power to the Bucky. If found faulty, check the wiring to both Bucky's before replacing the fuse. There may be loose connections, causing a short circuit. A film marker may have fallen onto electrical connections in the vertical Bucky.
    iii. Similar to the above, there may be a loose or broken connection to a particular Bucky. This is most likely to occur at either a plug or socket, or else at the Bucky terminals on the side of the Bucky.
    iv. The grid may have become dislodged, stopping operation of the grid drive motor. A common cause is poor attachment of lead numerals to the cassette. When the cassette and tray is inserted, these catch on the grid, dislodging it.
    v. Some older Bucky's have small motor drive gears. These gears may be damaged. As a result, although the motor operates, the grid does not oscillate or move to the expose position.
    vi. In the case of a vertical Bucky, a lead numeral may become dislodged, and fall into the drive motor area. This could prevent operation, or cause a short circuit.
    vii. For other problems with the Bucky, refer to **module 8.0 page 121**.
- Does the X-ray room have a safety 'Door closed' interlock? This may, depending on installation methods, interrupt preparation if the door is opened, or simply prevent an exposure. When the door is closed, can you hear the switch operate? If so, it is probably ok.
    i. If no indication is heard, check the switch operation with a multimeter.
    ii. **Ensure all power is turned off, including the room power isolation switch.**
    iii. Set the multimeter to the low ohms position, and check the meter shows minimum ohms with the probes connected together.
    iv. With the probes connected to the switch contacts, test the switch operation by opening and closing the door.
- Can you hear a relay operating on pressing the expose switch? If so, the switch and connecting cable are properly working.
- If the handswitch is suspect, then the switch and connecting cable should be checked with a multimeter set to the low ohms scale.
    i. **Ensure all power is turned off, including the room power isolation switch.**
    ii. If available, request an electrician or electronics technician to assist.
    iii. Use the method described in part two of this module. 'Check for a faulty preparation switch, or handswitch cable'. See **module 5.0, page 65**.

Part 7. No mA or kV

Refer first to the flow chart, Fig 6–7 page 84 'No exposure, part seven'.

On attempting to obtain an exposure, the film was blank, or very under exposed. The control indicates preparation is completed and ready for an exposure. The exposure indication operates on pressing the expose position of the handswitch.
There is no warning light or fault indication during the attempted exposure.

Check the cathode cable connection

A bad cable-end connection in the X-ray tube cathode-receptacle is a common cause of light or blank films. This occurs where the cable-end pins fit into the receptacle. For example: Sufficient current flows through the filament to satisfy the generator tests during preparation. However the filament temperature is too low for the required mA. A test exposure should be tried first on the other focal spot. However, if the cable-end centre pin has a bad connection, this can affect **both** focal spots.

Note. During an exposure, if there is low, or no mA, then the actual kV can rise above the set value of kV.

As a result, a microprocessor-controlled generator may display 'kV too high' as a fault message, instead of 'low mA'. Always select low mA and kV for test exposures.

- To check for a poor cable end connection.
  i. **Ensure all power is switched off, before any adjustment is made to the HT cables.**
  ii. To check, undo the cathode cable-end retaining ring nut. Partly withdraw the cable end about 2~4mm, then reinsert the cable end. Replace and firmly tighten the cable-end ring nut.
  iii. **Caution.** The cable end may have silicon grease as insulation instead of anti-corona silicon pads. Removal of the cable end by more than 2~4mm can reduce the effectiveness of the insulating grease. As a precaution, do not exceed 100kV until the cable end is re-sealed with fresh silicon grease. **See module 7.3 page 117.**
- Try a test exposure after adjusting the cathode cable end. If the test exposure is OK, there was a poor pin connection at the cable end. The cable-end pins need to be cleaned and adjusted. **See module 7.3 page 117.**
- If the test exposure still fails after adjusting the cathode cable-end, try exposing again on the other focal spot. If the exposure is now OK, the first focal spot may have a fault. This is caused by;
  i. The filament has a short circuit to the focus cup. Only part of the filament is heated.
  ii. The filament is able to pass the 'open circuit' safety test in preparation, but the actual mA on exposure is less than half the expected value. In this case, the X-ray tube requires replacement.
  iii. To inspect the filament, see **module 7.1 page 104**.
  iv. Unfortunately, this type of fault is not uncommon. The problem occurs when the filament bends, and touches the focus cup during preparation. This causes the filament to become welded to the focus cup, shorting out a section of the filament.
  v. Contact the service department for advice.

Has an inverter fuse failed?
If a trial exposure on the other focal spot also fails, this may be due to no high voltage. With high frequency generator systems, a common cause is an open circuit inverter fuse. This often fails due to fatigue, or a temporary fault with the inverter. Depending on the make or model of generator, there may be no warning signal, or message, to indicate high-voltage failure.

- ● *First ensure all power is turned off, including the room power isolation switch.*
- ● **Before attempting to test or replace the fuse, contact the service department for advice. As well as the fuse location, special test precautions can apply before and after replacement.**
- ● **Checking or replacing the fuse, should only be performed by an electrician, or electronics technician.**
- ● **See module 5.0 page 65.**

  i. Before attempting to test the fuse, measure the primary power filter capacitors, and ensure the residual voltage is at a safe level.
  ii. If the fuse is open circuit, check the inverter SCR or power transistors for a possible short circuit, before replacing the fuse.
  iii. After the fuse is replaced, examine the power circuit for charging the capacitors on initial power up. There is normally a resistor in series with each phase to limit the charging current. These resistors are shorted out by a contactor after a few seconds time delay. This circuit needs to be disabled so the resistors remain in operation.
  iv. On power up, monitor the voltage on the capacitors to ensure they are in fact charging normally. This ensures there is no undetected direct short circuit in the inverter.
  v. A short test exposure at low kV and mA should be attempted. If OK, then switch off, and restore the operation of the charge current-limit resistors to normal operation.

Other possible causes for no exposure
- A faulty fuse or connector in the generator control cabinet.
  i. *Ensure all power is turned off, including the room power isolation switch.*
  ii. With the aid of an electrician or electronics technician, look for an open circuit fuse, or tripped circuit breaker.
  iii. Look for loose plugs and sockets, or other connectors.
- Look for damage caused by rats eating the wiring. This can especially apply inside cable ducts, or cables to the HT transformer. Some species of rats can bite through cables, or a wiring harness, as if their teeth were cutting pliers.

Part 8. High-tension problems

Refer first to the flow chart, Fig 6–8 page 85, 'No exposure, part eight'.

On attempting to obtain an exposure, the film was blank, or very under exposed. The control indicated preparation was completed, and ready for an exposure. The exposure indication operates on pressing the expose position of the handswitch, but only a very short exposure results.

A warning light or error message is generated during, or at the end of exposure.

Record the exposure factors, focal spot, and technique in use. Include any fault codes or error messages.

Check for a high-tension fault

The cause of the problem may be due to:

- No kV was generated. The inverter fuse is open circuit in a high frequency generator. **See part seven.**
- A kV fault. Caused by a short circuit in the HT cable or cable end.
- A high voltage arc in the high-tension generator, cables, or X-ray tube housing.
- An unstable X-ray tube, due to gas. This can generate a kV or mA fault indication. Instability can occur if a high kV exposure is made, after using the tube for a long time at a medium or low kV output.
- A kV fault can be caused if mA is too low. For example, due to a bad connection in the cathode cable end, or a partial short circuit in the X-ray tube filament.
- The mA output has exceeded the correct value. For example, although 200 mA was selected for an exposure, the mA increased to over 300 mA. This could be due to a fault in the mA regulation circuit, a high-tension fault, or an unstable X-ray tube.

Check for a HT cable fault

HT cable faults normally occur at the X-ray tube end, due to the HT cable twisting as the tube is positioned. Arcing can also occur where the cable ends plug into the X-ray tube, due to a fault in the insulation grease.

- Was a 'bang' heard during the exposure?
- Check for a burning or acrid odour at the HT cable ends. If not sure, undo the retaining ring nut, 'sniff', and then replace the ring nut.
- If at all uncertain of the condition of the HT cable ends, refer to **module 7.3 page 117**.

- **Caution.** Do not make this test on a CD mobile, unless the capacitor is fully discharged. See **module 6.2 page 94**.
- If the HT cable or cable ends appear OK, then attempt a test exposure at low output.
    i. **Caution.** *If the fault occurred below 80 kV, do not test further without advice from the service department.*
    ii. *Select a low kV and mA output. A suggested exposure is 60 kV, 100 mA, for 0.1 seconds. Close the collimator and make a test exposure.*
    iii. *If the control is fitted with an mA or mAs meter, observe the meter carefully during the test exposure.*
    iv. *If the mA meter needle moves very quickly, or the mAs meter indicates 20% more than the expected value, this can indicate a high tension fault.*
    v. *Should the generator fail a test at 60 kV, stop. There is a possible fault at the HT cable ends, these should now be removed and carefully examined. See* **module 7.3 page 117**.
    vi. **Caution.** *Before removing the cable ends, contact the service department for advice. Include the exposure factors, focal spot, and any error codes or messages.*

Test the X-ray tube high-tension stability

If the test exposure at 60 kV is ok, then test the X-ray tube at higher kV.

- Do this by applying the seasoning procedure, described in **module 2.1 page 48**. When applying this procedure, take care to observe the HT cable at the X-ray tube, in case an arc occurs.
- Observe the mA meter (if available). An unstable tube can indicate an increase of mA when test exposures are first made, then return to the correct value after a few exposures at the same kV. Then when kV is increased, the first exposure may again show an increase of mA, returning to the correct value after the second or third exposure.
- In case any unusual occurrence takes place, immediately stop. HT cable or cable-end faults will become apparent at higher kV values. If not fully checked before, do so now. See **module 7.3 page 117**.
- **Caution.** *Before removing the cable ends, contact the service department for advice. Include the exposure factors, focal spot, and any error codes or messages.*

PART III. FAULT DIAGNOSIS AND REPAIR MODULES

- If the HT cable ends have been fully checked, then start the seasoning test again, starting at 10 kV below the level where a problem occurred. Should the problem still occur, then stop. Consult the service department for advice.

- Providing the X-ray tube passes the seasoning test, all is well.
- Make an entry in the logbook. Include a description of the problem, together with the tests and results.

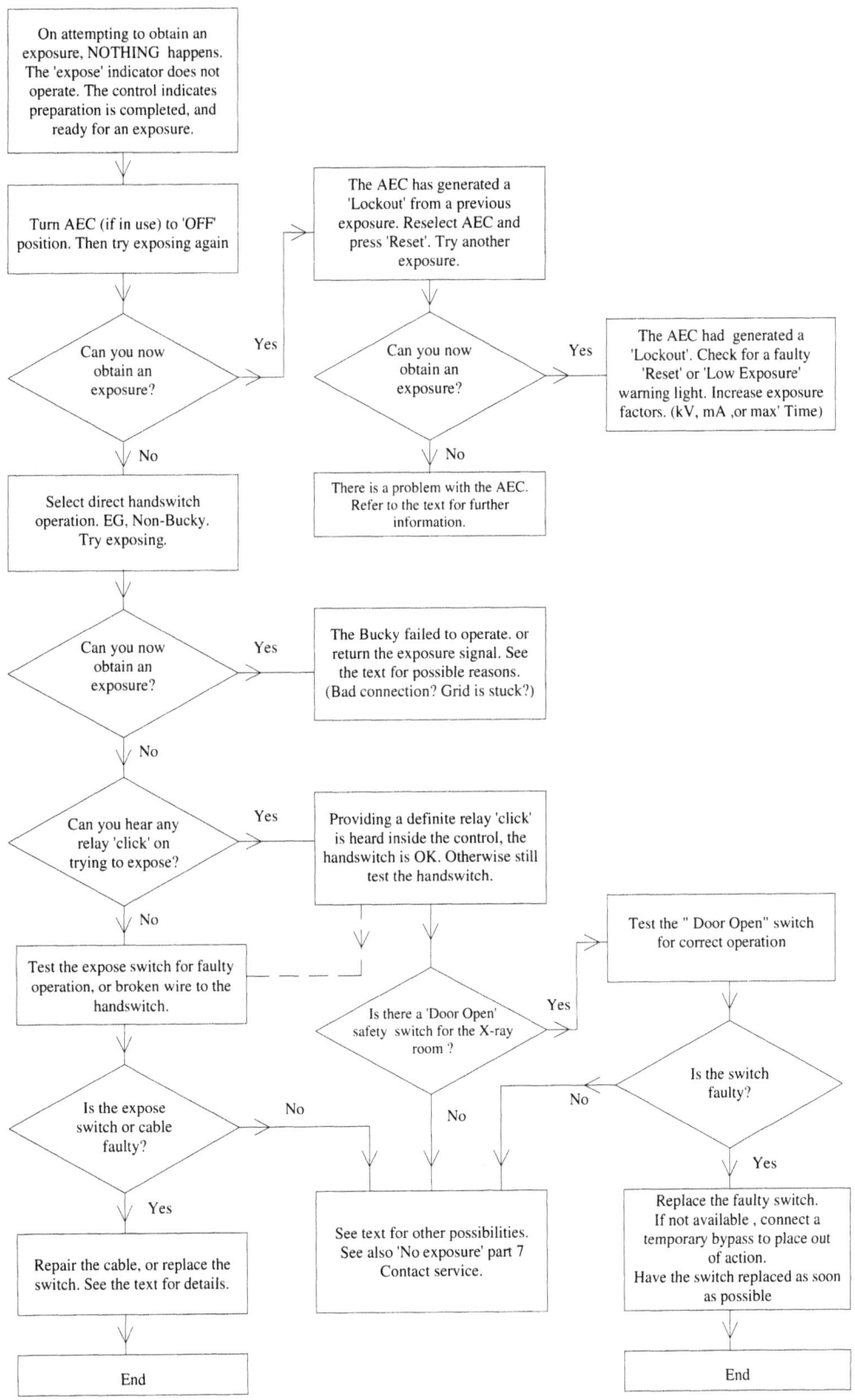

Fig 6–6. No exposure, part six

X-RAY EQUIPMENT MAINTENANCE AND REPAIRS WORKBOOK

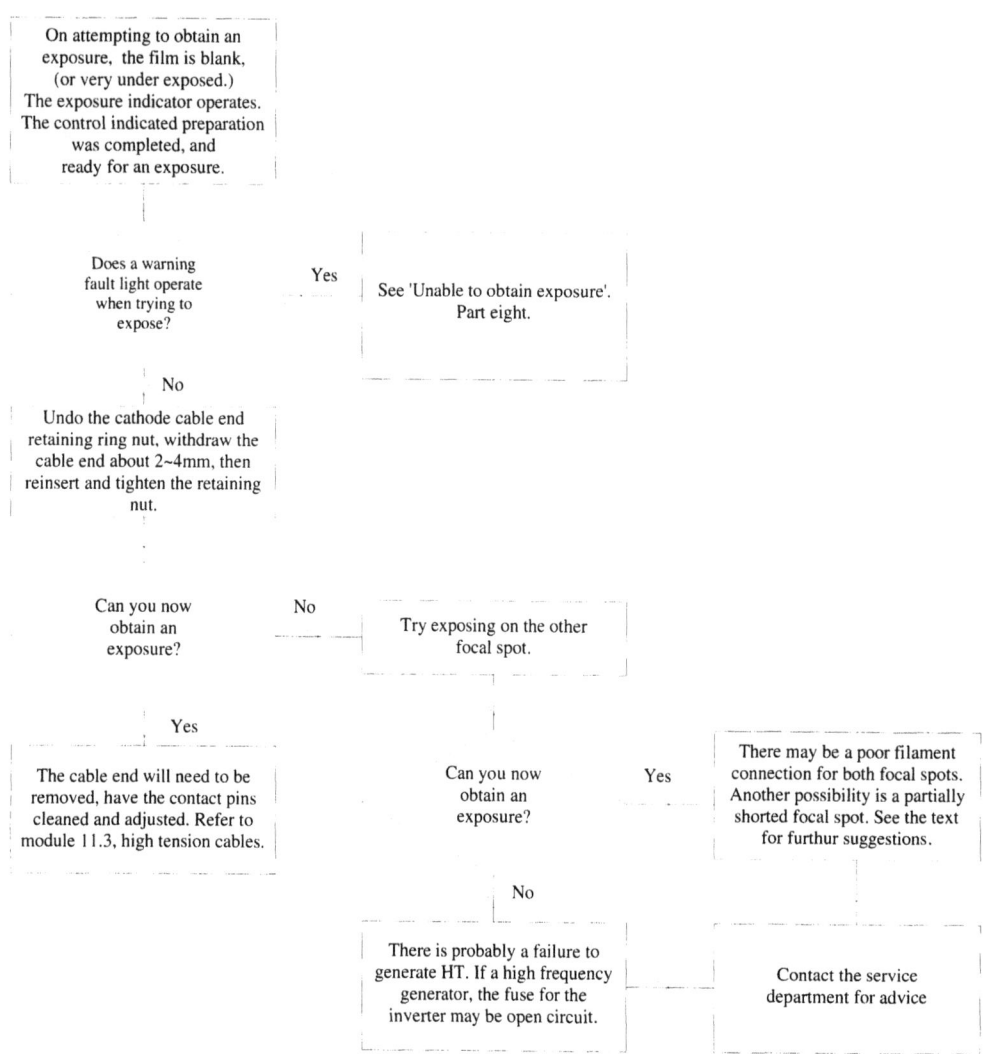

Fig 6–7. No exposure, part seven

PART III. FAULT DIAGNOSIS AND REPAIR MODULES

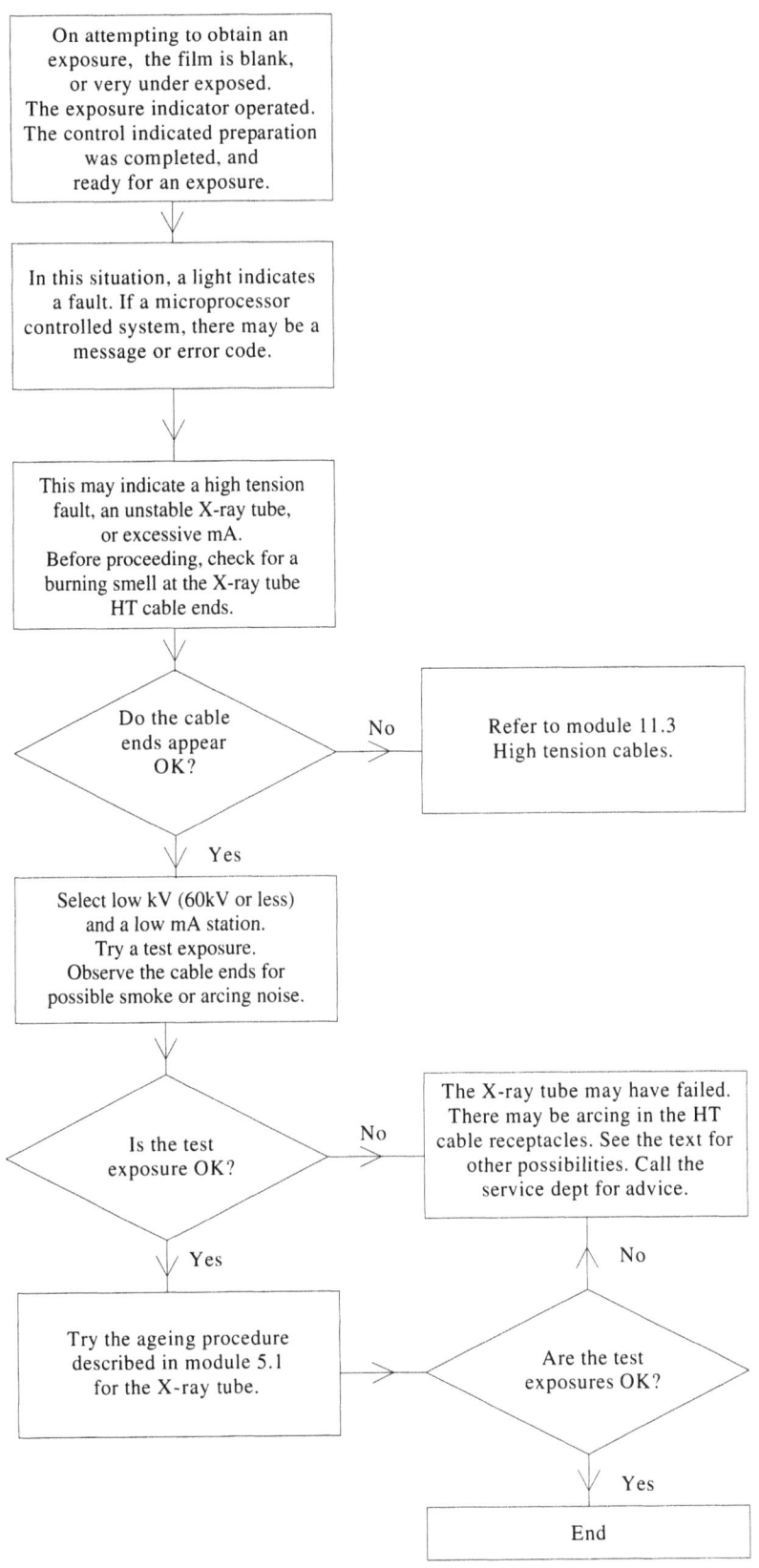

Fig 6–8. No exposure, part eight

TASK 9

No preparation
Part I

You have a patient on the table. After instructing him to 'take a deep breath and hold it', you press the preparation button. However, the 'ready to expose' signal does not occur.

What action should be taken first?
_____
_____

Make a list of your tests and observations.
_____
_____
_____
_____
_____
_____
_____

What do you think is the cause of the problem?
_____
_____
_____
_____

What action may be taken to correct the problem?
_____
_____
_____
_____
_____

Tutor's comments

Satisfactory/Unsatisfactory
_____

Signed _____ Date _____
Tutor

PART III. FAULT DIAGNOSIS AND REPAIR MODULES

TASK 10
No preparation
Part 2

The X-ray control does not provide the 'ready for exposure' signal after pressing the preparation switch. You have just completed a previous check for a similar report, and taken action that should have corrected the problem. However, while there are now 'signs of response' when the preparation button is pressed, the ready for exposure indication still cannot be obtained.

What was the previous observed problem, and action taken?

___

Make a list of your further tests and observations.

___

What do you think is the cause of this problem?

___

What possible actions may be taken to correct the problem?

___

Tutor's comments

Satisfactory/Unsatisfactory

Signed _____   Date _____
                Tutor

TASK 11
No exposure

You have positioned a patient on the upright Bucky to take a chest exposure. You place the generator into preparation, and press the exposure switch when the control indicates 'ready to expose'. However, no exposure occurs. After removing the patient, you investigate for a possible cause of the problem.

Make a list of possible reasons for this problem.

_____
_____
_____
_____
_____

Describe suitable tests to either confirm, or eliminate, possibilities from this list.

_____
_____
_____
_____
_____

Carry out these tests. What were the results?

_____
_____
_____
_____

What action is needed to correct the problem?

_____
_____
_____
_____
_____

Tutor's comments

_____

Satisfactory/Unsatisfactory

Signed _____ Date _____
Tutor

TASK 12
X-ray output linearity

You have a report of 'light' films occurring in room 2. This appears to happen only on selection of fine focus. Devise a test, using the stepwedge, to check the exposure linearity between the fine and broad focus selections of mA. _____
_____

Which exposure factors will be used for your generator? _____
_____

After carrying out the above test, you find the 100 mA station has reduced output. A comparison test of 50 mA and 200 mA indicates correct results. What is the most likely cause? _____
_____

What are two other possible reasons for light films on this 100 mA station? _____
_____

Give reasons why these possibilities are low, considering that 50 mA and 200 mA indicated normal results.
_____
_____
_____

Make another series of stepwedge tests. This time remain on the mA station that produced a low output. Make a series of four test strips, and increase kV by 2.0 kV for each exposure. What increase of kV was required to increase density by one step? _____

Your test was first carried out at 80 kV. You find that if kV is increased by 4.0 kV, the density steps on the 100 mA test strip are darker than the test trips for 50 or 200 mA stations. In terms of a 5% permitted kV error, would the test result be:

OK? _____ Marginal? _____ Outside acceptance? _____

Tutor's comments

Satisfactory/Unsatisfactory

Signed _____   Date _____
               Tutor

MODULE 6.1

Mobile or portable X-ray generators

Aim

The aim is to provide information related to servicing or repairing a mobile or portable X-ray generator. This information is additional to that provided for a fixed installation described in module 6.0 page 71.

Note. Capacitor discharge mobiles are discussed in module 6.2 page 94. Reference module page numbers refer to the title page.

Objectives

On completion of this module, the student will be aware of common problems with mobile or portable X-ray generators. When used together with the module 6.0 procedures, the student will be able locate a problem area, and carry out minor repairs. An electrician or electronics technician may provide added assistance where indicated in this module.

Contents

a. General precautions
b. Transport problems
c. The generator will not switch on
d. No preparation or no exposure
e. The generator appears to expose, but film is blank
f. Collimator light-beam alignment keeps changing
g. The magnetic locks sometimes do not work, or are weak in action
h. Problems with the motor drive

a. General precautions

- Before removing any covers, or testing any wires or connections:
  i. ***Ensure the system is switched off, and unplugged from the power point.***
  ii. Mobile high-frequency generators may be battery operated. The batteries in these are connected in series, and may have a total voltage of up to 240V DC. *Refer to the operating or installation manuals for the position of the battery isolation switch, and ensure this is switched off before removing the covers or testing wires and connectors.*
  iii. **Do not** attempt to replace the batteries if a mobile has more then two 12V batteries. Ask for assistance from an electrician or electronics technician.
- If removing a collimator or X-ray tube on a mobile generator.
  i. **Do not** rely on the vertical lock system.
  ii. Ensure the suspension system is at the limit of its maximum vertical travel.
  iii. Or, attach a rope so that the system cannot move upwards, once the weight of the collimator or X-ray tube is removed.
- **When replacing a motor drive battery.**
  i. Remove your wristwatch if it has a metallic band. The same precaution extends to any rings. While 12 volts, or even 24 volts, is too low to cause a serious shock, the battery can cause serious burns if a short circuit occurs across a watchband or ring.
  ii. **Disconnect first** the battery terminal that connects to the mobile body or framework. This prevents the danger of accidentally shorting the other battery terminal to the mobile body.
  iii. In the case of multiple battery systems, refer always to the operation or installation manuals. In some cases, the batteries will have the centre connections between two batteries connected to the mobile body. These connections should be removed first, and replaced last. It is advisable to

PART III. FAULT DIAGNOSIS AND REPAIR MODULES

request the aid of an electrician. If in any doubt, contact the service department for advice.

b. Transport problems

Mobile and portable equipment are subject to additional problems due to transport. Portable equipment may be dropped, while mobile generators may pass over severe bumps while travelling. There are also problems due to dust, or corrosion due to high humidity. This last can occur if the system is used in an air-conditioned area, then parked in a general area which is not air conditioned.

c. The generator will not switch on

- Check the power cable.
    i. Is the power point faulty? Check by plugging a lamp or other suitable item into the power point.
    ii. Is there a broken connection at the power-cable plug? Important, check also the earth connection.
    iii. The power cable may plug into a socket on a portable generator. In this case check the generator socket connections as well as the cable.
    iv. The power cable-reel for a mobile can have broken wires or faulty contacts in the mechanism. If faulty, have an electrician connect a temporary power cable directly into the mobile, while waiting for a replacement.
    v. If there is a bad connection to a power plug or socket, this should be checked and repaired by an electrician.
    vi. For information on locating bad connections, **see module 5.0 page 65**.
    vii. Moving the mobile while still plugged into a power point is not recommended.
- Check for a blown or faulty fuse. **See module 5.0 page 65**.
- The power on/off switch is faulty. Check for loose connections. Test the switch operation with a multimeter set to the low ohms range. Ask an electrician or electronics technician for assistance.

d. No preparation or no exposure

Note. The following checks are to be made with all power disconnected. If a battery operated high frequency system, ensure the battery circuit breaker or switch is opened.

- Look for possible blown fuses. **See module 5.0 page 65**.
- Check the hand switch, and handswitch cable. **See module 5.0 page 65**.
- Check all external plugs and sockets. **See module 5.0 page 65**.
- Look for dislodged components internally, or poor connections.
- Plugs and sockets can develop poor connections due to build up of oxides, or else slight corrosion of the plating. Check by unplugging and reconnecting the plugs and sockets.
    i. ***This procedure should only be carried out by an electrician or electronics technician.***
    ii. Printed circuit board edge connectors are subject to corrosion and poor connections, especially older types that are not gold plated. Later model generators use plugs and sockets for the printed circuit boards.
    iii. **Ensure the generator is unplugged from the power point.**
    iv. *Before removing a printed board, touch the main metal framework of the generator. This is to discharge any static electricity from the body. This is very important if working in a carpeted area, or there is low humidity.*
    v. Only remove and replace one board at a time. Take careful note of its position, so the board is not replaced the wrong way.
    vi. On removal of the printed board, clean the edge connectors with a cloth dampened with a little alcohol, or methylated spirits. After cleaning, do not touch the contacts with your fingers.
    vii. Printed circuit boards may become cracked, breaking some of the tracks. For example, if equipment was dropped. Examine with a magnifying glass for this possibility. If there are signs of corrosion, it is possible some of the tracks on the board have become open circuit. This can be a problem near the sea.
- Have any relays become dislodged? Move them slightly in their sockets in case of a bad socket connection.
- Bad contacts on control switches etc.
    i. A build up of dirt on switch contacts can cause excessive wear, and poor contact operation. In most cases, spraying with a suitable contact cleaner will restore normal operation.
    ii. An optimum spray is one designed to clean and lubricate. CRC 2.26 is recommended. WD-40 is less optimum, but may also be used. Your electrician may have a suitable contact cleaner.
    iii. Before spraying, cover adjacent areas with cloth or tissues to protect from unwanted spray.

iv. **Note. Never use any spray on electrical equipment while it is energized.**
- The X-ray tube head, or housing. (An X-ray 'tube head' combines the X-ray tube, and the high-tension transformer, in one housing.)
    i. Check all connections to the X-ray tube head.
    ii. For mobiles with a conventional X-ray tube housing, check the stator cable.
    iii. The HT cable may have a poor connection to the X-ray tube cathode. For test procedures, **see module 7.3 page 117.**

e. *The generator appears to expose, but film is blank*

- Ensure the collimator opened correctly. Look inside the collimator for loose or disconnected parts, due to vibration in transport.
- Was an unusual noise or arcing heard during the attempted exposure?
    i. With conventional X-ray tube housings, check the HT cable ends for arcing. **See module 7.3 page 117.**
    ii. If a rotating anode tube, does it slow down quickly when preparation is released? The tube may be broken, and the anode is now immersed in oil.
    iii. By rotating the X-ray tube head or housing, can you hear oil moving inside the housing? This can indicate a broken tube.
    iv. See also **module 7.1 page 104.**

f. *Collimator light-beam alignment keeps changing*

- Parts of the collimator may have become loose due to transport vibration. Remove the collimator cover, and check for loose sections.
- For adjustments to the collimator, **see module 7.2 page 110.**

g. *The magnetic locks sometimes do not work, or are weak in action*

There are two types of electromagnetic locks. Older types require power to operate. Later types have a permanent magnet, and require power to remove, or cancel, the magnetism.

- The lock sometimes operates, depending on the tube stand movement.
    i. There may be a poor connection to the lock coil or the lock switch.

    ii. The mechanical position of the lock coil has too large an air gap. Adjust the position for a smaller air gap, with the lock in the off position.
    iii. There is a build up of dirt, or oil, on the lock-coil pole-piece. A thin piece of cardboard soaked in methylated spirits is a good cleaning aid. Place the cardboard between the pole-piece and the brake plate. Energise the lock, while slowly pulling the cardboard out from between the lock and the brake plate. Repeat this a number of times, until fresh sections of cardboard show no smudges of dirt from this wiping action.
- The lock does not energise. (Old type.) Or does not release. (Permanent magnet type.) Check for the following.
    i. The lock switch may be faulty.
    ii. Look for an open circuit fuse.
    iii. A broken wire to the lock coil.
    iv. The lock coil itself may be open circuit. Test by disconnecting one wire, then with a multimeter set to a medium or high ohms scale, check for continuity of the lock coil. If necessary, request assistance from an electrician.

h. *Problems with the motor drive*

Many mobiles are fitted with battery-powered motor drives. A 12 volt car battery is most commonly fitted, and in some cases two batteries to provide 24 volts.

Caution
Battery powered high-frequency generators operate the motors from the same high voltage supply used to power the high-frequency inverter. **Due to the high voltages that may be present, do not remove the covers, or attempt any internal adjustment. This should only be attempted by an electrician or electronics technician, under instructions from the service department.**

- There is no motor drive in either direction.
    i. Units fitted with a drive control key switch. The switch may be faulty, or have a loose connection.
    ii. The fuse for battery power is open circuit.
    iii. The battery is discharged.
    iv. There is a bad connection to the battery terminal. Remove the battery terminals, and scrape corrosion from the posts or terminal clamps.
    v. **Caution.** See part a, general precautions, before disconnecting a battery.
    vi. The brake release bar on the handle operates a microswitch. The microswitch may need adjustment to operate correctly.

PART III. FAULT DIAGNOSIS AND REPAIR MODULES

vii. Check the electrical cables for loose plugs or sockets.

- There is no forward motor drive.
  i. The anti-crash bumper may be stuck in the operated position, or one of the switches operated by the bumper is damaged.

- There is only a low level of motor drive assistance.
  i. The motor may be in bedside mode. A control switch selects the change between bedside speed, and travel speed. The control switch could be faulty, or the knob has slipped, and indicates the wrong position.
  ii. Other mobiles require the tube stand to be placed in the travel position before permitting full power. In which case a microswitch to sense the tube stand position may need adjustment. Ask an electrician or electronics technician to check the switch, or microswitch operation.
  iii. New mobiles have microcomputer control of the motor power. Some systems allow the operator to change the amount of power assistance. Refer to the operation manual to adjust the level of motor power.
  iv. There may be a poor battery connection due to corrosion. Remove the battery terminal clamps, scrape of any corrosion, and reassemble.
  v. Caution. See part a, general precautions, before disconnecting a battery.

- The battery is not charged up.
  i. The battery was not charged overnight. Or the power point used was faulty. Check the power point with a lamp or similar item.
  ii. The power cord used for battery charging has a bad connection. Have an electrician repair the connection. (Do not move the mobile while plugged into a power point.)
  iii. There is an open circuit fuse, in the battery charge section of the mobile. To check a fuse, **see module 5.0 page 65**.
  iv. **Caution.** *If this is a battery operated high-frequency generator, this check should only be performed by an electrician, or electronics technician.*

- After charging, the battery soon looses power.
  i. Has the electrolyte level of the battery been checked? (This may not be possible with sealed or low maintenance batteries.)
  ii. The battery may have a shorted cell, or the cells are sulphated, in this case a new battery is required. Before replacing the battery, have the battery tested at a garage, or by an auto electrician.
  iii. **Caution.** See part a, general precautions, before disconnecting a battery.

MODULE 6.2
Capacitor discharge mobile

Aim

The aim is to provide information related to servicing or repairing a capacitor discharge (CD) mobile. This information is additional to that provided for a fixed installation described in module 6.0 page 71, and mobile generators described in module 6.1 page 90. (**Note:** Reference module page numbers refer to the title page.)

Objectives

On completion of this module, the student will be aware of common problems with capacitor discharge mobile generators. When used together with modules 6.0 and 6.1 procedures, the student will be able locate a problem area, and carry out minor repairs. The student will also be aware of special precautions when dealing with the capacitor high-voltage system, where repairs must only be attempted by an electrician or electronics technician.

Contents

a. CD mobile operation modes
b. General precautions
c. High-tension cable precautions
d. The mobile does not switch on
e. Unable to charge the capacitor
f. On charging the capacitor, there is a loud 'bang'
g. Unable to obtain preparation
h. On going into preparation, the capacitor discharges
i. No exposure
j. The kV does not adjust to a lower value
k. Apparently there is an exposure, but the film is blank
l. On exposing, the exposure continues till the capacitor is fully discharged
m. Problems with the motor drive

a. CD operation modes

Unlike the fixed installation, or standard mobile generator, the CD mobile has special modes of operation. This list is provided as a reminder.

- High voltage is applied to the X-ray tube continuously while the capacitor is charged. This includes the kV remaining after the exposure.
- When not in preparation, the X-ray tube filament has no pre-heating. There is also a high negative voltage applied between the cathode cup or grid, and the filament. Despite these precautions, a very small electron emission does occur. This is called 'dark current'.
- To prevent X-ray emission due to dark current, the collimator has an additional lead shutter. This shutter blocks all X-ray emission. The shutter moves out only when the mobile begins preparation, or else just before the X-ray exposure.
- When in preparation, the exposure is prevented by the negative voltage applied to the X-ray tube grid. During an exposure, this voltage is removed, allowing full emission from the cathode. At the end of the exposure, the negative voltage is again connected to the grid. This shuts off the electron beam to the anode, ending the exposure.
- Earlier CD mobiles set the exposure as a percentage of kV drop during an exposure. Later systems have an mAs timer.
- The capacitor discharges during an exposure, at the rate of 1 kV per mAs. As a result, the X-ray output of a CD mobile is not linear. A 20 mAs exposure will not give twice the output of a 10 mAs exposure.

b. General precautions

- **Before removing any covers, ensure the mobile is switched off, and unplugged from the power point.**
- If removing a collimator or X-ray tube.
  i. Do **not** rely on the vertical lock system.
  ii. Ensure the suspension system is at the limit of its maximum vertical travel.

iii. Or, attach a rope so that the system cannot move upwards, once the weight of the collimator or X-ray tube is removed.
- When replacing a motor drive battery.
  i. Do **not** attempt to replace the batteries if a mobile has more then two 12 V batteries. Ask for assistance from an electrician or electronics technician.
  ii. Remove your wristwatch if it has a metallic band. The same precaution extends to any rings. While 12 volts, or even 24 volts, is too low to cause a serious shock, the battery can cause serious burns if a short circuit occurs across a watchband or ring.
  iii. Disconnect **first** the battery terminal that connects to the mobile body or framework. This prevents the danger of accidentally shorting the other battery terminal to the mobile body.
  iv. If two batteries are fitted, refer always to the operation or installation manuals. In some cases, the batteries will have the centre connection between the two batteries connected to the mobile body. These connections should be removed first, and replaced last. It is advisable to request the aid of an electrician. If in any doubt, contact the service department for advice.

c. High-tension cable precautions

Special care is required due to dangerous high voltage stored in the capacitors. Several of the tests described involve removing or exchanging the high-tension cables. This requires taking care to ensure the capacitor is fully discharged, before any attempt is made to remove or adjust the cable ends. *Only an electrician, or electronics technician, should attempt this procedure.*

- If possible, make an exposure to fully discharge the capacitor, or reset the kV to the minimum level possible.
- Wait until the kV has dropped to below 5 kV, as indicated on the panel meter. If no indication of capacitor high-voltage by a panel meter, wait overnight before proceeding.
- **Switch off, and unplug the power cord.**
- Open the control panel, and locate the two manual capacitor-discharge control knobs. These may be operated by lifting and rotating, and must stay in the discharge position. **If uncertain of their operation, refer to the installation or service manual for the mobile. Otherwise contact the service department for advice.**
- Before operating the discharge knobs, you may observe two neon lamps glowing. On operation of the discharge knobs, both of these lamps should turn off. **If one lamp continues to glow, contact the service department for advice before proceeding.**
  i. Undo the ring nut holding the cable end in position at the X-ray tube receptacle.
  ii. **Do not** remove the cable end, but first inspect the safety-shield metal braid for damage. Twisting of the cable may have caused it to break, and become disconnected from the cable end.
  iii. If the shield does not appear damaged, then re-tighten the ring nut so the cable shield is properly grounded.
  iv. Undo the ring nut holding the high-tension cable in the high-tension tank receptacle. On withdrawing the cable end, **do not touch** the pins, but first short them to the side of the receptacle. This is to ensure any possible charge in the high-tension cable is completely shorted out.
  v. In case the high-tension cable shield appears damaged at the X-ray tube end, then remove the cable end from the X-ray tube **first**. Again, touch the end pins to side of the receptacle to discharge any residual high voltage. This will include any charge in the capacitor as well as the high-tension cable. Now undo the ring nut, and remove the cable end from the high-tension tank.
  vi. The above precautions are in case the knobs for discharging the capacitors have not operated correctly. In part (iv) only a small spark will occur if the capacitor was not discharged, but in part (v) take care, as there may be a very big spark. Normally, there should be no spark at all. In case there is, then contact the service department before proceeding further.
- When the high-tension cable is removed or replaced, the cable ends must be cleaned and resealed. **See module 7.3 page 117.**
- After replacing or adjusting the high-tension cable, return the discharge knobs to their normal position.

d. The mobile does not switch on

- Is the power cable faulty?
  i. Is the power point faulty? Check by plugging a lamp or other suitable item into the power point.
  ii. Is there a broken connection at the power-cable plug? Important, check also the earth connection.

iii. The power cable-reel may have broken wires or faulty contacts in the mechanism. If faulty, have an electrician connect a temporary power cable directly into the mobile, while waiting for a replacement.
iv. If there is a bad connection to a power plug or socket, this should be checked and repaired by an electrician.
v. For information on locating bad connections, **see module 5.0 page 65**.
vi. Moving the mobile while still plugged into a power point is not recommended.
- Check for a blown or faulty fuse. **See module 5.0 page 65.**
- The power on/off switch is faulty. Check for loose connections. Test the switch operation with a multimeter set to the low ohms range. Ask an electrician or electronics technician for assistance.

e. Unable to charge the capacitor

On pressing the charge button, nothing happens. Otherwise all appears normal.

- With power switched off, and the mobile unplugged from the power point, check the following;
  i. Some CD mobiles have the capacitor charge switch mounted on the hand switch. There may be a broken wire or a faulty switch.
  ii. To check the hand switch or cable, **see module 5.0 page 65**.
  iii. Open the control panel, and check for an open circuit fuse. At this time also check internally for loose plugs or sockets, and or loose connections.
  iv. Check the wiring to the collimator for broken or loose connections.
  v. Remove the collimator cover. Locate the dark-current shutter mechanism. Check that the shutter is in the correct position, and the shutter microswitch operates correctly.
  vi. Have the manual capacitor-discharge knobs been left in the discharge position, after a high-tension cable was adjusted? Or, is a microswitch, operated by these knobs, sticking? This will indicate the manual discharger is still operated, and prevent the capacitor charging.

f. On charging the capacitor, there is a loud 'bang'

This indicates a high-tension fault. This could be the capacitor, but may instead be in a cable end, or perhaps the X-ray tube.

> *These tests should only be performed by an electrician or electronics technician.*
> **Refer to (b), 'High-tension precautions', before proceeding.**

- Undo the ring-nuts retaining the high-tension cable-ends in the X-ray tube receptacles. Is there a strong odour from one of the cable ends?
  i. If there is, either the cable end is faulty, or there has been an arc-over in the receptacle.
  ii. **With the capacitor fully discharged and safe**, withdraw the suspect cable end. (See part b. High-tension cable precautions.)
  iii. After taking the precaution of shorting the pins to ground, examine the cable end and receptacle for traces of carbon. An easy way to find contamination of grease etc is to wipe with a paper tissue.
  iv. If there is no indication of arcing inside the receptacle, but there is a strong odour from the high-tension cable as it enters the cable end, the cable is faulty and requires replacement.
  v. In case there are signs of arcing in the receptacle, then the receptacle and cable end must be carefully cleaned, and re-sealed. **See module 7.3 page 117.**
  vi. After a cable end has been withdrawn, it will need to be re-sealed with fresh grease, or have new anti-corona pads fitted. **See module 7.3 page 117.**
- Do the high-tension cables or X-ray tube receptacles appear OK? The capacitors or X-ray tube could have a short circuit. Contact the service department for assistance. Provide full details of the fault, tests made, and the results.

g. Unable to obtain preparation

- Is the capacitor fully charged? Check by pressing the charge button. Or else by increasing the preset kV, in which case the charge mode should operate.
- There may be a faulty preparation switch, or a faulty cable from the handswitch. To check the switch or cable, **see module 5.0 page 65**.
- Check the wiring to the X-ray tube and collimator, especially if it has been subject to pulling. This includes;
  i. Stator cable and connections.
  ii. Connections to the thermal overload switch, in the tube housing.

PART III. FAULT DIAGNOSIS AND REPAIR MODULES

- Check for a loose plug or socket. **See module 5.0 page 65.**
- Look for a loose or open circuit fuse. **See module 5.0 page 65.**
- Contact the service department for assistance. Provide full details of the fault, tests made, and the results.

h. On going into preparation, the capacitor discharges

The negative control voltage applied to the X-ray tube grid is missing. As the filament heats up, this results in an uncontrolled exposure, and may fully discharge the capacitor. A common cause is a faulty high-tension cathode cable, with an internal short circuit between the grid wire, and one of the filament wires. To check for a possible fault in the high-tension cable, the cables can be exchanged between anode and cathode. (Of course, there is a possibility the cables were previously exchanged, but no record kept of this event.)

> ***This procedure should only be performed by an electrician, or electronics technician.***
>
> ***Refer to (b), 'High-tension precautions', before proceeding.***

- Mark each cable end before removal, to ensure they are correctly exchanged after being re-inserted.
- Ensure the capacitor is fully discharged, power is disconnected, and the manual discharge knobs are in position.
- Undo the ring nuts' holding the cable ends in position, and withdraw the cable ends from the high-tension transformer. Take note, which receptacle is positive, and which is negative. The X-ray tube cathode must connect to the negative receptacle.
- Remove the high-tension cable ends from the X-ray tube housing.
- For cleaning and resealing the cable ends on reinsertion, see **module 7.3 page 117**.
- Both high-tension cables have now been disconnected. Now reconnect the cables to the X-ray tube and generator. The previous cathode cable is connected to the anode, and the previous anode cable is connected to the cathode.
- Take care to connect the anode cable to the positive receptacle, and the cathode cable to the negative receptacle of the high-tension transformer.
- Power up. Recharge the capacitor and make a test exposure.
- If all is well, attach a label to the previous cathode cable. This should indicate it has an internal short circuit, and is suitable as an anode cable only.
- If the problem still occurs, contact the service department for assistance. Provide full details of the fault, tests made, and the results.

i. No exposure

- Does the mobile have a Bucky connection option? Try exposing with the Bucky option switched off or bypassed.
- Is the capacitor fully charged? In some designs, if the capacitor is not fully charged to the required value, this will prevent preparation. In other systems it may instead prevent a radiographic exposure.
- Check the operation of the handswitch, or a possible broken wire in the handswitch cable. To check the switch or cable, **see module 5.0 page 65**.
- Check the wiring and connections to the collimator.
    i. Remove the collimator cover and check for correct operation of the dark-current shutter. This should move out of the way, either during preparation, or just before an exposure.
    ii. Check the microswitch operated by this shutter. If sticking, or not fully operated, this may be the cause.

j. The kV does not adjust to a lower value

For example, after charging the capacitor, you decide to use less kV for the exposure. However, on trying to reset to a lower kV, nothing immediately happens, although the capacitor voltage may very slowly drop in value.

To reset the kV down to the new setting, the generator makes a low mA exposure. Radiation is prevented from leaving the collimator by the dark-current shutter. The shutter also operates a safety microswitch. This prevents the low mA exposure if the shutter is not closed.

- Switch the power off, and unplug the power cord from the power point.
- Remove the collimator cover.
- Check the collimator dark-current shutter for correct operation.
- Check if the shutter microswitch has operated correctly, and there are no bad connections or broken wires to the collimator.
- If the shutter and microswitch appear correct, contact the service department for advice. Provide full details of the fault, tests made, and the results.

k. Apparently there is an exposure, but the film is blank

On exposing, the capacitor voltage drops the expected amount of kV. For example, the kV dropped by 10 kV after a 10 mAs exposure.

- In this situation, the dark-current shutter has failed to open properly. Remove the collimator covers and check the operation of the shutter, and its associated microswitch.

l. On exposing, the exposure continued till the capacitor was fully discharged

- The X-ray tube may have become unstable. Try a test exposure at a much lower kV setting. If successful, then try the X-ray tube seasoning procedure. **See module 2.1 page 48.**
- If a low kV test exposure shows the same fault, contact the service department for assistance. Provide full details of the fault, tests made, and the results.

m. Problems with the motor drive

Some mobiles are fitted with battery-powered motor drives. A 12 volt car battery is most commonly fitted. In some cases two batteries are used, to provide 24 volts.

Whenever possible, refer first to the operation or installation manual. If in doubt contact the service department, or request the assistance of an electrician.

- There is no motor drive in either direction.
  i. Units fitted with a drive control key switch. The switch may be faulty, or have a loose connection.
  ii. The fuse for battery power is open circuit. **See module 5.0 page 65.**
  iii. The battery is discharged.
  iv. There is a bad connection to the battery terminal. Remove the battery terminals, and scrape corrosion from the posts or terminal clamps.
  v. Caution. See part b, general precautions, before disconnecting a battery.
  vi. The brake release bar on the handle operates a microswitch. The microswitch may need adjustment to operate correctly.
  vii. Check the electrical cables for loose plugs or sockets.
- There is no forward motor drive.
  i. The anti-crash bumper may be stuck in the operated position, or one of the switches operated by the bumper is damaged.
- There is only a low level of motor drive assistance.
  i. The motor may be in bedside mode. A control switch selects the change between bedside speed, and travel speed. The control switch could be faulty, or the knob has slipped, and indicates the wrong position.
  ii. Other mobiles require the tube stand to be placed in the travel position before permitting full power. In which case a microswitch to sense the tube stand position may need adjustment. Ask an electrician or electronics technician to check the switch, or microswitch operation.
  iii. There may be a poor battery connection due to corrosion. Remove the battery terminal clamps, scrape of any corrosion, and reassemble.
  iv. Caution. See part b, general precautions, before disconnecting a battery.
- The battery is not charged up.
  i. The battery was not charged overnight. Or the power point used was faulty. Check the power point with a lamp or similar item.
  ii. There is an open circuit fuse in the battery charge section of the mobile. **See module 5.0 page 65.**
- After charging, the battery soon looses power.
  i. Has the electrolyte level of the battery been checked? (This may not be possible with sealed or low maintenance batteries.)
  ii. The battery may have a shorted cell, or the cells are sulphated, in this case a new battery is required. Before replacing the battery, have the battery tested at a garage, or by an auto electrician.
  iii. **Caution.** See part b, 'general precautions', before disconnecting a battery.

PART III. FAULT DIAGNOSIS AND REPAIR MODULES

MODULE 7.0
X-ray tube stand

Aim

The aim is to provide information for repairing or adjusting the tube stand. This is additional information to the maintenance procedures, provided in **module 2.0 page 44**. Procedures for electrical tests are provided in **module 5.0 page 65**.
(**Note:** Reference module page numbers refer to the title page.)

Objectives

On completion of this module, the student will be aware of common problems with the X-ray tube stand. When used together with the module 5.0 procedures, the student will be able identify a problem area, and carry out minor repairs. An electrician or electronics technician may provide added assistance where indicated in this module.

Task 13. 'Bucky tabletop and tube-stand centre', should be attempted on completion of this module.

Contents

a. General precautions
b. The vertical movement is not balanced
c. When power is turned off, the tube stand starts moving
d. The tube stand is not centred to the vertical Bucky
e. Check the tube stand centre stop position
f. Mechanical centre-stop adjustments
g. Electrically operated centre-stop adjustments
h. An electromagnetic lock fails to operate
i. A group of locks fail to operate

a. General precautions

- Electrical safety.
  i. In most installations the tube-stand power will come from the generator, but in some installations, switching off the generator does not remove power from the tube stand.
  ii. **Before removing any covers, ensure the generator is switched off, and the room power isolation switch is also turned off.**
  iii. This also applies if testing wiring connections, or electrical components.
- If removing an X-ray tube, or collimator.
  i. **See module 7.1 page 104, and module 7.2 page 110.**
  ii. Ask an electrician or electronics technician for assistance.
  iii. Do **not** rely on the vertical lock system.
  iv. Attach a rope so that the system cannot move upwards, once the weight of the collimator or X-ray tube is removed.
  v. The X-ray tube is heavy. Removal or replacement requires two people.
  vi. Make a diagram of electrical connections. Attach labels to wires or high-tension cables. This is to ensure correct connection when an X-ray tube or collimator is replaced.
  vii. Place all screws or other small parts in a box, so they are not lost.
- Do not place a ladder against a tube stand. The tube stand may suddenly move.
- An adjustment to any tube-stand bearing requires skill, and good mechanical knowledge. When a problem is identified, request the service department to make the required adjustments.

b. The vertical movement is not balanced

For example, with the vertical locks off, or if power is turned off, the tube carriage tends to move down, or up. This problem may have occurred after fitting a replacement X-ray tube.

- Most tube stands have a system of 'trim weights'. Adding or removing these weights balances the vertical suspension.
  i. With ceiling suspensions, these weights may be positioned inside the cross arm.
  ii. Floor ceiling tube stands allow for trim weights to be attached to the main counterbalance weight. To gain access, a panel is removed from either behind the tube stand, or from one side of the tube stand.
  iii. If added trim weights are required, these may be formed from lead sheet, available from a builder's hardware shop.
- Some ceiling mounted tube stands require a spring to be added or removed to achieve balance. The service department should make this adjustment.
- Floor ceiling tube stands may have a large spring instead of a counterweight. The variable ratio pulley at the top of the tube stand can identify this method. Final counterbalance may still be achieved using trim weights attached to the cross-arm. Otherwise contact the service department.

c. When power is turned off, the tube stand starts moving

- A common reason is the support method of the HT cables. Providing the tube stand movement is not restricted, arrange for added or more suitable HT cable support.
- With a floor ceiling stand, this may be due to a floor that is not level. It may be possible to improve by adding shims under the floor rail. Check the floor rails with a spirit level.
- With a ceiling mounted system, the ceiling rails may not be level. This may be due to incorrect initial installation. There is a possibility the ceiling attachment points have shifted, or a problem with the building. Check the rails with a spirit level. Depending on the age or style of construction, have the installation checked by a building inspector.

d. The tube stand is not centred to the vertical Bucky

In this situation, the x-ray tube may appear to be correctly centred to the table centre. However, when the X-ray tube is rotated, it is not centred to the vertical Bucky.

- The tube stand may not be vertical.
  i. A floor-ceiling tube stand may be checked with an accurate spirit level. A more accurate check is to use a plumb bob.
  ii. A ceiling mounted tube stand can only be checked with a plumb bob. With the tube stand first at lower, then at maximum height, the plumb bob should deviate by only a few millimetres.
  iii. A ceiling mounted tube stand may need adjustment of the gantry-rail bearings.
  iv. The floor-ceiling tube stand has a ceiling or wall mounted guide rail. Check the guide-rail bearing assembly. This may be loose or incorrectly adjusted.
- The cross arm may not be horizontal.
  i. This can be a common fault with some floor ceiling tube stands. Check the cross arm with a spirit level.
  ii. Look for broken or loose bearings, especially with the cross-arm bearings inside the vertical movement.
  iii. *Adjustments to any tube-stand bearings require specialized knowledge. When a problem is identified, request the service department to make the required adjustments.*
- Is the light-field vertical alignment correct?
  i. Bring the collimator down onto the tabletop. If necessary adjust the X-ray tube rotation in the trunnion rings, so it sits 'flat' on the tabletop.
  ii. Raise the tube a small amount. With the collimator light switched on, place a marker in the centre of the light field.
  iii. Raise the collimator to the normal operating height. The centre of the light field should stay on the marker. If not, rotate the tube a small amount in the trunnion rings, and repeat this test.
  iv. Alignment is correct when the light field does not shift, as the X-ray tube is raised or lowered.

After checking the first three items, now centre the light field to the centre of the tabletop. **Caution**, do not use the cross arm centre stop as a guide, as this may also need adjustment.

- If the Bucky table has lateral movement of the tabletop, the centre stop position of the tabletop should first be checked.
  i. Move the tabletop to the centre position.
  ii. Place a cassette in the Bucky, with a marker positioned on the centre of the cassette.
  iii. Place another marker on the centre of the tabletop.
  iv. Make a low kV and mAs exposure. Process the film. The markers should have the same position on the film.

v. If required, adjust the position of the tabletop centring device. **See module 8.0 page 121.**

Rotate the X-ray tube to face the vertical Bucky. Move the tube stand close to the Bucky. The light field should be centred to the Bucky. Next, move the tube stand away from the Bucky. Check the position of the light field.

- Although the light field is not centred to the vertical Bucky, it does not shift as the tube stand moves away from the Bucky.
    i. Was the vertical Bucky positioned correctly during installation? Attach a string to the far end of the tabletop, positioned at the centre. Take the other end of the string to the centre of the vertical Bucky.
    ii. When the string is tightened, it should remain centred along the full length of the tabletop.
    iii. Is the tube stand movement parallel to the tabletop? The light field should remain centred while the tube stand is moved from the table foot end to the table head end. If the light moves off centre, this could indicate either the tube stand or the Bucky table was incorrectly installed

- The light beam shifts off centre, as the tube stand is moved away from the vertical Bucky.
    i. The tube stand cross-arm may not be horizontal. Check with a spirit level.
    ii. Many tube stands allow rotation of the cross arm. (In some cases, the entire tube stand rotates.) The rotation index-plate may be loose, or not correctly centred.
    iii. Common rotation angles are −90 degrees, centre, and +90 degrees. A lock pin is inserted into a slotted index-plate to hold the rotation position. The index-plate and locking pin may be worn, or incorrectly adjusted.

e. Check the tube stand centre-stop position

- If the Bucky table has lateral movement of the tabletop, the centre stop position of the tabletop should be checked.
    i. Move the tabletop to the centre position.
    ii. Place a cassette in the Bucky, with a marker positioned on the centre of the cassette.
    iii. Place another marker on the centre of the tabletop.
    iv. Make a low kV and mAs exposure. Process the film. The markers should have the same position on the film.
    v. If required, adjust the position of the tabletop centring device. **See module 8.0 page 121.**

- Is the tube stand centred to the Bucky table?
    i. Make this test, after checking the light-field vertical alignment described in part 'd'.
    ii. Move the Bucky tabletop to the centre position.
    iii. Switch on the collimator lamp.
    iv. Test the lateral centre-stop position of the tube stand. With the cross-arm retracted, move the X-ray tube out towards the centre stop position. The light field should be centred to the tabletop.
    v. Repeat the test, start with the cross-arm extended, then move inwards to the table centre. The light field should again be centred to the tabletop.
    vi. The actual centre stop position can depend on how quickly the X-ray tube is moved across the table, and the method used to indicate the stop position. Depending on the design, accurate centring may require moving the X-ray tube from one direction only.

f. Mechanical centre-stop adjustments

- A steel ball, pushed by a spring, clicks into a slot when the cross arm is centred. This holds the cross arm in position.
- If the spring is weak, it is difficult to feel when the centre position is reached.
- There is usually a screw provided, to adjust the spring tension. Adjust this screw to provide the best 'feel' when centring the X-ray tube.
- The centre-stop position is adjusted by changing the position of the mechanical system on the cross arm.
- To adjust the position, see the directions provided in the installation manual. Otherwise contact the service department for advice.

g. Electrically operated centre-stop adjustments

Caution: Before making any electrical tests, ensure the generator is switched off, and the room power isolation switch is also turned off. An electrician or electronics technician should carry out electrical tests or adjustments. **See module 5.0 page 65.**

A number of different electrical centre-stop sensors have been developed. These operate the lateral lock when in position.

- A microswitch, operated by a cam. In normal operation, you may hear a small 'click' as the switch passes over the cam. The position of the cam con-

trols the centre stop position. A problem may be caused by:
  i. The cam height is too small. As a result, there is insufficient pressure on the microswitch for reliable operation.
  ii. The cam or microswitch has become loose, and the microswitch does not operate.
  iii. A broken wire or connection to the microswitch.
  iv. A ceiling tube stand may have incorrect adjustment of the lateral movement bearings. This may cause the cam to move away from the switch, so it does not operate. In some cases, it may instead move too close, damaging the switch. A close visual inspection can indicate if this is a problem.
- A vane operated sensor. A vane passes through a small slot in the sensor. The position of the vane controls the centre stop position. A problem may be caused by;
  i. The vane is positioned too high, and does not fully enter the sensor. This can cause unreliable operation.
  ii. The vane is missing.
  iii. The vane or sensor is damaged. Check by visual observation.
  iv. A broken wire or connection to the sensor.
- An optical sensor, operated by reflected light. This system requires a white or silvered reflector, mounted opposite the sensor at the stop position. A problem may be caused by;
  i. The reflector is a small piece of foil, with a self-adhesive backing. Due to poor adhesive, this may have become dislodged.
  ii. The sensor is not close enough to the reflector for reliable operation.
  iii. The reflector is dirty, or there is dirt on the sensor.
  iv. A broken wire or connection to the sensor.

h. An electromagnetic lock fails to operate

This may be due to a faulty lock coil. Other reasons may be a faulty switch, or a broken connection due to a cable being pulled. **See module 5.0 page 65.**

First ensure the generator is switched off, and the room power isolation switch is also turned off.

- The lock may have too large an air gap. This can also cause erratic or slow operation. Most locks have slotted mounting plates. Adjust by undoing the screws a small amount, adjust the lock position, and retighten the screws.
- The lock may only partially release. In this case it may be too close to the surface. Again, adjust its position. In some cases, the lock has residual magnetism. This can be a design problem with some tube stands. Contact the service department for advice.
- Permanent magnet locks have become popular. These ensure the locks remain on when power is removed.
- Some ceiling suspended tube-stands use a solenoid operated 'piston', attached via a lever to a brake pad. A spring maintains brake operation, until the solenoid operation pulls the pad away from the surface. If the stroke is too long, the piston fails to pull inside the solenoid, and the lock does not release. Adjustment is by a screw thread fitted with a locknut.
- In some cases, there is a fuse specific to the failed lock. Before checking fuses, ensure all power is turned off.
- **See Module 5.0 page 65.**

i. A group of locks fail to operate

First ensure the generator is switched off, and the room power isolation switch is also turned off.

- Look for an open circuit fuse at the tube stand.
- In most installations, the tube stand obtains power from the generator. This may involve several different voltage supplies. Check at the generator and at the high-tension transformer for an open circuit fuse.
- **See Module 5.0 page 65.**
- Check where cables enter the tube stand or control panel. If the cables are pulled during the tube stand movements, a wire may have broken from a terminal strip.

PART III. FAULT DIAGNOSIS AND REPAIR MODULES

TASK 13

Bucky tabletop and tube-stand centre

A number of films have appeared which show incorrect lateral centring. You decide to verify the accuracy of the X-ray tube and table centre-stop.

1. Design a method to verify the tabletop is accurately centred over the Bucky. Note: this should include possible errors due to the film position in the cassette.

_____

_____

_____

Carry out this test; include moving the tabletop to centre position from either direction.

Is the tabletop correctly centred to the cassette? _____

If not, is the cassette tray correctly centred in the Bucky? _____

_____

Is the crosshair on the collimator faceplate correctly centred to the 'closed' position of the collimator leaves?

_____

With the X-ray tube positioned close to the tabletop, is the crosshair aligned to the tabletop centre? Include moving the X-ray tube to centre position from either direction. _____

_____

As the X-ray tube is raised from the tabletop, does the crosshair move away from the centre mark?

_____

What adjustment might be made so that the crosshair position does not move as the X-ray tube is raised from the tabletop? _____

_____

If this adjustment is performed, will it affect the centre-stop position of the X-ray tube cross-arm?

_____

_____

Tutor's comments

Satisfactory/Unsatisfactory

Signed _____ Date _____
        Tutor

MODULE 7.1

X-ray tube

Aim

The aim is to provide information for testing or replacing the X-ray tube. Different failure modes are examined. This module is an extension of module 2.1 page 48. Reference should also be made to module 7.3 page 117.
(**Note:** Reference module page numbers refer to the title page.)

Objectives

On completion of this module, the student will be aware of common problems with the X-ray tube, together with the test procedures. This includes removal or replacement of the X-ray tube, together with assistance from an electrician or electronics technician.

Contents

a. General precautions
b. X-ray tube failure modes
c. Inspection of the anode or filament
d. Focal spot performance
e. Oil leaks
f. Removal of the X-ray tube
g. X-ray tube transport
h. Re-installation of the X-ray tube

a. General precautions

- **Before disconnecting any wires, or removing the high-tension cables, always ensure power is turned off and unplugged from the power point. If the equipment is part of a fixed installation, besides switching the generator power off, ensure the isolation power switch for the room is also switched off.**
- Mobile high-frequency generators may be battery operated. The batteries in these are connected in series, and can have a total voltage of up to 240V DC. *Refer to the operating or installation manuals for the position of the battery isolation switch, and ensure this is switched off before removing the covers, or testing wires and connectors.*
- If removing a collimator or X-ray tube from a tube stand.
    i. Do not rely on the vertical lock system.
    ii. An X-ray tube is heavy. Two people are required for removal or installation.
    iii. Use a rope to prevent the system moving upwards, when a collimator or X-ray tube is removed.
    iv. Provide a container to hold all small parts, or screws. Protect against loss.
- If removing an X-ray tube from a capacitor discharge mobile, observe the high-tension precautions described in **module 7.3 page 117**.

b. X-ray tube failure modes

- The X-ray tube is unstable. A common cause is gas, which causes very high current to flow during an exposure. Unstable operation is usually corrected by 'seasoning'. This is described in **module 2.1 page 48**.
- Attempts to improve the performance by seasoning are not successful. This can be due to;
    i. The glass has developed micro-fine cracks. With the collimator removed, this will be observed as a fine 'crazing' effect on the output window, or port. These cracks indicate the glass is punctured. As a result, the tube is gassy.

ii. The bearings have seized, so X-ray exposures are hitting a stationary anode.
   iii. In both cases, the tube requires replacement.
- Arcing at the HT cable ends or in the receptacles.
   i. You have may have observed smoke at one receptacle. Or an actual spark.
   ii. The X-ray control might generate 'mA overload' or 'kV fault' signals. This depends on the design of the X-ray control and the severity of the arcing.
   iii. You have noticed there is a strong odour at the suspect cable end or receptacle.
   iv. High-tension cable problems are discussed in **module 7.3 page 117**.
- The bearings have become very noisy. In many cases a tube with noisy bearings can still have a useful life. However, budget for a replacement if the anode slows down quickly once preparation is released. This can indicate a failure in the near future, and is especially the case if the anode slows down while still in preparation.
- Poor X-ray resolution.
   i. The anode is badly worn.
   ii. The anode is cracked or distorted, so that the focal spot wobbles as the anode rotates.
   iii. Heavy metal deposits on the output window. This causes excessive hardening or filtration of the X-ray output. In this case, it will not be possible to observe the anode or cathode after removing the collimator. Metal deposits can also lead to a micro arc through the glass, causing the tube to become unstable or gassy.
   iv. See 'inspection of the anode or filament' in part 'c'.
- A filament is open circuit. To test, use a multimeter set to the low ohms scale. There should be a very low resistance between any two pins in the cathode receptacle. An exception is the X-ray tube for some mobiles, which may have only one focal spot.
- A filament has a partial short circuit. This is due to a section of the filament touching the cathode focus cup, and then welding itself to the cup. Unfortunately, this is not a rare occurrence.
   i. The generator will indicate sudden low mA output, while films will not only appear underexposed, but may also have poor contrast due to an increase of kV. With high frequency, or microprocessor-controlled generators, a kV fault signal may be generated.
   ii. Checking the other focal spot will indicate normal operation.
   iii. Attempting to re-calibrate mA output will indicate a rapidly increasing filament drive current, especially at higher mA output. In addition, correction of mA at medium to low kV calibration points becomes very difficult. (Space charge compensation).
   iv. In some cases it is possible to see a faulty filament, after the collimator is removed. See 'inspection of the anode or filament'.

c. Inspection of the anode or filament

> *Note. This technique must not be attempted with a capacitor discharge mobile, due to high voltage that may be stored in the capacitor.*

- **Removing the collimator.**
   i. Where possible, refer first to the installation manual of the collimator. If in doubt, contact the service department for instruction. Two people are recommended, to hold the assembly in position as it is removed or replaced.
   ii. Rotate the X-ray tube so it is aimed at the ceiling. Adjust the height close to the tabletop.
   iii. Secure the vertical movement of the tube stand so it cannot move upwards once the collimator is removed. **Do not rely on the magnetic lock system, this can slip, or not operate when power is switched off.**
   iv. Examine the connecting cables to the collimator, and the tube-stand operation panel. Is there sufficient length? Undo any cable ties if required, to allow cables to hang freely.
   v. **Ensure all power is off. Turn of power at the room isolation switch. Do not rely on the generator power switch, as some installations allow direct power to the tube stand.**
   vi. To avoid pulling on cables once the collimator is removed, place a box of a suitable height on the tabletop. The collimator can rest on this box when removed from the tube head. This may include the tube-stand control panel.
   vii. If the cables are short, disconnection is required. Make a careful diagram of connection terminals before disconnecting, and ensure the wires have a suitable label or mark. Ensure any attached labels will not fall off. Any exposed terminals attached to wires must be covered with insulation tape. (Power may need to be re-applied to see inside the X-ray tube.)
   viii. Undo the retaining screws holding the collimator to the tube housing, and place the colli-

mator on the tabletop, or box. Take care on removal, as part of the collimator can extend into the tube-housing port. In some cases, the tube-stand control panel will be detached at the same time. Assistance may be required to hold or place components as they are removed.

ix. *Have a container ready to receive screws etc. These are easily lost.*

x. The tube-housing port may have an aluminium filter, and a lead 'proximal' diaphragm fitted. The latter is often in the form of cone extending into the port, and the filter is then placed under this lead diaphragm. The proximal diaphragm is held in place by a spring clip, or else by two or four very small screws. Before removing the lead diaphragm, make a mark so it can be replaced in the same position.

xi. **Caution:** do not remove the larger screws holding the port assembly in place. Air will enter the housing, or oil will leak out. If this happens, the tube housing needs to be reprocessed.

- **Inspecting the anode.**
  ii. With the collimator, proximal diaphragm, and any added filter removed, it should now be possible to see the anode and filament.
  iii. Often observation of the anode may be made using a torch. However, any metal evaporation on the glass acts as a mirror, and prevents observation.
  iv. Most generators have a filament pre-heat circuit, which will light up the filament, and allow observation of the anode. Ensure any disconnected wires have their ends taped up, and then switch on the generator.
  v. **Note.** If the glass has heavy metal deposits, this technique may only yield limited results. In this case, the future reliability of the tube is not good.
  vi. To observe the anode for defects, the anode needs to slowly rotate.
  vii. To rotate the anode, press the preparation button on the handswitch. This should be very brief, so that the anode only just starts spinning. ***Do not expose.*** *As a safety precaution, preset minimum kV and time, and a low mA station.*
  viii. As the anode slows down, carefully observe the track area. Look for the following.
    —Anode wobble, this indicates possible cracking, and poor focal spot performance.
    —Stationary hits. These appear as small melted areas of the anode, as if hit by a small arc welder.
    —Worn anode. This appears as a fine crazed pattern, like coarse sandpaper.
    —Overloaded anode. This has a fine orange peel pattern.
    —Smudged areas. This often occurs during manufacture. However, if the tube is unstable, this can be an indication of gas.

- **Inspecting the filament.**
  ix. When the generator is switched on, the pre-heat circuit will light up the selected focal spot.
  x. Select fine focus, and then broad focus. The broad focus will appear a little longer and larger in diameter.
  xi. (An exception to the above will occur with a fluoroscopy table. In this case, the fine focus remains selected at all times, unless in preparation for radiography).
  xii. If there is a partial short in the broad focus, then the broad focus will appear shorter in length than the fine focus. Careful observation can sometimes see a short length of filament that is not heated.

- **Are there fine cracks in the glass?**
  i. These can appear over the anode or cathode area. A minor case may appear as a single fine line, like a single strand of spider web. More severe cases can appear as a fine crazed pattern.
  ii. These marks are due to high voltage discharge through the glass. This condition occurs more often with metal deposits on the glass, which increases the possibility of arcing in this area.
  iii. The presence of these marks, together with a suspect unstable or arcing tube, means the tube is gassy and will need replacement. In this case, seasoning is not effective.

- **Replacing the collimator.**
  i. Re-assembly is in the reverse order as the dismantle process.
  ii. Take care that any added aluminium filters are returned to their previous position, and the proximal diaphragm is correctly aligned.
  iii. After reassembly the collimator will need realignment. Please refer to **module 7.2 page 110**.

d. Focal spot performance

Focal spot performance can be tested using a 'Star pattern' gauge.

In use, the gauge is positioned in the centre of the X-ray beam, close to the focal spot. This gives a magnified view of the star pattern. Part of this pattern is blurred. The diameter of the blurred area is used to calculate the size of the focal spot.

There are several versions of star patterns. Use the directions enclosed with the pattern, which includes a formula specific to the supplied star pattern.

A star test pattern may be obtained as a loan item from the service department, or from the physics department of a major hospital.

When measuring the focal spot, take the following precautions.

- Use a low value of kV. (60~70 kV)
- Use a medium value of mA suitable for the focal spot under examination. **Note.** As mA is increased, the focal spot will also increase in size.
- Use non-screened film. Or else a cassette with detail screens.
- Exposure time should be more than 0.04 seconds. This allows at least two rotations of the anode, in case anode wobble is degrading the focal spot.
- If the test result is too light, increase the time or mA station.
- If the result is too dark, consider increasing the FFD or reducing kV.
- If the outer blurred area is too large in diameter to measure easily, then reduce the magnification, and adjust mA and kV to suit.

e. *Oil leaks*

Oil leaks should always be reported. Even a very slow oil leak has the possibility of letting air into the housing. An air bubble in the wrong position can lead to arcing, and possible destruction of the X-ray tube.

- Oil leaks may be seen at either end of the tube housing, or at the collimator, if a crack or faulty seal occurs in the housing port.
- An X-ray tube with oil leaks will need to be repaired by the service department. In some situations, the service department may supply a loan unit, while the faulty housing is repaired.
- To locate where the oil leak occurs, first thoroughly wipe clean with a paper tissue. Leave overnight, and then test by wiping with a fresh tissue next morning. This will indicate the origin, and assist in repairs when the tube is returned to the service department.
- Occasionally, when a tube is returned after a repair, an apparent oil leak might appear. This can be caused by a small amount of spilt oil around such areas as the external terminal strip etc. A few drips may initially occur, and then no further symptoms appear. If drips continue after a few days, then this needs to be reported, and have the tube returned for further attention.

f. *Removal of the X-ray tube*

Due to the presence of an oil leak, or to have a new insert installed, the X-ray tube and housing is required at the service department.

- Preparation for removal.
  i. Where possible, refer first to the installation manual of the tube stand. If in doubt, contact the service department for instruction.
  ii. **Two people are recommended to assist in removing or installing the X-ray tube assembly. This is a heavy object.**
  iii. Rotate the X-ray tube so it is aimed at the ceiling. Adjust the height close to the tabletop.
  iv. Secure the vertical movement of the tube stand so it cannot move upwards once the collimator is removed. **Do NOT rely on the magnetic lock system, this can slip, or not operate when power is switched off.**
  v. Examine the connecting cables to the collimator, and the tube-stand operation panel. Undo any cable ties or clamps, to allow cables to hang freely.
  vi. Carefully mark the anode and cathode cables.
  vii. **Hint.** The stator cable normally enters at the anode end of the housing. (In some cases, it enters at the centre. Be careful in this situation)
  viii. **Ensure all power is off. Turn of power at the room isolation switch. Do not rely on the generator power switch, as some installations allow direct power to the tube stand.**
  xi. If removing an X-ray tube from a capacitor discharge mobile, observe the high-tension precautions described in **module 7.3 page 117**.
  x. Undo the ring nuts holding the HT cable ends, and withdraw the cable ends from the housing. As they are withdrawn touch the end pins to the tube stand. This is to discharge any residual high-tension that may be present. When they are withdrawn, wrap the cable ends in cloth or paper towel to protect from damage.
  xi. To avoid pulling on cables once the collimator is removed, place a box of a suitable height on the tabletop. The collimator can rest on this box when removed from the tube head. This could also include the tube-stand control panel.
  xii. Disconnection of all wires to the tube housing is required. Make a careful diagram of con-

nection positions before removing, and ensure the wires have a suitable label or mark. Ensure any attached labels will not fall off. **Any exposed terminals attached to wires must be covered with insulation tape.** (This is in case power is turned back on)

xiii. Undo the retaining screws holding the collimator to the tube housing, and place the collimator on the tabletop, or box. Take care when removing, as part of the collimator may extend into the tube housing port. In some installations, the tube-stand control panel will be detached at the same time. Assistance will be required to hold or place components as they are removed.

xiv. Have a container ready to receive screws etc. These are easily lost.

- Removal of the X-ray tube housing from the tube stand.
  i. Check again, that vertical movement of the tube stand is properly secured.
  ii. After the collimator, high-tension cables, and stator cables have been removed, examine carefully the shape of the trunnion mounting rings. With the X-ray tube aimed at the ceiling, the bottom section of the rings should be able to hold the housing in place, after the top section is removed. Sometimes the assembly is installed in the reverse direction, so then the tube needs to face the tabletop.
  iii. With the housing in the required position, undo the top half's of the trunnion rings, taking special note if any washers have been inserted between where the trunnion rings are fastened together. (These are sometimes fitted to adjust the trunnion rings, in case they are too tight a fit for the tube housing.)
  iv. The tube housing may now be lifted up out of the rings. This is a heavy object. Two people should assist in this process.

g. X-ray tube transport

The X-ray tube housing offers no protection to the X-ray tube if it is bumped or dropped. Incorrect packaging for transport can easily result in a broken tube, due to the weight of the anode.

- Before sending the tube to the service department, take careful note of all housing and X-ray tube details. Include serial numbers.
- Attach full documentation to the housing. This should give a full description of the problem to be rectified. There should also be full contact details, such as hospital address, phone number, person to contact, etc. Include an order number or other authorisation if required.
- Include a request for suitable silicon grease, and or anti-corona silicon pads to be supplied when the tube is returned.
- Before looking for suitable boxes etc, contact the service department. They may be able to send a suitable size box and packing material.
- Select a box size about twice that of the housing. Pack the housing in the centre, using material to cushion any bumps. For example, shredded polystyrene foam. Make a mark on the box to indicate the anode end.
- Place this box in another box about twice the size of the first box. Fill the space between the two boxes with suitable cushion packing.
- Position the second box so that the X-ray tube is vertical, and the anode end is towards the bottom of the box.
- Attach very large labels with an arrow to indicate 'This side up' on the sides of the box. Attach another label on the top to indicate 'Top side'. Attach 'Fragile, do not drop' labels on all sides.
- Take care both the service department address and the hospital return address is protected, eg, inside a transparent plastic cover. If sending to another country, be sure to provide suitable information for customs etc.
- Ensure you have a full copy of the shipping details. Also phone the service department and notify them of the method of transport etc. If the X-ray tube is sent to another country, enclose copies of the required customs forms.

h. Reinstallation of the X-ray tube

Caution: If a new tube insert or assembly is supplied, a complete mA re-calibration is required. This should be performed by the service department.

The X-ray tube is re-installed in the reverse order of the instructions for removal. Eg, first it is mounted in the trunnion a ring, then the collimator is attached, and finally the wiring and HT cables. These precautions should be observed.

- Check the HT receptacles. If there is any grease residue, this must all be removed, and the receptacles left in a polished condition. Even if apparently clean, still wipe them carefully with fresh paper tissues. This is to remove any possible moisture. Do not touch the inside with the fingers, or scrape the sides with a metal object. (This may leave very slight traces of metal behind).

PART III. FAULT DIAGNOSIS AND REPAIR MODULES

- If the housing has been repaired, or a new insert fitted, check carefully for any areas of oil residue, and wipe away with paper tissues.
- Before installing the collimator, ensure the proximal diaphragm and aluminium filters are in place.
- Reconnect the wires, using the diagram previously made when the tube was removed. If a new tube and housing is supplied, and the connection points appear different, contact the service department before connecting any wires.
- When inserting the high-tension cable ends, use the instructions provided in **module 7.3 page 117**.
- Important. After the cable ends have been inserted, and the ring nut fully tightened, retighten a few hours later, and again next day.
- The collimator will need re-alignment. Please refer to **module 7.2 page 110**.
- Before making a test exposure, just enter preparation only. Listen carefully to the tube anode as it rotates. Is this the normal anode rotation sound?
- *Providing the original tube insert and housing has been returned*, then the mA calibration should be the same. Select a low mA position, set 60 kV, and 0.1 second exposure time. Make a test exposure. If any problem occurs, **STOP**. Contact the service department for advice.
- The X-ray tube should now be seasoned. Use the technique described in **module 2.1 page 48**.
- Keep the boxes and packing the X-ray tube assembly arrived in for future use.

MODULE 7.2
X-ray collimator

Aim

The aim is to provide information and procedures related to adjusting the X-ray collimator. This is in addition to the collimator maintenance, provided in module 2.2 page 50.
(**Note:** Reference module page numbers refer to the title page.)

Objectives

On completion of this module, the student will be aware of common problems with the collimator and their solutions. Adjustments and repairs may be carried out. Some tests for the collimator lamp can be made with the help of an electrician, or electronics technician.

Task 14. 'Help! No spare globe for the collimator', should be attempted on completion of this module.

Contents

a. General precautions
b. The light field has insufficient brightness
c. Changing the collimator globe
d. A wrong collimator globe
e. This collimator was not designed to rotate
f. Centring of the collimator lamp
g. Centring of the X-ray beam
h. The light field is larger than the X-ray field
i. The Bucky centre light
j. The X-ray field fades out on one side of the film
k. The collimator blades close, after adjusting the field size
l. The collimator lamp fails to operate
m. The globe has failed, and there is no spare globe

Equipment required

- Basic tool kit.
- X-ray alignment template.*
- 24/30 cm cassette.
- Spare collimator globe.
- Cloth, for cleaning.
- Detergent.

* The template is described in appendix 'B' page 169.

a. General precautions

- **Before disconnecting any wires, or removing a cover, always ensure power is turned off and unplugged from the power point. If the equipment is part of a fixed installation, besides switching the generator power off, ensure the isolation power switch for the room is also switched off.**
- Whenever changing a collimator lamp, ensure all power is turned off.
- If removing a cover, position the collimator close to the tabletop. Secure the tube stand so it cannot move upwards once the cover is removed.
- The timer switch may be fastened to the cover. Place a small box or pillow within reach to support the cover when removed, so the connecting wires are not pulled.

b. The light field has insufficient brightness

This may be due to several causes, some of which can combine to give an overall drop in light level.

- Dirt or dust builds up on the inside of the transparent exit cover of the collimator.
  i. To clean, it will be necessary to include removal of the collimator outer cover. Before removal, ensure all power is turned off.
  ii. Clean with a soft rag and mild detergent. Wipe off any residual detergent.

PART III. FAULT DIAGNOSIS AND REPAIR MODULES

iii. The above also applies if cleaning the mirror. Take care not to scratch the surface, or change the position of the mirror.
- The globe has metal evaporation on the inside of the globe. Fit a new globe.
- The voltage supply to the lamp is too low. This is due to a supply voltage that has not allowed for voltage drop, due to wiring resistance. As an example, when the lamp is switched on, the voltage can drop by 2~5 volts. This is a common problem when installed with long connecting cables.
  i. **Ask an electrician, or electronics technician, to measure and adjust the lamp voltage.**
  ii. Check if the lamp has the correct voltage. To do this, set the multimeter to a convenient AC voltage range. For example, 25 V or 100 V AC. Remove the lamp covers, and place the meter probes on the lamp terminals. Look away from the lamp while switching the light on, and then measure the operating voltage. This might be only 7~8 V for a 12 V lamp, or perhaps 17~20 V for a 24 V lamp.
  iii. A number of systems have a transformer, with a selection of output voltages. The required voltage is selected by changing a connection on a terminal strip.
  iv. During installation this may be set at the lamp voltage. For example, set at 12 V output for a 12 V lamp. This is incorrect, as it does not allow for voltage loss in the connecting cable.
  v. A correct installation may even set the voltage as high as 16 V for a 12 V lamp. This compensates for voltage drop due to cable resistance, when the lamp is switched on.
  vi. If the test voltage is low, eg, below 10 V for a 12 V lamp, then contact the service department for instructions to adjust the voltage supply. Include the make and model of the collimator and the generator.
  vii. In some other situations, there may be spare conductors in the cable. These may be placed in parallel with the existing wires for the lamp, to reduce the voltage drop. This is best left to a technician from the service department to carry out.
  viii. **Note.** While an 8 V operating voltage for a 12 V lamp is too low, increasing to the full 12 V will give a shorter lamp life. A compromise between brightness and life for a 12 V lamp is 10~11 V.

c. Changing the collimator globe

Before attempting to replace a collimator globe, ensure all power is turned off.

When replacing the globe, take care not to touch the glass with the fingers. This especially applies to quartz iodide globes, as slight oil or perspiration from the fingers will cause premature failure. Use a paper tissue to hold the globe.

d. A wrong collimator globe

Two versions of a quartz iodide globe appear very similar. If the wrong version is installed, there is a large error between the light field, and the X-ray field.

- The correct globe has longer connecting pins. OR, the filament is placed further towards the tip of the lamp. Both are correct, in that the filament is the same distance from the rear end of the pins.
- The incorrect version has shorter pins, so that the distance between the filament and the rear end of the pins is smaller.
- In an emergency, the short pin version may be used. Insert the lamp sufficiently to make good contact in the socket, however do not push it all the way in. There will still be an error in the light beam to X-ray alignment, so obtain the correct version as soon as possible.

e. This collimator was not designed to rotate

Older installations may have a collimator of European origin. With this collimator, four adjustable metal 'fingers' attach the collimator to a circular flange, or plate. There is no other adjustment. Correct adjustment is with the fingers tightened, so the collimator does not rotate.

However, in some installations these fingers are not tightened, allowing the collimator to be rotated.

- The collimator can only be aligned correctly to the focal spot in one position. When it is rotated, correct alignment to the light beam may be lost, especially if the light beam has also been adjusted. See 'Centring of the X-ray beam', part 'g'.
- Rotating the collimator can cause wear to the metal fingers. As the wear increases, the collimator may 'wobble' when pointed at a wall Bucky. In a severe wear case, the top of the metal finger breaks off. Replacement metal fingers are difficult to obtain.
- As the adjusting screws were not tightened, these can vibrate to a more open position. This will cause

erratic collimation. In a severe situation, the collimator may even detach from the flange, and fall off.
- Assuming the collimator **must** rotate, please take the following precautions.
    i. Apply a very thin layer of oil to the upper surface of the mounting ring, to reduce wear.
    ii. After the metal fingers have been adjusted, apply a dab of nail polish to the outer threads of the adjusting screws. This will help prevent them from unwinding.

f. Centring of the collimator lamp

There are two methods of aligning the light beam to the X-ray field. One is to move the position of the lamp, and the other is to adjust the position of the collimator relative to the X-ray beam.

With a collimator that can rotate, it is essential to adjust in the correct sequence, otherwise alignment is correct only in one position.

- Bring the collimator down until it touches the tabletop, and adjust the tube rotation so the collimator face is flat against the tabletop. Now raise the collimator to its normal working height.
- Rotate the collimator clockwise 90 degrees.
- Place a used X-ray film on the table top as a template. A suggested size is 24 × 30 cm. Switch the collimator lamp on, and adjust the collimator so the light field just covers the film.
- Next, rotate the collimator anti-clockwise 180 degrees. With the lamp switched on, look for any error in alignment. This should be less than 2.0 mm in any direction.
- Before making any adjustment, check to see the correct lamp is fitted. If in doubt, contact the service department to obtain positive identification.
- If adjusting the lamp position, adjust so the error is reduced by 50%. Then adjust the film position to the light, and test again with the collimator rotated 180 degrees to the previous position.
- Can the mirror be adjusted?
    i. With most collimators, the mirror is fixed in position. Attempting to move the mirror against the clamping screws can distort or break it, requiring a replacement. (If the mirror is distorted, the sides of the light field are at an angle, and not parallel).
    ii. The exception is where there is a spring-tensioned adjustment screw. This may be found on some mobiles or portable units. In this case, the lamp may be adjusted sideways, and the mirror rotation replaces the vertical adjustment of the lamp.

g. Centring of the X-ray beam

Attempting to adjust the collimator to the X-ray beam should only be attempted after checking that the light beam is correctly centred. This especially applies when the collimator can rotate.

- An X-ray alignment template is required. A suitable design is shown in **appendix B page 169**.
- Place the X-ray alignment template on a 24/30 cm cassette.
- Collimate the light beam to the outer 20 by 26 cm rectangle.
- Make a low kV and mAs exposure.
- Develop the film.
- **Does the alignment meet the required compliance?**
  Two versions are provided as an example only. The actual compliance requirement will depend on individual country regulations.
    i. *The X-ray field edges should not deviate by more than 2% of the distance between the plane of the light field and the focal spot.*
      [a1] + [a2] ≤ 0.02 × S.
      [b1] + [b2] ≤ 0.02 × S.
      Where S is the distance from the focal spot, a1 and a2 are the two sides on one axis, and b1 and b2 are the two sides of the other axis.
      For example; at a FFD of 100 cm, if the two vertical edges of the light field were displaced by 10 mm, this would be at the limit of acceptance. *If only one edge was displaced, then 2.0 cm is at the limit of acceptance.*
    ii. Another version has a different requirement.
      The total misalignment of any edge of the light field with the respective edge of the irradiated field must not exceed 1% of the distance between the plane of the light field and the focal spot.
      For example; at a FFD of 100 cm, the maximum displacement of **any** edge should be less than 10 cm.
- In case the X-ray field is off-centre by more than the permitted amount, re-centring is required.
- **To adjust a non-rotating collimator.**
    i. Refer where possible to the installation or operation manuals for the collimator. If necessary, contact the service department for information specific to your version of collimator.
    ii. Locate the adjusting screws for the metal fingers. These usually require an Allen key for adjustment.
    iii. By slackening off one finger, then tightening the opposite finger, the collimator will move relative to the X-ray field. Only a small adjustment,

of about one turn of the screw, is sufficient to move the X-ray field several millimetres.
iv. Adjust the collimator to move in the same direction you require the X-ray field to move.
v. After adjusting, make another test film and compare the results.
vi. If necessary, make further adjustments until compliance is achieved.
vii. Tighten carefully all four fingers, ensuring the collimator position does not change.
viii. Make a final test film for verification.

- **To adjust a rotating collimator.**
    i. Refer where possible to the installation or operation manuals for the collimator. If necessary, contact the service department for information specific to your version of collimator.
    ii. With a true rotating collimator, the ring or bearing on which it rotates is repositioned relative to the X-ray beam. (Unfortunately some simplified systems may not have this facility, and so can only be correct in one position).
    iii. It is necessary to locate the screws that clamp this ring in position. These may require a small spanner. Undo these screws a small amount, so the ring can just be moved.
    iv. With some rotating collimators, once the rotation ring is free to move, adjusting screws similar to the fixed-collimator, are used for alignment. Otherwise gently tap the collimator into position, moving it only a part of a millimetre at a time.
    v. Adjust the position and test in the same fashion as the fixed-collimator.
    vi. Repeat the above test, with the collimator rotated 90 degrees clockwise, and 90 degrees counter clockwise.
    vii. Ensure the ring clamping screws are correctly tightened when alignment is satisfied. Make a final test film for verification.

h. The light field is larger than the X-ray field

Most collimators depend on a standard distance between the X-ray tube housing and focal spot. If a manufacturer supplies non-standard tube housing, this distance may be incorrect.

- The collimator is required to be positioned at a specific distance from the focal spot. Some collimators are supplied with shims. These can be added or subtracted to make the required adjustment. Check with the service department for this possibility.
- A common reason is the method of installation.

i. The bracket for the tube-stand command panel is placed between the collimator and the X-ray tube port. This increases the collimator to focal spot distance. As a result the X-ray field becomes smaller.
ii. In some cases it may be possible to have a mounting block machined to reduce this added distance, or have shims removed. Otherwise the mounting method of the control panel will need to be changed to correct the situation.

i. The Bucky centre light

Collimators have been fitted with a number of methods to indicate Bucky centre. Two versions are discussed here.

- One method is a fixed slot, immediately below the collimator lamp. If the lamp is not correctly adjusted, then the light shines at an angle through this slot, creating an error. This is usually corrected by re-alignment of the collimator. See 'Centring of the collimator lamp', part 'f'.
- Other versions may have a small focussing lens, attached to a slit in the collimator cover. Adjusting the lens can shift the position of the light beam.

j. The X-ray field fades out on one side of the film

If the fade out occurs towards the anode side of the film, when selecting a large format, this is probably due to the 'heel affect' of the X-ray tube anode.

Otherwise, it may be due to the following.

- The collimator is not centred to the focal spot, and the lamp has been adjusted to align the light beam to the X-ray field. Test by making sure the light field remains centred as the collimator is rotated. See 'Centring of the collimator lamp'.
- The collimator primary-beam shutter, or blade, is touching the side of the X-ray tube port, or 'throat'. The cause is due to incorrect centring of the collimator.
(This problem depends on the collimator design, and how the collimator is attached to the X-ray tube.)
- The lead proximal-diaphragm has been incorrectly fitted inside the tube port. For example, after replacement of the X-ray tube.
- The collimator lead-shutters, or blades, are out of adjustment. This may be either the middle blade, or the bottom blade.
    i. The shutters are coupled to the field size knob by a thin stainless-steel cable.

ii. The cable may be loose, or has slipped where it attaches to the shutters. To check, remove the collimator cover, and make a careful comparison of both sets of collimator blades. The lead strip, on the bottom blades, can be adjusted in most collimators.
- In some cases there may be a slow fade off towards the cathode side of the film. This is more noticeable at lower kV levels. If this is an older, or hard worked X-ray tube, then it might be due to metal deposits on the glass. To check, you will need to remove the collimator. **See module 7.1 page 104.**

k. The collimator blades close, after adjusting the field size

A collimator has an internal 'brake' or 'clutch'. If this becomes loose, the springs, fitted between the collimator blades, cause the blades to close.

Two common methods are described here.

To adjust, it is necessary to remove the collimator cover. Ensure power is turned off, and the tube stand is prevented from moving vertically.

- Japanese origin.
  i. The shutter control knob is attached to a round shaft. This shaft, which controls the opening of the blades, passes through a cylinder attached to the inside front of the collimator.
  ii. A small nylon pad forms the brake action. This is pressed firmly against the shaft that passes through the cylinder.
  iii. A screw, attached to the cylinder, controls the amount of pressure. As the screw is turned clockwise, the pressure of the nylon pad against the shaft increases.
  iv. To adjust, first undo the locknut on the screw. Then turn the adjusting screw about a quarter turn clockwise. Retighten the locknut, and test the feel of the control knob.
  v. Repeat the above action so the knob is firm to turn, without being over tight.
  vi. Check and adjust the other shutter control knob.
  vii. Replace the cover.
- European origin.
  i. The shutter control knob is attached to a round shaft. This shaft, which controls the opening of the blades, passes through to the rear of the collimator.
  ii. At the rear of the collimator, a circular disc is attached to the end of the shaft.
  iii. There are two screws on the outer side of this disc. These adjust the pressure of a wide spring washer on the disc.
  iv. The screws are adjusted so the control knob is firm to adjust, but not over tight.
  v. These screws tend to become loose. After they are adjusted, clean around the screw heads with alcohol. Then paint the immediate area with nail polish. This will help retain the screws in position.
- For other collimators, contact the service department for advice regarding adjustment of the brake and its location.

l. The collimator lamp fails to operate

The most common cause of failure is, of course, a burnt out globe. In case this is not the reason, then check the following. **Ask an electrician, or electronics technician, for assistance.**

- The lamp timer switch. Mechanical types are prone to failure, and to a lesser degree, electronic versions.
  i. Ensure all power is switched off, while removing the cover to gain access to the timer.
     Note. Some tests for the timer will require power after the cover is removed.
  ii. Check the internal wiring, looking for loose connections.
  iii. Mechanical timers have only two terminals. Operate the timer, and with a multimeter set to low ohms range, check the timer-switch contacts for continuity. **See module 5.0 page 65.**
  iv. An alternate test is to set the multimeter to AC volts, and look for voltage across the terminals. This should be 12~15V for a 12V lamp. When the timer is operated, there should be no voltage across the terminals.
  v. Electronic timers have several connections. It is necessary to trace out the wiring and locate the two terminals that switch the power to the lamp. Look for voltage across these terminals. This should be 12~15V for a 12V lamp. When the timer is operated, there should be no voltage across the terminals.
  vi. In case the timer is faulty, a temporary repair is to remove the timer and replace it with a standard on-off switch. Order a new timer and replace the temporary switch at the first opportunity.
- No power to the collimator. Check the connecting cable for broken connections, especially if the collimator is a rotating version.
- There may be a faulty fuse in the collimator power circuit. Contact the service department for the location of this fuse.

m. The globe has failed, and there is no spare globe

- The collimator should have a scale on the front. This indicates the field aperture as the blades are opened. The accuracy of this scale has, hopefully, been checked at the last routine service. However, let us assume this has not happened.
    i. Lower the collimator so it is touching the tabletop. Adjust any angulation or rotation, so it sits flat on the tabletop.
    ii. Adjust the tube stand so the collimator is at the table centre.
    iii. Place a ruler on the tabletop, end-on against the centre of the collimator. Use this as a guide to assist in centring the Bucky to the collimator.
    iv. Raise the tube to the normal working height. Adjust the X-ray control for a low kV and mAs exposure.
    v. Place a 24 × 30 cm cassette in the Bucky. Adjust the collimator to the film size using the scale on the collimator. Make a test exposure.
    vi. If the test exposure shows all the film was exposed, then repeat the above test, this time reducing the collimator aperture a small amount. If all is well, the film should now have a border around all four sides.
    vii. Continue the above test with the most commonly used sizes of films and orientation. Adjust the position of the control knob on the collimator shaft to obtain a correct indication. Or, place a mark on the collimator front to indicate the required opening for the different films.
    viii. A similar test to the above is required for the wall Bucky.
- To estimate the position of the anatomy under examination. AP view.
    i. With the patient on the tabletop, bring the collimator close to the area under examination. View the position both from the head, or foot, end of the table, as well as the side of the table.
    ii. To estimate the area to be covered, place a sheet of film on the patient, centred directly under the collimator. This is to simulate the previous appearance with the light beam. The actual area will be about 10% less.
    iii. Where possible, protect other immediate areas of the patient by masking with lead rubber strips.
    iv. Raise the X-ray tube to its normal height, and set the aperture size using the scale on the front of the collimator.
- To assist in Bucky centring during an examination.
    i. Attach a length of string to the front of the collimator side, positioned at the centre. Attach a small weight at the end to act as a plumb bob.
    ii. Move the X-ray tube across the table, so the plumb bob is over the Bucky tray. Centre the Bucky to the X-ray tube, and then return the X-ray tube back to the table centre position.
    iii. Coil up the string etc on the X-ray tube when not in use.
- To estimate the position of the anatomy under examination. Oblique view.
    i. A simple method is to use a long ruler, or similar object, resting against the upper or lower side of the collimator. This is extended towards the patient. By alternating the ruler on the upper and lower side of the collimator, a reasonably accurate positioning of the X-ray field may be made.
    ii. A torch may be used. The torch should be the type that has a focussed spotlight, and a flat bottom end.
    iii. Hold the torch bottom end against the centre of the collimator faceplate. Switch the torch on, and place a marker in the centre of the light beam on the tabletop.
    iv. Rotate the torch, and check that the light beam stays in position. This test indicates the torch is suitable for use.
    v. Now place the patient on the table. Adjust the X-ray tube to the required angle. Place the torch on the collimator front as before, and use the torch light to indicate the X-ray beam centre.
    vi. As before, a sheet of film may be used to estimate the area to be covered.
    vii. Set the collimator aperture, using the scale on the front of the collimator.
    viii. Use lead rubber strips to protect the patient.

TASK 14

Help! No spare globe for the collimator

The collimator globe has failed. On checking supplies, there is no spare globe. You are required to continue processing patients, while waiting for a new globe.
Please refer to module 11.2 for an outline of suggested techniques.
Note; for this exercise, the collimator lamp must not be used.

Using a 24/30 cm film, test the accuracy of the collimator scales. Is the accuracy adequate? Does the scale need to be reset?

_____
_____

Make a suggestion for other methods to achieve patient positioning, with an AP view.

_____
_____
_____

Can this method be adapted for a wall Bucky?

_____
_____
_____

With a water phantom to simulate the patient, try the methods suggested for an oblique view. Will this give the required accuracy?

_____
_____

Discuss other problems that may arise if the lamp fails. Suggest a technique that may be used.

_____
_____
_____

Tutor's comments

Satisfactory/Unsatisfactory

Signed _____   Date _____
            Tutor

MODULE 7.3
HT cable

Aim

The aim is to provide information related to the high-tension (HT) cable. This includes repairing common faults, and procedures for replacing the HT cable. Information in this module also applies to module 7.1 page 104.
(**Note:** Reference module page numbers refer to the title page.)

Objectives

On completion of this module, the student will be aware of common problems with the high-tension cable, and their symptoms. With the assistance of an electrician or electronics technician, a number of corrective actions can be carried out.
 This includes:

- Repairs to eliminate high-tension arcing.
- Correcting bad cathode cable connections to the X-ray tube filament.
- Removal and reinsertion of the cable ends, when the X-ray tube is replaced.
- Replacement of the HT cable.

Contents

a. Safety precautions
b. High-tension failure of the HT cable
c. Damage to the cable electrical safety shield
d. Arcing in the X-ray tube receptacle
e. Burnt pins on the cathode cable end
f. Caution on removing HT cable ends
g. HT cable replacement
h. HT cable fault with CD mobiles
i. Preparation prior to inserting the HT cable end
j. Inserting the HT cable end
k. Need for re-calibration

a. Safety precautions

- **Do not attempt repairs or replacement of the HT cables by yourself. Ask an electrician, or an electronics technician, for assistance.**
- **Before disconnecting any wires, or removing the HT cables, always ensure power is turned off and unplugged from the power point. If the equipment is part of a fixed installation, besides switching the generator power off, ensure the isolation power switch for the room is also switched off.**
- Mobile high-frequency generators may be battery operated. The batteries in these are connected in series, and can have a total voltage of up to 240 V DC. *Refer to the operating or installation manuals for the position of the battery isolation switch, and ensure this is switched off, before removing the covers, or testing wires and connectors.*
- **If removing a HT cable from a capacitor discharge mobile, observe the high-tension precautions described in module 6.2 page 94.**
- **Whenever a HT cable is removed from a receptacle, immediately short the cable-end pins to ground. This is to remove any residual high voltage in the cable. The same precaution applies before applying grease or any other handling of the cable end. Failure to take this precaution could cause a severe electrical shock.**

b. High-tension failure of the HT cable

The HT cable tends to fail at the cable ends. This is due to the added flexing, or twisting, as the X-ray tube is rotated and repositioned. This is often due to poor support of the HT cable. Failure is usually accompanied with a pungent, or acrid, smell.

- A metal 'cuff' often hides the actual failure point. This cuff helps support the cable end where it enters the tube housing. If suspicious of the HT cable, then undo the retaining ring nut, and slide the cuff out of the way. Then inspect again for an unusual smell. Make a comparison with the other cable end.

- If a HT cable is suspect, test by replacing the cable.
    i. This may be a spare cable from an old installation, or else a loan cable sent from the service department.
    ii. Observe carefully the procedures and precautions in this module, before replacing a cable.
- See 'Arcing in the X-ray tube receptacle', part 'd'.

c. Damage to the cable electrical safety shield

The HT cable is fitted with a wire mesh safety shield. This is just below the outer insulation. If there is a failure of the cable insulation, the shield conducts the high-voltage spark to ground. The safety shield is soldered to the metal flange of the cable end. Twisting of the HT cable can cause the wire strands to break.

In some cases, the shield is found completely disconnected. This can be very dangerous. See 'Caution on removing high-tension cable ends', part 'f'.

- To inspect, undo the cable-end retaining ring-nut. Slide back the cable support cuff. Check for broken strands.
- In some cases, the shield connection is wrapped in insulation tape. This form of construction is weak, and is prone to have damage to the shield. Unwrap the tape to inspect for broken strands. If ok, then re-wrap using fresh tape.
- If there are broken strands, and especially if this is extensive, a repair should be attempted. An electrician or electronics technician should perform this repair, after obtaining advice from the service department.
    i. A soldering iron is required. 75~100 watt is optimum. This is to allow quick soldering to the cable end without spreading excessive heat.
    ii. You will need some fine multi-strand 'hook up' wire. (The type needed for general electronics wiring). Or if possible, the braided shield from a length of co-axial cable. If using hook-up wire, remove the insulation from the wire.
    iii. Gather the broken strands of the shield wire. If necessary, remove a little of the cable outer insulating sheath. Twist together to make four bunches, spaced around the cable end.
    iv. Solder to one end of the hook-up wire. Take the hook-up wire a full turn clockwise around the cable end, then solder to the cable end.
    v. Repeat this with the other three bunches, alternating the direction around the cable end. Eg, anticlockwise, then clockwise, and finally anticlockwise.
    vi. Use insulation tape to cover the repaired shield connection.

- **Please note.** In some cases it may be claimed that the system is safe, providing the shield is connected to ground at the transformer end. This is not correct. Besides possible danger, this can upset the performance of high-frequency X-ray generators, and create interference in other equipment.

d. Arcing in the X-ray tube receptacle

This is a common cause of failure. Arcing can be caused by a number of reasons. There may be poor quality or dried-out insulating grease. The grease may have been incorrectly applied. If the cable end is loose, this will create air gaps, and eventual arcing. Later systems use silicon rubber anti-corona insulating pads. Unless care is taken installing these pads, arcing will occur. Finally, a fault can occur inside the cable end itself. In affect, a fault in the HT cable, but not externally apparent until the cable end is removed.

- *See 'Caution on removing HT cable ends', and 'Preparation prior to inserting the cable end'.*
- With the cable end withdrawn, look for possible carbon tracks on the cable end, or in the receptacle.
- Where there is grease in the receptacle, wipe the grease with a fresh paper tissue. If arcing occurs in the grease, this will show up as carbon deposits on the paper. Old grease may have a yellow colour, but this does not indicate arcing.
- Examine the cable end carefully for signs of swelling or cracking. This would indicate arcing. In this case a replacement HT cable is required.
- Wipe out all grease from the receptacle and the cable end. With a torch examine the receptacle carefully for signs of arcing.
- If silicon rubber anti-corona pads are fitted, these may remain attached to the cable end. More often they will remain in the receptacle.
    i. Pads are often used without grease. However, wipe the inside of the receptacle and the cable-end with a fresh paper tissue. If it appears dirty, this is a sign of arcing.
    ii. Examine the pads for possible hairline black marks, which indicate arcing.
    iii. The pads should be replaced after being disturbed.
    iv. In case a replacement pad is not immediately available, then they may be returned to service. Take care not to touch them directly with the fingers. (Use a paper tissue.) Have the pads replaced at the first opportunity.

e. Burnt pins on the cathode cable end

The cathode filament current may be between 4.0~5.5 amps. If the pins on the cable end do not make good contact inside the receptacle, they will become burnt. This produces added resistance to the filament circuit, and reduced filament heating.

In the case where light or no exposures occur, this may be due to poor contact of the cathode cable-end pins.

Do not make the following test with a capacitor discharge mobile. Please refer to 'High tension precautions' in module 6.2 page 94.

Undo the ring nut sufficiently to withdraw the cable end about 2~4mm. Then reinsert the cable end and tighten the ring nut. If this action restores or improves the X-ray output, then the cable-end pins are suspect.

The cable end should now be fully withdrawn and examined.

- **See 'Caution on removing HT cable ends', and 'Preparation prior to inserting the cable end'.**
- Examine the cable-end pins. Look for a pin that shows burn marks, or pitting.
- Clean the pin with fine emery cloth or sand paper.
- If the cable-end pins are burnt, then the pin sockets of the housing receptacle will also need cleaning. One method is to use a wire coat hanger, with one end filed flat. This can be used to scrape the sides of the socket into which the cable-end pins fit.
- Most HT cable-end pins are solid brass, split in two halves. These tend to close together, and make a less secure fit in the receptacle. The pins may be carefully spread apart, so the air gap in the middle is parallel.
- **Caution;** these pins are brittle. Do not try to spread them apart using a screwdriver. The best tool is a utility knife with a retractable blade. This blade is just slightly thicker than the required gap. Push the blade into the gap very carefully, so the gap becomes almost, or just, parallel.
- **Help.** A pin is broken. All is not lost. However, you will need to exchange the anode and cathode cables. (At the HT transformer as well as the X-ray tube)
- Before attempting to reinsert the cable end, ensure it is thoroughly cleaned. This is especially important after handling the end, as small traces of perspiration or fingerprints etc may be left behind. See 'Inserting the HT cable end'.

f. Caution on removing HT cable ends

- **If in a capacitor discharge mobile, please refer to 'High tension precautions' in module 6.2 page 94.**
- When replacing a cable, remove the cable-end first at the X-ray tube, and then at the high-tension transformer. This especially applies if the safety earth shield is damaged at the X-ray tube end.
- **As the cable end is withdrawn, touch the endpins to the screw-thread side of the housing receptacle.** This is to short out any residual high voltage in the HT cable. This especially applies if high voltage was generated, but with no mA.
- The cable end may be a very tight fit. Do not try tugging on the HT cable to remove it. Instead, use two screwdrivers, one on each side of the cable-end flange, to lift, or ease, the cable end from the receptacle.

g. HT cable replacement

- If the anode HT cable is replaced with a different type or length, in most cases this makes little difference to the performance, especially if the difference in length is less than 10~15%.
- An exception may be with some medium frequency inverter systems, which have an adjustment for different lengths of HT cables. Check with the service department for this possibility.
- In the case of the cathode cable, a different length or type can change the mA calibration. Providing the anode cable is the same type and length as the failed cathode cable, then exchange cables, so the replacement is used for the anode side. This will avoid the requirement for immediate mA recalibration.

h. HT cable fault with capacitor discharge mobiles

The CD mobile cathode-cable can develop a short circuit between the internal control-grid wire, and the filament wires. In most cases, the anode cable will be in good condition, and can be exchanged for the cathode cable. mA calibration is not critical, and will not need to be adjusted. If exchanging cables, attach an internal notice to indicate a change has been made. This will avoid future frustration in case another change is attempted.

Before attempting to remove or replace a CD mobile cable, please refer to 'High tension precautions' in module 6.2 page 94. Otherwise replacement is the same as for a standard system.

i. Preparation prior to inserting the HT cable end

- Old grease on the cable end, and in the receptacle, should be removed.
- Thoroughly clean the receptacle and cable end. When cleaning the receptacle, use paper tissues wrapped around a wood or plastic rod.
- If necessary, a hydrocarbon cleaning solvent may be used. Be sure to remove all residues. When cleaning a cable-end, avoid touching with your hand, this can leave unwanted perspiration or skin oil. After cleaning, the cable end or receptacle should have a high polished appearance.
- Examine the cable-end pins of the cathode cable. If of the split pin version, check the pins are not bent together, and the gap is parallel. See **'Burnt pins on the cathode cable end'**, for tips on adjusting.
- Most HT cable ends have a rubber sealing-ring. This is placed over the cable end, up against the flange. Ensure this is fitted correctly before applying grease.
- Fresh insulating grease is now required, or a silicon anti-corona disk.
  i. Grease is applied using a wooden or plastic spatula. For example, a tongue depressor. Do not apply or smooth the grease with a finger. First wrap a paper tissue around the finger, to avoid directly touching the grease.
  ii. Apply the grease to about 70% of the length of the cable-end, starting at the pin end. The depth of grease at the pin end should be about 2~3mm, tapering off at the 70% point. The application of grease is not critical, as any irregular area will flow around the sides of the cable end, as it is inserted.
  iii. A layer of about 1~2mm may also be applied to the front of the pin end, between the pins.
- If a silicon rubber anti-corona disk is used.
  i. The disk should be supplied in a sealed package. Handle the disk with a pair of clean tweezers, or else by a paper tissue.
  ii. Place the disk in position on the cable end, with the pins passing through the disk.
  iii. A small layer of silicon grease may be placed around the sides of the cable end. This is an option. If in doubt, check with the service department for advice.

j. Inserting the HT cable end

- The cable end has a 'key' at the flange end. This fits into a 'notch' in the receptacle. Before inserting the cable end, check the rotational position, so these two areas will be aligned on insertion.
- On inserting the cable end, try to keep it aligned in the centre of the receptacle. This ensures an even distribution of the grease.
- As it becomes fully inserted, rotate the end a little to align the cable-end pins into the receptacle sockets.
- A very firm continuous pressure is often required. This is due to pockets of air in front of the grease, as well as the viscosity of the grease itself.
- Once the cable-end pins are properly inserted into the receptacle sockets, it should now be possible to attach the cable-end retaining ring-nut and cable support cuff.
- Tighten the ring nut fully. Then check again every few minutes until it can no longer be even partially rotated. This should be checked again over the next few days.
- Attach cable ties to support the HT cable in position.

k. Need for recalibration

Note. Replacement of the **cathode** cable can alter the mA calibration. While replacement with an identical type and length may have very little affect on the calibration, this should still be checked.

In case the **cathode** cable is a different length or type, this may have a large affect on the mA calibration, depending on the design of the generator. ***Before attempting any calibration, check first with the service department for the recommended procedure.***

PART III. FAULT DIAGNOSIS AND REPAIR MODULES

MODULE 8.0
Bucky and Bucky table

Aim

The aim is to provide information and procedures related to problems with the Bucky and Bucky table. This is an addition to the maintenance procedures, provided in module 3.0 page 53.
(**Note:** Reference module page numbers refer to the title page.)

Objectives

On completion of this module, the student will be aware of common problems with the Bucky and Bucky table, together with their solutions. Adjustments and repairs may be carried out. Some tests or repairs will require the assistance of an electrician or electronics technician.

Task 15.'A film exhibits grid lines', should be attempted on completion of this module.

Contents

a. General precautions
b. Grid lines sometimes appear on the film
c. Grid lines appear on all films
d. The film is dark in the centre, but fades out to either side
e. No exposure on a selected Bucky
f. The cassette tray does not hold the cassette properly in the vertical Bucky
g. The Bucky lock does not operate
h. The table magnetic locks slip, or are unreliable
i. The table auto-centre does not operate, or is not accurate
j. Noisy tabletop movement
k. Elevating Bucky-table problems

a. General precautions

Please take the following precautions.

- **Before testing any fuses, or removing a cover, always switch the generator power off, and ensure the isolation power switch for the room is also switched off.**
- Test procedures for fuses or wiring, are described in **module 5.0 page 65**.
- When removing the cover from a vertical Bucky, make sure the Bucky cannot move upwards when the cover is removed. For example, attach a rope to hold it in position, or remove the cover with the Bucky set to maximum height.
- In most cases removal of the Bucky cover is a simple operation. However, where possible refer to an installation manual. This will indicate if there is any special procedure for removing or installing the cover.
- In the case of a Bucky table, removal of the tabletop may be required. The method used depends on the table design.
    i. Most tabletops may be removed once the screws holding the 'profile rails' in position are removed. Before attempting to remove these screws, make sure the screw slots are not blocked with dirt, and use a screwdriver that has a good fit and is not blunt.
    ii. In other cases, removal of the tabletop end-stops will allow the tabletop to extend to over one end. As the tabletop is moved past the end-stop position, the tabletop will disengage from the far-end bearings. Have a chair or other suitable object ready to support the tabletop.
    iii. In some cases, power will need to be switched on to release the table locks.
    iv. To avoid unexpected problems, make sure at least one person is available to assist.
    v. On replacement of the tabletop, check and manually reseat the locks to allow the tabletop to pass over them.
    vi. Ensure the end stops are securely replaced.
- Keep all screws, or other small parts in a container, to avoid loss.

b. Grid lines sometimes appear on the film

There are several possibilities.

- The grid starts oscillating as soon as the X-ray control is placed in preparation.
  i. This form of operation does not synchronise the exposure to the grid position. As a result, an exposure can commence when the grid has reached the end of travel, and is reversing direction.
  ii. Test by looking for grid movement when in preparation, before an exposure. If this is the cause of the problem, re-installation of the Bucky wiring is required. Consult the service department for advice.
- The Bucky has a slow moving grid. In this case, there is insufficient grid movement when performing short exposure times. A replacement Bucky of a later design is required.
- Some Bucky's have a rotating cam to operate the grid.
  i. The grid moves quickly at first, then slower until the full 'in and out' cycle is completed. This is repeated till the exposure is completed.
  ii. The Bucky has an adjustment to ensure the exposure commences at the point of maximum speed. If incorrectly adjusted, then the exposure could commence before that point, when the grid is at minimum speed.
  iii. This is indicated if grid lines occur on short exposure times. If adjustment is required, request the service department to adjust the Bucky.
- In the case of mammography, many Bucky's have a speed adjustment.
  i. Optimum adjustment is for the grid to reach 75% of its stroke during an average exposure. If grid lines occur on short exposures, then increase the speed. Or, if grid lines occur during long exposures, then reduce the speed.
  ii. This may be a screwdriver adjustment at the back of the Bucky, or it may be an internal adjustment. In that case, the service department must make the adjustment.
- The grid movement may be hitting an obstruction, and while moving far enough to permit an exposure, then stops moving. Or, in some cases, the grid drive may be sticking. This can be indicated if short exposure times are ok, but long exposure times have grid lines.
  i. Close the collimator, and direct the X-ray tube away from the Bucky.
  ii. Select minimum kV, the lowest mA station, and a long exposure time.
  iii. Remove the cassette tray, so the grid may be clearly observed.
  iv. Have an assistant make an exposure, using the Bucky under test. Observe the grid movement, looking for signs of hesitation. Does it tend to stop before reversing?
  v. If a problem is indicated, then remove the Bucky cover, or tabletop, and examine the mechanism while it is moving. Look for film markers causing an obstruction.
  vi. Some motors can have damaged gears, with missing teeth. In this case, repair kits may be available.
  vii. Apply a small amount of oil to moving surfaces.

c. Grid lines appear on all films

- The wrong Bucky was selected. Is the selection switch correctly labelled?
- Listen for a Bucky sound during a test exposure. Does the selected Bucky operate? Does it sound normal?
- A common cause is a dislodged grid. For example, the grid has fallen from the grid frame, or holder. In this situation, although the frame moves, permitting an exposure, the grid itself does not move. To remount the grid in the frame, removal of the Bucky cover, or tabletop, may be required.

d. The film is dark in the centre, but fades out to either side

- The grid was removed, and then reinserted upside down.
- Is the grid focal distance within the range you are using?
  i. Can the grid be removed? Look for a label that provides the focal length of the grid.
  ii. If it is difficult to remove the grid, try a test exposure after changing the FFD. Make a direct low kV and mAs exposure, without a patient.
- Is the vertical Bucky correctly aligned to the X-ray tube?
  i. For example, is the Bucky at an angle to the x-ray beam?
  ii. This can occur if the vertical Bucky is not correctly installed. The Bucky may be mounted against a wall, which is not at an angle of ninety degrees to the tube-stand.
- Image fade off to one side may be a problem due to the X-ray tube or collimator. To check, make a direct exposure to a cassette on the tabletop.

i. A direct exposure at low kV will exhibit some fade off towards the cathode side of the tube. On large films, a larger fade-off can occur towards the anode side of the film, due to the anode heel-affect.
ii. To test at the kV values normally used, a suitable filter is necessary. For example, 1.0 mm of copper in front of the collimator. Or else 10 cm of water in a plastic container.
iii. Set a similar kV to that used when observing the problem. Adjust the mAs to achieve a suitable density.
iv. If the fade off is still present, the collimator may be out of alignment. For example, the collimator is incorrectly centred to the X-ray tube, and the light beam was adjusted to compensate. **See module 7.2 page 110.**
v. There may be excessive metal deposits on the X-ray tube glass. You will need to remove the collimator to check. Contact the service department for advice before attempting removal. **See module 7.1 page 104.**

e. No exposure on a selected Bucky

In this situation, a test exposure using direct non-Bucky radiography is successful.

- Close the collimator, and move the tube away from the Bucky.
- Select minimum kV, a low mA station, and a medium time setting.
- When trying to make a Bucky exposure;
  i. Listen carefully at the Bucky for any sound. With the cassette tray removed, see if the grid moves.
  ii. If no sign of any grid movement, check the Bucky cable for a possible loose connection or broken wire. **See module 5.0 page 65.**
  iii. Check for a possible blown fuse. To locate the fuse, **see module 5.0 page 65**, or consult the service department.
- When trying to make a Bucky exposure, the Bucky starts to operate, but there is no exposure.
  i. Check the Bucky cable for a possible loose or broken connection. **See module 5.0 page 65.**
  ii. Remove the Bucky cover, or the tabletop.
  iii. Look for any object that could be blocking the grid movement, such as a lost film marker. This can happen with a wall Bucky, or a Bucky with a fluoroscopy table.
  iv. On attempting an exposure, does the grid drive motor operate? Look for damaged fibre gears, or a broken drive cord.
  v. Does the grid manage a full 'stroke'? As the grid moves from the 'rest' to the 'expose' position, a microswitch is operated. This microswitch allows the exposure to commence. Check the microswitch for correct operation. **See module 5.0 page 65.**

f. The cassette tray does not hold the cassette properly in the vertical Bucky

- In some cassette tray designs, the amount of 'grip' is insufficient. To prevent the cassette slipping down, the manufacturer supplies small wood blocks with an attached magnet. In other designs, a metal support is provided, which fits into a series of holes.
- The rubber grips attached to the tray jaws become smooth, allowing the cassette to slip. The rubber grips can be improved by cleaning with lighter fluid, or a similar hydrocarbon.
- The jaws may not be closing fully. Check and adjust the position of the clamping knob on the shaft.

g. The Bucky lock does not operate

- Look for a faulty switch, or broken connection, either to the switch or the magnetic lock coil. The lock coil may be open circuit. Test with a multimeter set to medium 'ohms' scale. **See module 5.0 page 65.**
- The lock coil may have too large an air gap. Adjust it closer to the operating surface.

h. The table magnetic-locks slip, or are unreliable

- **Before testing any fuses, or removing a cover, always switch the generator power off, and ensure the isolation power switch for the room is also switched off.**
- Test procedures for fuses or wiring, are described in **module 5.0 page 65.**
- No locks operate.
  i. A fuse could be open circuit. Before replacing, check the wiring to the lock coils or switches, and look for possible damaged insulation.
  ii. A foot switch is faulty, or has a bad cable connection.
- A specific lock fails to operate.
  i. If other locks are operating, it is unlikely to be an open circuit fuse. However, still check, as the lock may not be the same type, and has a separate fuse.

    ii. There may be an open circuit lock coil, or winding.
    iii. To check the lock coil, first ensure all power is switched off. Disconnect the lock from the table. Use a multimeter set on a medium ohms scale to check the resistance of the lock coil. This may measure around 500 to 2000 ohms depending on design. If unsure, check against a similar lock coil in the table, or contact the service department.
    iv. Check the wiring for possible loose or broken connections.
    v. Check the control switch.
- The table lateral-movement lock is weak.
    i. Check the number of locks installed for lateral operation. Some tables were supplied with only one lock. In this case, apart from cleaning the front of the lock, little improvement can be made.
    ii. Where there are two locks, one of the lock coils may have failed. Watch the locks when they are switched on and off. If only one lock moves, the other lock might have an open circuit winding. In some cases, you may find the suspect lock cool to touch.
    iii. **Note.** Locks may be positioned either at one end, or at both ends of the table.
    iv. To check the lock coil, first ensure all power is switched off. Disconnect the lock from the table. Use a multimeter set on a medium ohms scale to check the resistance of the lock coil. This may measure around 500 to 2000 ohms depending on design. If unsure, check against a similar lock coil in the table, or contact the service department.
    v. Check the wiring for possible loose or broken connections to the suspect lock coil.
- The locks make a rattling or buzzing noise. Check the mounting of the locks. Is the lock parallel to the operating surface? Is there a large air gap when not switched on? Another possibility is dirt on the top, or face, of the lock.

i. The table auto-centre does not operate, or is not accurate

- To test the table lateral centre-position.
    i. Place a cassette in the Bucky, with a marker positioned on the centre of the cassette.
    ii. Place another marker on the centre of the tabletop.
    iii. Move the tabletop to the lateral centre-position.
    iv. Make a low kV and mAs exposure. Process the film and check if both markers are in the same position.

- Some tables have a mechanical centre stop. A spring tensioned steel ball clicks into a slot when the table is centred.
    i. The spring can have a tension adjustment screw. Tighten the screw to obtain a firm stopping action.
    ii. Are the screws holding the mechanical centre stop loose?
    iii. To adjust the stop position, undo the screws a small amount, and push the centre stop to the required position. Tighten the screws to prevent movement. Make a test film to confirm the table is centred.
- Other tables may switch on the electromagnetic locks when in the centre position. This is usually by a cam passing over a microswitch. (Later designs may use electronic sensors, such as optoelectronics.)
    i. If the microswitch is positioned away from the cam, unreliable operation can result. If positioned too close, then poor centring action results. For example, the stopping position becomes wide.
    ii. Centre position is adjusted by moving the cam, or else the microswitch.
- The centre microswitch, or the auto-centre selection switch can have faulty contacts. Ensure power is turned off. Check with a multimeter set to low ohms scale to test the switch.
- For other possibilities, contact the service department for advice.

j. Noisy tabletop movement

A 'clunking' noise is heard as the tabletop is moved.

- This may be due to a faulty bearing, or it may be caused by dirt in the bearing tracks, or on the rim of the bearings. Spray the bearings and bearing track with light aerosol oil, and wipe down with a rag.
- Watch the bearings as the tabletop is moved. A faulty bearing may have a cracked or missing rim. In some cases, the bearing does not rotate, and the table is stiff to move.
- Contact the service department for a replacement bearing, plus advice for replacing the bearing.

k. Elevating Bucky-table problems

- The tabletop will rise up, but not move down.
    i. Many tables have been damaged after being brought down onto a chair or patient stool. Two common safety devices are now used.

PART III. FAULT DIAGNOSIS AND REPAIR MODULES

ii. A pressure-pad is installed on the floor, positioned at both ends of the table. Pressure on these pads activates a relay, which stops the downward movement of the table. Is there an object pushing on the pad, or has the pad become damaged?

iii. A sensitive microswitch is installed in the middle of the longitudinal bearing tracks. This may require adjustment. Contact the service department for advice.

- The table does not stop at the operating height.
   i. A cam-operated microswitch is used to switch off the motor once the operating height is reached.
   ii. The position of the cam or microswitch may require adjustment.

- The motor does not operate.
   i. On some tables, occasional failure of the motor power fuse occurs. *Before attempting to replace the fuse, ensure all power is turned off.*
   ii. When replacing the fuse, use a delay or slow-blow type. **See module 5.0 page 65.**
   iii. For location of the fuse, refer to the parts or installation manuals.
   iv. If the fuse continues to fail, contact the service department for advice.

TASK 15

A film exhibits grid lines

After taking a chest X-ray, you notice prominent grid lines on the film.

Make a list of possible reasons for this problem.

_____
_____
_____
_____
_____

Describe suitable tests to either confirm, or eliminate, possibilities from this list.

_____
_____
_____
_____
_____

Carry out these tests. What were the results?

_____
_____
_____

What action is needed to correct the problem?

_____
_____
_____
_____
_____

Tutor's comments

Satisfactory/Unsatisfactory

Signed _____ Date _____
            Tutor

PART III. FAULT DIAGNOSIS AND REPAIR MODULES

MODULE 8.1
Tomography attachment

Aim

The aim is to provide information and procedures related to problems with a tomography system. This may be a tomography attachment, fitted to a standard tube-stand, or integrated with a Bucky table. This is an addition to the maintenance procedures, provided in module 3.1 page 55.
(**Note:** Reference module page numbers refer to the title page.)

Objectives

On completion of this module, the student will be aware of common problems with the tomography attachment, including operator error, or problems with a safety interlock. Adjustments and repairs may be carried out. Some tests or repairs will require the assistance of an electrician or electronics technician.

Contents

a. General precautions
b. Failure to operate
c. The tomography image has poor definition
d. The required exposure time is difficult to estimate

a. General precautions

Please take the following precautions.

- **Before testing any fuses, or removing a cover, always switch the generator power off, and ensure the isolation power switch for the room is also switched off.**
- When investigating a possible bad connection, open circuit fuse, or faulty switch, refer to the procedures in **module 5.0 page 65**.
- Older tomography attachments have either limited, or no safety interlocks. Do not operate the motor without the fulcrum pole.

b. Failure to operate

This may be due to incorrect set-up, operation of a safety interlock, or an open circuit fuse to the motor. Where sections of the attachment are connected via plugs and sockets, these need to be checked.

- Incorrect set-up may cause operation of a safety interlock. This helps guard against operator error. When all else fails, then read the operation manual.
- Current tomography attachments can have a number of safety interlocks, depending on the design, and integration with the tube stand.
- Older systems depend more on correct set-up, such as ensuring the rotation and longitudinal tube-stand locks, and Bucky lock, is off. This requires care by the operator before using the system.
- Typical interlocks for a tomography attachment can include:
  i. The fulcrum pole interlock. If the fulcrum pole is not attached, the tomography system does not 'know' the stop or start positions. If energized, the motor would drive the tube-stand to the end of the tube-stand track. The interlock usually consists of a microswitch. This is operated when the fulcrum pole is attached to the tube-stand cross-arm. The actuating lever for this microswitch may be damaged, or out of adjustment. Listen for a small 'click' as the fulcrum pole is placed in position.

ii. An exception to the above is where the motor has an arm that engages with a floor, or ceiling, slotted guide plate. This system can only operate over the distance controlled by rotation of the arm. However, later systems do have an interlock to check if the fulcrum pole is fitted. This prevents a wrong exposure.
iii. X-ray tube height. If not correct, then tomography calibration is incorrect. This is measured by a cam-operated microswitch, or in some cases by an optoelectronic sensor and reflective strip. Operation of the height sensor occurs only over a narrow distance. Try moving the tube-stand up or down a small amount.
iv. Tube-stand and Bucky locks are often automatically turned off or on by selection of tomography operation. However, in some systems this is not the case, and failure to switch off the longitudinal lock can prevent the motor from moving the tube stand.

- The tomography motor has a high 'inrush' current on start up. An open circuit fuse is not uncommon.
  i. **Before testing or replacing a fuse, ensure all power is turned off. See module 5.0 page 65.**
  ii. Look for any cables that might be damaged, causing a short circuit.
  iii. The fuse could be positioned close to the motor system, or else in the tomography control cabinet. Use a delay, or slow-blow, fuse as a replacement.
  iv. If the fuse fails shortly after replacement, contact the service department for advice.
- The fulcrum tower has a number of contacts controlled by rotation of the fulcrum. These may be cam operated switches, or else a metal-strip with a sliding contact, or 'commutator'. Selection of the appropriate section controls the start-stop position of the motor, and the tomographic angle.
  i. There may be broken connections in the cable plug and socket. Another possibility is a broken microswitch, or commutator brush, inside the fulcrum tower. **Before investigating, ensure all power is turned off.**
  ii. An exception is where the tomographic angle and start-stop position is controlled by a series of cams coupled directly to the drive motor. In this case, the tower will only have a motor and light for setting the fulcrum height.

c. The tomography image has poor definition

During a tomographic scan, the film in Bucky must retain the correct alignment with the X-ray tube focal spot. This requires a smooth movement when travelling through the actual exposure area.

- To evaluate the actual performance, the tomographic resolution test piece described in appendix B page 169, section is recommended. When operated at the correct height, a clear image of the central paper clips should be obtained.
  i. To avoid over exposing the film, use a low mA station, and low kV. If the film is still too dark, then insert a sheet of paper between one intensifying screen and the film.
  ii. Repeat this test for different combinations of speeds and angles.
- If a good image is not obtained with the test piece, then check the following.
  i. Uncouple the fulcrum pole. Otherwise remain in tomographic set-up mode.
  ii. Check that the tube rotation lock is off. The tube should be able to rotate smoothly, and remain balanced as it is rotated.
  iii. Check the Bucky movement. The Bucky lock should be fully off, and the Bucky should move smoothly in the table.
  iv. Is the longitudinal lock released? With some systems, it will be necessary to exit tomographic mode, and then check the lock releases correctly.
  v. Is the fulcrum tower securely mounted?
  vi. Is the fulcrum pole in good condition? The fulcrum pole should not bend or twist.
  vii. With the fulcrum pole in position, but not in tomographic mode, push the tube-stand across the floor. Look for any sudden stiffness, or jerking. Look for dirt in the guide rails.
  viii. A wire cable is used to pull some systems. Is the cable firm, and not slipping on the motor drive pulley?
  ix. The same applies to units with a belt drive. A loose belt can stretch, and give an uneven start to the movement.
  x. Some motors have a drive wheel pressed against the floor. Depending on the floor surface, the wheel slips and drives unevenly. To stop the wheel slipping, glue a strip of material that has a rough surface, to the floor. For example, the type that is fitted to the steps of a staircase.

d. The required exposure time is difficult to estimate

- A microprocessor controlled tomography unit, integrated with the X-ray control, may directly set the

PART III. FAULT DIAGNOSIS AND REPAIR MODULES

exposure time. Otherwise the operator must select a minimum exposure time, called the 'backup' time.
- The backup time must be longer than the actual exposure time, which is controlled by the combination of tomography speed and angle. The tomograph operation manual should indicate the actual exposure times. A minimum backup time of 5~10% longer is recommended.
- If the operation manual indicates times for 60 hz operation only, a correction factor is needed for 50 hz operation. Multiply the 60 hz tomographic times by the conversion factor of 1.20.
- If no information is available, contact the service department. They may need to measure the actual exposure times, and then make a suitable reference chart.

MODULE 9.0
Fluoroscopy table

Aim

A fluoroscopy table can range from the most basic version to a highly sophisticated remote control microprocessor system. Many problems are caused when 'standard' operating procedures are changed. For example, switches and selections may be set to a different position.

The fluoroscopy table has a large number of safety interlocks. These can be activated by operator error. Other possibilities are problems with the X-ray tube, the X-ray control, and the TV imaging system.

The information and procedures provided in this module are for basic tables fitted with an under-table tube. Some suggestions will however be common to all versions.

(**Note:** Reference module page numbers refer to the title page.)

Objectives

On completion of this module, the student will be aware of common problems with the fluoroscopy table, including operator error, or problems with a safety interlock. Adjustments and repairs may be carried out. If assistance is required from the service department, an accurate description of the problem can be provided.

Note: Some tests or repairs will require the assistance of an electrician or electronics technician.

Contents

a. General precautions
b. No fluoroscopy exposure, X-ray control checks
c. No fluoroscopy exposure, table interlock checks
d. No fluoroscopy exposure, table electrical checks
e. The X-ray control indicates fluoroscopy is operating, but no image
f. No radiography exposure
g. Artefacts on the film
h. Manual collimation has unwanted beam limitation
i. X-ray beam alignment is incorrect
j. Table movements do not operate
k. Table locks do not operate

a. General precautions

Please take the following precautions.

- **Before testing any fuses, or removing a cover, always switch the generator power off, and ensure the isolation power switch for the room is also switched off.**
- Test procedures for fuses, switches, or wiring; are described in **module 5.0 page 65**.
- Any dismantling of the table to test connections, switches, or broken wires; should be performed by an electrician, or electronics technician.
- If removing a cover, or dismantling any section, place the screws in a container to avoid loss.

b. No fluoroscopy exposure, X-ray control checks

- Has a correct technique been selected?
- Check the fluoroscopy timer. Has it timed out?
- Does the X-ray control have automatic regulation of fluoroscopy output? If so, turn this off, and try a manual setting of mA and kV.
- Is there a fault indication at the control on trying an exposure?
    i. In some cases, if actual fluoroscopy mA is too high, the exposure immediately stops. A fault condition may only illuminate while attempting an exposure, but disappear once the exposure attempt is released. In other cases, it may be necessary to switch 'off' then 'on' again to reset the safety interlock.
    ii. Reduce the fluoroscopic mA and kV setting, close the table collimator, and try another fluoroscopy exposure.
    iii. If at a lower mA setting, fluoroscopy is now ok, then try slowly advancing the mA knob during a test exposure. Watch the mA meter. Look for

instability, as this could indicate a gassy tube, or perhaps a high-tension fault. An 'over mA' fault may occur when mA reaches 6.0 mA, but in some cases be as low as 4.0 mA, depending on equipment design.

  iv. Some controls have no manual selection of mA; instead the value of mA is controlled by fluoroscopy kV. In this case, advance the kV control, as in part (v). Observe the mA meter for excessive mA or instability.
  v. If excessive mA is not the cause of the problem, then slowly advance the kV control. Again, watch the mA meter, looking for instability. Should a problem occur as kV is increased, this could indicate gas in the X-ray tube, or an arc where the high-tension cable-end enters the X-ray-tube receptacle. **See module 7.1 page 104, and module 7.3 page 117.**

c. No fluoroscopy exposure table, interlock checks

The fluoroscopy table can have a number of interlocks for radiation safety. Some of the possibilities discussed depend on individual table design, and may not be present in your table.

- Is the correct technique selected at the table?
  i. For example, a remote controlled table may be in tomographic mode, or a collimator key switch may have been turned to manual operation.
  ii. The table may have a separate fluoroscopic timer. Has this timed out?
  iii. Some designs have a fluoroscopy-preparation switch. This must be pressed to place the table in operation, *after* selecting fluoroscopy operation at the X-ray control. (The switch has a fluoroscopy symbol.) With this system, the table is automatically deselected if the technique is changed at the control, and must be reselected each time prior to use.
- Are there any warning lights or fault codes displayed on the table? Refer to the operating manual, or contact the service department for further information.
- Is the image intensifier, or fluorescent screen correctly mounted, and not loose?
  i. This especially applies if the image system is removed from the table, to park the serial-changer out of the way.
  ii. The safety interlock is a small microswitch. The microswitch actuator may have become bent, or out of adjustment, when the image system was repositioned, or not operated if the image system clamps are loose.

- Is the serial-changer fully positioned in the operating position?
  i. When the serial-changer is brought forward from the parked position to the operating position, a microswitch is operated, permitting fluoroscopy operation.
  ii. Locate the position of this microswitch, and check if it has operated correctly. It may be possible to hear a small 'click' as the serial-changer is moved into position.
  iii. Some designs uncouple the undertable tube carriage to allow parking of the serial-changer. As the serial-changer is brought forward, the tube carriage should lock back in position. Check to make sure this has happened. Again, a safety interlock microswitch needs to be operated. This microswitch may need adjustment.
- Is there a cassette incorrectly positioned?
  i. For example, in the 'load/unload' position.
  ii. Manually operated serial-changers have a microswitch, which prevents fluoroscopy unless the cassette carriage is fully retracted. For example, when the cassette carriage is brought to the radiography position, fluoroscopy is immediately switched off. Further movement operates another microswitch, which sends the preparation request to the X-ray control. Check the operation of these microswitches.
- Is the Bucky parked at the foot-end of the table?
  i. This can apply to tables where a radiation shield covers the Bucky-slot when the Bucky is parked. Check the safety microswitch operated by this shield. It may be possible to hear a small 'click' as the shield is opened and closed.
- Some older tables were designed to enable the undertable tube to be used with a wall Bucky. These have a microswitch to ensure the tube is correctly positioned for fluoroscopy. This rarely has a problem, but should be checked.

d. No fluoroscopy exposure, table electrical checks

Note. Test procedures for fuses, switches, or wiring; are described in **module 5.0 page 65**.

- The footswitch is a common cause of failure. The connecting cable can have broken connections, either at the footswitch, or where it connects to the table. In addition, the cable itself may have a broken internal wire.
  i. **Before removing any covers, or make any measurements, ensure all power is disconnected.**

ii. With a multimeter set to the low ohms position, test the continuity of the cable and footswitch contacts. This test should be made where the cable enters the table.

iii. With the footswitch operated, a low resistance reading of less than five ohms should be obtained.

- The connecting cables from the serial-changer to the table foot, or to the table serial-changer 'tower', may have broken or loose connections. This especially applies if the cables are pulled or stretched during operation. Broken internal wires in the cables can also occur. If suspect, the individual wires can be checked by measuring across both ends, in a similar fashion to the footswitch. **This procedure must only be attempted by an electrician or electronics technician.**

- Have there been any indications of rats? Some rats enjoy biting wires. This especially applies to the internal wiring of a table. Damage can appear similar to cutting wires with a pair of scissors.

e. The X-ray control indicates fluoroscopy is operating, but no image

On attempting fluoroscopy, the X-ray control indicates a fluoroscopic exposure is operating, and the mA meter indicates a normal value of mA.

- Has the TV been properly switched on and adjusted? **See module 9.1 page 135.**
- Is the collimator closed?
  i. Try operating the collimator controls to the half open position.
  ii. To ensure the collimator is in fact open, place a cassette face down on the tabletop. Make a one or two second fluoroscopy exposure, and process the film.
  iii. If the film is blank, there can be a problem with the collimation control. Contact the service department for advice.
  iv. If the film is exposed, there is a problem with the TV image system. **See module 9.1 page 135.**
- If operating under automatic regulation, change to manual operation. Check that a suitable value of mA and kV can be obtained. If mA is too low, there may be a poor filament connection at the cathode cable end into the X-ray receptacle. **See module 7.3 page 117.**

f. No radiography exposure

The safety interlocks and conditions that prevent fluoroscopy will also prevent radiography. So providing fluoroscopy operates correctly, we can consider the requirements of most interlocks satisfied. However, there are some additional requirements for radiography.

- Has the cassette has already been exposed? Try another cassette.
- Has a cassette of the correct size for the required format been inserted?
  i. Try a different size cassette or format selection.
  ii. If a different size cassette allows operation, there is either a problem with the internal recognition of the cassette size, or else the cassette is not compatible with the table. For example, trying to use an imperial dimension cassette, in a table designed for metric sized cassettes.
  iii. Contact the service department for advice.
- Does the cassette move forward into the expose position? (This assumes a motor drive cassette carriage.)
  i. Try ejecting and reinserting the cassette. If the cassette does not eject, it may have been incorrectly inserted, and fallen out of position. This can cause the carriage to become jammed.
  ii. In case of a cassette jam, try lifting the cassette using a long **wooden or plastic** ruler. This may then allow the cassette to be driven out. Otherwise it will be necessary to gain access by either removing the serial-changer cover, or removing the image intensifier. **Make sure all power is switched off before removing the cover.**
- As the cassette moves forward into the expose position, does the X-ray control go into preparation mode, and then indicate 'ready for exposure'?
  i. Check for normal operation of the X-ray control by a test exposure on the over-table tube.
  ii. A hand operated cassette carriage operates a microswitch when it moves towards the expose position. This microswitch produces the preparation request for the X-ray control. There may be two microswitches close together. As the carriage moves forward, listen carefully for a 'click' from each microswitch. A small adjustment may be required to obtain correct actuation.
  ii. Listen to the X-ray tube. Can you hear anode rotation on the preparation request? If not, check the stator cable for possible damage. For example, it may have been caught up in the undertable mechanism, or have a broken internal wire where it enters the serial-changer longitudinal carriage. Ask an electrician or electronics technician for assistance.

iv. There may be a problem with the large focus. Try radiography on the fine focus. If preparation is now OK, there could be a poor filament connection with the cathode cable-end, in the X-ray tube receptacle. In this case affecting the large focus only. **See module 7.3 page 117.**
- The cassette has moved into the 'expose' position. The X-ray control indicates 'ready for exposure' however, an exposure cannot be made.
    i. Check for normal operation of the X-ray control by a test exposure on the over-table tube.
    ii. Some table designs prevent an exposure if the motorized cassette carriage has stopped outside the correct expose position. Try either a 'full format' exposure, or else a 'split' format exposure. If the change of format allows an exposure, the carriage drive requires adjustment. Contact the service department for advice.
    iii. The serial-changer has additional 'close to film' shutters that are operated when selecting split formats. If these shutters are not in their correct position, this can cause prevent an exposure. This problem can occur with motor driven shutters after a cassette jam occurs, or due to lack of maintenance. Track lubrication is required.
    iv. With a manually operated cassette carriage, a microswitch is operated when the carriage reaches the 'stop' position. On some tables this assembly can become loose and require adjustment. Before attempting any disassembly, contact the service department for advice.
    v. See also 'No fluoroscopy exposure, table electrical checks' regarding the possibility of broken wires and rat damage etc.

g. Artefacts on the film

Film artefacts are due to several possibilities.

- The cassettes were incorrectly stored in the room, and have been subject to scattered radiation. Remember, one minute of fluoroscopy at 2.0 mA and 110 kV is equal to four 30 mAs exposures at 110 kV.
- Poor calibration of the automatic collimation can mean the X-ray field is much wider than the size required for fluoroscopy. This can allow radiation to penetrate the lead shield, designed to protect the film while the cassette is waiting in its 'garage' or parked position. This can cause intermittent bar patterns on the film, depending on the type of examination, fluoroscopy kV levels, and exposure duration.
    i. To test, attach a 35 × 35 cm directly beneath the serial-changer. Apply a few seconds of fluoroscopy, then process the film
    ii. The fluoroscopy pattern on the film should be about 5~10% less than the stated diameter of the image intensifier.
    iii. A problem was experienced similar to the above with some earlier designs of remote controlled tables. In this case, it was possible to obtain fluoroscopy with the collimator key switch in the manual position. The problems disappeared after a design change, so that fluoroscopy was only permitted under automatic beam limitation.
    iv. Some models of fluoroscopy tables, although collimation **was** correct, still required additional lead shielding. For example, the problem was due to scatter. Discuss this possibility with the service department.
- Radio-opaque contrast solutions find their way into unusual places. Barium deposits are easy to see. Media used for an IVP is less easy to see. If in doubt clean all surfaces, including under the serial changer. Look also under the tabletop, and on top of the undertable tube collimator.
- Another cause can be wiring cables moved out of position under the tabletop. A frequent cause is due to the Bucky not parked fully at the table end.

h. Manual collimation has unwanted beam limitation

In many examinations, such as a barium swallow, the radiologist will desire to cone in horizontally to optimize the image. When the film is developed, it is found the top and bottom areas of the film are not exposed. This affect is due to the collimator field size required for fluoroscopy. For example, the X-ray field must not exceed the diameter of the image intensifier field of view.

Many later designs of tables have an added facility, called 'semi automatic collimation'. In this mode the lateral collimation remains in the position set during fluoroscopy, while the vertical collimation opens to the film size during radiography.

It is sometimes possible to convert an older table, depending on make and model, to also have semiautomatic collimation. Discuss this with your service department.

i. X-ray beam alignment is incorrect

- Remove the bottom table cover.
- Check if the collimator is loose on the X-ray tube.
- The X-ray beam is shifted to one side of the image. Lateral shift of the image.
    i. In most cases, the tube may have shifted in the trunnion mounting rings.
    ii. Release the trunnion ring clamp, rotate the tube position only a small amount and test again.
    iii. The most critical position is with the spot filmer at maximum height from the tabletop.
    iv. Check with the table horizontal and vertical. If necessary, a small adjustment between the two positions may be needed.
- The X-ray beam has shifted vertically.
    i. In many cases the tube and trunnion assembly is mounted to the table via a 'spigot'. A bump or vibration of the table can cause this to rotate slightly.
    ii. Locate the clamping screws for the spigot, undo them just a small amount. Then rotate the assembly so the beam is realigned. Tighten the screws, and re-check the alignment.
    iii. Small changes in alignment are easily seen with the spot filmer at maximum height from the tabletop.
    iv. Check with the table horizontal and vertical. If necessary, a compromise adjustment for the two positions may be required.
- If not able to adjust the X-ray tube position, or if not sure of the procedure to suit your table, contact the service department for advice.

j. Table movements do not operate

- Has an emergency stop switch been activated? The warning lamp may have failed. Some switch designs, once pushed in, require the knob to be rotated to release. Remote controlled tables can have two switches, one at the control desk, and the other at the table body. Check both switches.
- Is the vertical compression lock activated? As a safety precaution this can disable tabletop movements. In some alternate designs, the compression lock is automatically released when moving the tabletop.
- Has a patient protection device been operated? In some tables this is a light beam. Collisions with a patient trolley can cause this to be misaligned, or else there is dirt on the optical system. Look also for table drapes in the wrong position, which can block the light beam.
- The tabletop will only move in one direction. This usually means a limit switch has been operated. A common cause is a faulty microswitch. See also 'No fluoroscopy exposure, table electrical checks' in case of broken wires, or rat damage.
- Is there an open circuit fuse? Depending on table design, this may only affect one motor, or else a group of motors. *Always ensure power is turned off before checking or attempting a replacement.* **See module 5.0 page 65.** *If unsure, contact the service department for the fuse location, and to verify the correct rating and type.*
- The table will not tilt vertically.
    i. If the table will not tilt in either direction, there may be an open circuit fuse. **Always ensure power is turned off, before checking or attempting to replace a fuse. See module 5.0 page 65.**
    ii. The table may have reached its maximum angle in one direction, and be unable to return. (Perhaps it is in Trendelenburg position) This could be caused by operation of the anti-crash safety interlock. This may be a bar, or metal flap at the table end. This can be damaged and stick in the operated position. Some systems use a pressure mat on the floor. Look for an object trapped between the table base and the mat.

k. Table locks do not operate

- Has the compression lock been activated? Depending on the table design, this will release the serial-changer longitudinal and lateral locks.
- Is there a problem with the wiring? See 'No fluoroscopy exposure, table electrical checks' for possible broken wires and rat damage.
- The lock may have too large an air gap. With the power switched off, adjust the lock so this gap is at a minimum.
- A lock coil, or winding, may be open circuit.
    i. **An electrician, or electronics technician, should test the lock coil.**
    ii. Before testing, first ensure the lock activation switch is off. Then ensure all power is turned off.
    iii. Disconnect one of the lock coil connections.
    iv. With a multimeter set to medium ohms position, test the lock coil for continuity.
    v. Depending on design, the lock coil should measure well below 20,000 ohms. If unsure of the typical value to expect, contact the service department.

MODULE 9.1
Fluoroscopy TV

Aim

The aim is to provide information and adjustment procedures, related to problems with a TV imaging system.

The basic TV imaging system consists of the image intensifier (II), the TV camera and monitor, with possibly a videocassette recorder (VCR). Systems with greater complexity, such as DSA and electronic radiography, are not included.

Older TV cameras use camera tubes such as 'Vidicon', 'Chalnicon', or similar device. Some adjustments for these cameras are included in this module. (**Note:** Reference module page numbers refer to the title page.)

Objectives

On completion of this module, the student will be aware of common problems with the TV imaging system. This includes adjustments to the monitor, and tests to locate the cause of poor image quality. The VCR is included in these objectives.

Contents

a. General precautions
b. No image on the TV monitor
c. The image is not sharp
d. The picture has no detail in bright areas
e. Is it possible to connect a VCR
f. The VCR recording is the wrong shape on another monitor
g. Is the image intensifier faulty
h. The image rotates as the fluoroscopy table is tilted

a. General precautions

Please take the following precautions.

- **Before testing any fuses, or removing a cover, always switch the generator power off, and ensure the isolation power switch for the room is also switched off.**
- *Do not attempt any internal adjustment of a TV monitor. Ask an electronics technician to assist if internal adjustment of a TV monitor is required. Dangerous voltages can exist for some time after the monitor is switched off.*
- The TV monitor may not use a standard power voltage. Damage to the monitor will occur, if connected to the wrong voltage.
- Test procedures for fuses, switches, or wiring; are described in **module 5.0 page 65**.
- If removing a cover, or dismantling any section, place the screws in a container to avoid loss.
- Current TV cameras now use a 'charge coupled device' (CCD) instead of a camera tube. CCD cameras have very good stability and reliability. Adjustments to a CCD camera are complex, and should only be attempted by a qualified technician.
- If in doubt of any adjustment described in this module, contact the service department before proceeding.

b. No image on the TV monitor

This most common help request to the service department can have a large variety of causes. Many are due to operator error. Whenever a request is made to the service department, accurate reporting of the problem will save time.

- Has the TV been properly turned on and adjusted?
  i. Check the position of the brightness and contrast controls. Adjust the brightness control and check if the picture tube lights up.
  ii. Some monitors have two video inputs. Check that the selection switch is in the right position.

iii. Is a VCR fitted? This may not be switched on. Or, the video connecting cables have been wrongly connected. This occurs if the VCR was used at another location, and then returned.
iv. If a VCR is fitted, this may be incorrectly set up. Check the input settings. To make a positive check, disconnect the VCR, and connect the video cable from the TV camera directly into the monitor.

- Is the image intensifier receiving a correct fluoroscopic exposure? For tests of the collimator, table interlocks, foot switch and generator, please see **module 9.0 page 130**.
- Check the video cable from the TV camera to the monitor. This cable is sometimes pulled partway out of its connector, disconnecting the centre wire, or else causing the connecting pin to pull out.
  i. A quick test is to disconnect the video cable. In most cases, this will cause a change in the monitor brightness level.
  ii. The above test can also indicate if a video signal is coming from the TV camera. For example, has the camera been switched on?
- For further tests, see 'Is the image intensifier faulty?'

c. The image is not sharp

Besides the possibility of poor 'system focus', this can also be caused by problems with the video cable, or a faulty picture tube.

System focus includes electronic focus of the image intensifier and TV camera, and optical focus of the TV camera. (A CCD camera does not have an electronic focus adjustment.)

> Focus adjustments are sometimes attempted without checking for other reasons first. They are also attempted without a focus test tool, which makes it difficult to find the optimum position. Unfortunately, the focus adjustment is often the first adjustment that is 'fiddled' with.

Many image intensifiers have multiple adjustments; these must be carried out in the right sequence. For these reasons, always consult the service department first, before attempting any focus adjustments.

- Is the picture tube, or monitor, faulty?
  i. The simplest test, if available, is to try another monitor in the same position.
  ii. Adjust the brightness control for a medium setting. Examine the picture tube closely. The scanning lines, or raster, should be clearly visible. In some monitor designs, a focus control may be available either from the front panel, or the rear of the monitor.
  iii. Does the monitor focus become blurred at medium brightness levels, but appears ok at a minimal brightness setting? This indicates a worn picture tube. Replacement is required.
  iv. Does the monitor take a long time to 'warm up'? For example, at first the available brightness is low, and brighter areas of an image merge together. This is an indication of low electron emission, from the picture tube cathode. Picture tube replacement is required.
- Is the 75 ohm video-cable termination switch set correctly?
  i. This is a common error. It is often found that if this switch is turned 'off', or unterminated; the picture appears brighter and has more contrast. However, in many cases this will cause a loss of fine detail. The correct position is 'on', or terminated; except when there are two or more monitors. In this case, monitors in the middle should have the termination switch turned off, while the last, or end, monitor has the termination switch turned on. (The final monitor will have only one video cable connection.)
  ii. Has a 'T' connector been used to connect another monitor or VCR? This is incorrect, as proper termination of the video cable cannot be obtained, together with possible loss of fine detail.
- The video, or coaxial, cable has a woven metal shield under the first layer of insulation. This is connected to ground by the video connector. Is the shield pulled out from the connector? This can also give rise to interference patterns on the monitor, as well as a loss of picture sharpness.
- Is the TV camera electronic focus correct?
  i. This does not apply to CCD cameras, and only to some cameras that have an accessible focus control. This adjustment is normally very stable.
  ii. Contact the service department to locate the position of the focus adjustment.
  iii. Tape a line-pair gauge directly under the serial changer, as close as possible to the image intensifier. If the gauge is not available, then use the 'focus aid' described in appendix B page 169.
  iv. Set a minimum fluoroscopic kV and mA level, just sufficient to obtain a good image.
  v. With fluoroscopy 'on', adjust the focus control for best results. This should be better than 12

LP/mm for a 9" image intensifier. (A typical result might be 14 LP/mm with a CCD camera, and 16 LP/mm with a vidicon camera.)

- Is the TV camera optical focus correct?
  i. Optical focus is normally very stable, however sometimes the image intensifier moves a slight amount in the housing, or the camera tube and deflection assembly moves a small amount in the TV camera head.
  ii. Older cameras may have a screwdriver operated focus control at the rear of the camera head. (This is sometimes 'fiddled' with.) In other cases it is necessary to remove a cover plate to obtain access to the lens.
  iii. If directly adjusting the lens, first make a mark on the adjustment ring so the lens can be returned, if needed, to its previous position. In most cases, it is necessary to undo a 'locking' screw before an adjustment is possible. Some lenses also have an adjustable 'iris'. Take care not to accidentally adjust this instead of the focus.
  iv. Tape a line-pair gauge directly under the serial changer, as close as possible to the image intensifier. If the gauge is not available, then use the focus aid described in appendix B, page 169.
  v. Set a minimum fluoroscopic kV and mA level, just sufficient to obtain a good image.
  vi. With fluoroscopy 'on', adjust the lens focus for best results. Or adjust the focus control at the rear of the camera head with a screwdriver.
  vii. This should be better than 12 LP/mm for a 9" image intensifier. (A typical result might be 14 LP/mm with a CCD camera, and 16 LP/mm with a vidicon camera.)
- Is the image intensifier focus correct?
  i. Current image intensifiers have very stable focus adjustments. Older systems may occasionally require adjustment.
  ii. If you have a multi-field, or dual-field image intensifier, check the resolution first on all fields. If all fields indicate poor resolution, this will indicate the problem is elsewhere, or else a component failure in the intensifier power supply.
  iii. Older designs have a single external focus adjustment. This is a screwdriver adjustment, positioned towards the top of the image intensifier. *Do not attempt adjustment of the image intensifier focus, if there is more than one adjustment. In this case, contact the service department for advice.*
  iv. Tape a line-pair gauge directly under the serial changer, as close as possible to the image intensifier. If the gauge is not available, then use the focus aid described in appendix B page 169.
  v. Set a minimum fluoroscopic kV and mA level, just sufficient to obtain a good image.
  vi. With fluoroscopy 'on', adjust the image intensifier for best focus.
  vii. The combined TV and II focus should be better than 12 LP/mm for a 9" image intensifier. (A typical result might be 14 LP/mm with a CCD camera, and 16 LP/mm with a 'vidicon' camera.)

d. The picture has no detail in bright areas

- Is the automatic fluoroscopy system set at too high a level?
  i. Change over to manual operation, select a lower kV or mA and observe if this corrects the problem.
  ii. If there is now a good image, remain on manual operation, and contact the service department to have the automatic fluoroscopy system adjusted.
- Is the TV monitor correctly adjusted? If the monitor contrast is set too high, in some monitors this will cause bright parts of the image to merge together, or to appear 'flat'. In other cases, an overbright image will appear smeared, or wiped, horizontally across the picture tube.
- Does the monitor take a long time to 'warm up'? For example, at first the available brightness is low, and brighter areas of an image merge together. This is an indication of low electron emission. Picture tube replacement is required.
- TV cameras that use a camera tube instead of a CCD have a 'beam current adjustment'.
  i. If beam current is low, bright areas of the image lack contrast and merge together. If beam current is too low, the image will just appear white, with no detail. In this case, the image may become clear for a very short time, immediately after fluoroscopy is switched off.
  ii. If beam current is too high, this will result in poor image quality and reduced focus.
  iii. Beam current is a common adjustment for older TV cameras. Contact the service department to locate the position of this adjustment, and any required precautions before adjusting. Ask an electronics technician to make this adjustment.

e. Is it possible to connect a VCR?

- The videocassette recorder (VCR) must operate to the same scanning format as your imaging system.

In most cases, the X-ray TV system will use the same standard as domestic TV, allowing a domestic VCR to be used.
  i. A 525 line, 60 hz system requires a NTSC compatible VCR
  ii. A 625 line, 50 hz system requires a PAL compatible VCR.
- In case your imaging system operates at a higher line rate, such as 1049 or 1249 lines, then connecting a VCR is not possible, unless there is an alternate output from the TV camera, at the standard line rate.
- Playback of a recording will be the same aspect ratio used by the X-ray TV camera and monitor. The standard aspect ratio is 4:3, however, some systems, including CCD cameras, use a 1:1 aspect ratio. If a recording is played back on a monitor adjusted for a 4:3 aspect ratio, the image will be stretched horizontally, and appear 'egg shaped'
- The VCR must have direct video input and output connections. A simple installation connects the video lead from the TV camera to the video input of the VCR, and the VCR video output to the monitor. In most cases, this will require the VCR to be always switched on during fluoroscopy.

f. The VCR recording is the wrong shape on another monitor

This is a common complaint when the recording is made from an X-ray TV using a 1:1 aspect ratio, instead of the domestic 4:3 aspect ratio.

- To see if the system is adjusted for a 1:1 aspect ratio, adjust the monitor brightness so the scanning lines, or raster, is clearly visible. A 4:3 aspect ratio will show scanning lines extended fully across the screen, while a 1:1 aspect ratio will show a small blank area at both sides of the picture tube.
- Apart from systems deliberately set to a 1:1 aspect ratio for a CCD TV camera, it is possible the TV camera and monitor is incorrectly adjusted.
  i. For example, if the original size and shape of the image from the camera was incorrectly adjusted, and then the monitor was adjusted to give the required size and shape.
  ii. This is an incorrect set-up, as the monitor should first be adjusted to the required aspect ratio and size, then finally the TV camera adjusted to suit the monitor.
  iii. Fortunately, in most cases the TV camera and monitor can be realigned for use with other monitors. This possibility should be discussed with your service department.
  iv. **Note.** This adjustment is not available if a CCD camera is used.

g. Is the image intensifier faulty?

There is no X-ray image on the monitor. The monitor adjustments appear correct, and the X-ray control indicates a normal fluoroscopic exposure.

- Is radiation entering the image intensifier?
  i. Is the collimator closed? The collimator may have a fault.
  ii. Place a cassette on the table, underneath the image intensifier.
  iii. Make a 2~3 second fluoroscopy exposure, and develop the film.
  iv. The film should be very dark.
  v. If the film is unexposed, the collimator could have a fault. Contact the service department for advice.
- Is the TV camera faulty?
  i. Check the camera is switched on, and the video cable is not disconnected or damaged.
  ii. **The following test should only be performed on advice from the service department.**
  iii. Remove the TV camera from the image intensifier. Switch on the camera power. Point the camera around the room, but do **not** aim the camera at any bright light. An 'off focus' image should be obtained of objects in the room.
  iv. Some systems, especially those with 'last image hold', will require fluoroscopy to be 'on' during the above test. First ensure the collimator is closed, and the fluoroscopy kV is adjusted to its lowest setting.
- There is a flickering background illumination, even without fluoroscopy.
  i. This is a possible electrical discharge or instability in the image intensifier.
  ii. With a dual or multi-field intensifier, selecting another field size can alter the appearance of this illumination. This indicates a fault in the image intensifier, or image intensifier power supply. If the image intensifier loses focus on selecting a different field size, the power supply is faulty.
  iii. **The following test should only be performed on advice, and instructions, from the service department.**
  iv. Remove the TV camera. Cover the camera lens.
  v. Switch the power back on.
  vi. Close the fluoroscopy collimator. **Do not** make a fluoroscopy exposure.

vii. Turn off the room lights. Now look directly into the image intensifier lens.
viii. If any 'glow' pattern is observed, the image intensifier is faulty.
ix. Repeat this test for each image intensifier field size.

- The image has a bright central area during fluoroscopy.
  i. This is an indication of 'gas' inside the image intensifier. This may occur if the image intensifier has not been used for some time. In this case, leave the system switched on overnight, and test again next day.
  ii. To test, place a piece of lead, of about one third of the image intensifier diameter, up against the serial changer. Position the lead test piece in the middle of the image. This prevents radiation entering the centre of the image intensifier.
  iii. With fluoroscopy 'on', the area covered by the lead should show very little illumination. If there is gas, this will be seen as a bright area in the centre, where radiation is blocked by the lead test piece. The Image intensifier will need replacement.

- Is the image intensifier 'worn out'? This is a common question with older systems, especially if there is poor penetration with a 'noisy' or 'snowy' image.
  i. The conversion gain of an image intensifier drops with age as well as use. However in most cases, adjusting the lens aperture on the TV camera can compensate for reduced brightness.
  ii. A measurement is made of the radiation value, required to produce a standard video level, from the TV camera.
  iii. The radiation is set to the required value, and the lens iris is adjusted to obtain the required video level. If the required video level is not obtained, then the image intensifier may need replacement.
  iv. Before replacing the image intensifier, the TV camera should also be checked.
  v. This applies if the TV camera uses a 'Vidicon' camera tube. The 'target' voltage may need adjustment.
  vi. The above test and adjustment should be requested from your service department before considering a replacement system.

h. The image rotates as the fluoroscopy table is tilted

This affect is caused by an interaction with the earth's magnetic field. It depends not only on the local field strength inside the hospital, but also the orientation as the table is tilted.

Standard image intensifiers have a magnetic shield in the housing to reduce this effect. However they are not shielded at the entrance plane. (This is where the X-ray radiation enters the II.)

Image intensifiers intended for high performance digital image systems might be fitted with a 'Mu-Metal' magnetic shield, to cover the entrance plane into the II. While this virtually eliminates the rotation problem, the added filtration reduces the conversion efficiency of the image intensifier.

MODULE 10.0
Automatic exposure control (AEC)

Aim

The aim is to provide a series of tests for an automatic exposure control (AEC). These tests are aimed at determining whether there is a problem due to the AEC, or caused by other reasons.

Objectives

A correctly installed and adjusted AEC can produce excellent results. However, correct use of the system is still required to obtain the desired performance. On completion of this module, the student will be aware of AEC operation requirements, and will be able to perform basic tests for the AEC performance. These tests will provide important information when requesting advice, or else attendance by the service department.

Contents

a. AEC operation precautions
b. AEC film density test
c. AEC incorrect operation
d. When a request is made for service

a. AEC operation precautions

- Incorrect selection of a technique not covered by AEC operation.
  For example:
  The wall Bucky may be fitted with a selection of three separate measuring fields or chambers, while the table Bucky has a central field only. Good design of an AEC should prevent selection of an incorrect field. Unfortunately, some AEC systems, in particular those fitted outside the X-ray control as an accessory, may allow selection of an invalid field position.
- Some AEC systems have a 'film sensitivity' control in addition to the film density control. While this gives greater flexibility, the film sensitivity control is often never used. Instead it is left at a standard setting.
  i. The use of the film sensitivity selection control can be forgotten. This includes the original setting. If the setting position is accidentally changed, then a sudden large change in film density occurs. This results in an unnecessary service call.
  ii. In some cases, due to bad calibration, the film sensitivity control may need to be reset when changing from the table Bucky to the vertical Bucky. This can easily result in operator error, and should be corrected wherever possible.
- X-ray output for the exposure is too low. For example, too low a kV, or insufficient mAs. In this situation, the generator timer terminates the exposure, and not the AEC. A light film results.
  i. The AEC system, depending on design, may prevent further exposures until a 'reset' button is pressed. There may be only a small indicator lamp to indicate this condition, which is sometimes overlooked.
  ii. Other AEC systems may provide a short audible signal. In some cases this is for only a few seconds, and is easily ignored. As a result another radiograph is made without adjusting

the exposure setting. A complaint is made about 'occasional light films'
- X-ray output is too high. As a result the AEC cannot terminate the exposure quickly enough.
  i. The 'minimum switch-off time' is more of a problem on older single or three-phase generators; especially those fitted with mechanical exposure contactors. This can cause unreliable results if AEC exposure times below 0.02~0.03 seconds are attempted.
  ii. Modern high-frequency systems are able to respond quickly to the AEC exposure-stop signal, so this is not a problem. However it is still good practise to keep exposure times above 0.005 seconds. This is due to the energy stored in the HT cable, which will slightly extend the actual exposure time.
  iii. If AEC exposure times are close to the minimum switch-off time, reduce kV or mA for the next exposure.
- Selection of a less optimum chamber. For example, use of a chamber centred behind the spine, instead of the left or right chambers for a chest or lung exposure.
- Combining two or more chambers. This depends on the method of AEC operation. While the systems appear similar, they can deliver different results.
  i. In one system, the chambers are combined together, and the **average** output of the chambers controls the exposure.
  ii. In the other system, each chamber has separate control of the exposure. When chambers are combined, only the chamber receiving the **higher level of radiation** controls the exposure. (This is called the 'OR' technique by one manufacturer)
- Incorrect collimation. If collimating to a smaller area, part of the measuring chamber is also coned off. A dark exposure results.
- Bad patient positioning. In this case, radiation passes through a relatively thin portion of the anatomy, compared to the main item of interest. In some cases, the measuring chamber may receive a portion of direct radiation. In either case a light exposure will result.
- The AEC chambers are sensitive to soft, or scattered, radiation.
  Although the grid removes most of this radiation, the remainder still has an effect on the exposure. In addition, with the chamber in front of the cassette, this soft radiation is filtered from entering the cassette.
  i. The AEC is calibrated for a patient to be positioned against the Bucky. If there is an air gap, this reduces the amount of soft radiation entering the chamber, and the film becomes darker.
  ii. The above problem is much greater if there is no grid in front of the AEC chamber.
  iii. An extreme example is found on older fluoroscopic tables that have the chamber mounted in front, instead of behind the grid. In this case, an air gap of only about 6 cm may double the exposure time. The doctor should keep the spot filmer close to the patient at all times, while using the AEC.
- In case the cassettes, film, or intensifying screens are changed, the AEC will need recalibration.

b. AEC film density test

The AEC calibration may require calibration, or there may be a fault.

This test allows the performance to be verified, and indicate if an individual chamber has a problem.

- A test phantom is required. This can be a plastic bucket with water, or else a large flat-sided plastic bottle. An empty plastic container used for bulk detergent is ideal.
- Place a plastic bucket filled with water to a height of 18 cm on the tabletop, positioned over the middle of the Bucky.
- Place a 24/30 cm cassette in the Bucky. Set X-ray tube height to 100 cm.
- Select exposure factors of 90 kV, 100~250 mA, and backup time of 0.5 sec. (Some X-ray controls allow kV adjustment only, mA and backup time is automatically set by the AEC)
- Select the AEC central chamber only. Set the density control to the middle, or '0' position.
- Make a radiographic exposure. Note the actual exposure time, or alternately the mAs value. This depends on the meter indications provided.
- In case a warning signal indicates an incorrect exposure, reduce the level of water to about 10 cm. If the next exposure still indicates a problem, stop testing that chamber. Contact the service department for advice.
- Process the film. Film density should be in the region of 1.4 to 1.6. If a densitometer is not available, compare the film density to a previously exposed reference film. (The actual value of film density often depends on individual doctor preferences.)

- For under-table Bucky's fitted with three fields, repeat the above test for each chamber. Adjust the phantom so it is positioned above the selected chamber.
- A similar exposure time and film density should be obtained with each chamber. In case there is a difference, expose another film, after adjusting the density control by one step. If more than one step is required to obtain a similar density, then calibration of individual chambers is out of tolerance.
- If any chambers fail this test, do not use that chamber until the problem is corrected. Contact the service department; advise them of the problem, and the tests carried out

A similar test can be applied to the vertical Bucky. A flat-sided plastic container with water is required. If an empty bulk detergent container is not available, then two empty fixer or processor bottles can be used, placing them side by side. Rinse the bottles before filling with water.

c. AEC incorrect operation

These tests are only provided as a general guide. This is due to the great variety of AEC systems in use, most of which require specialized instructions and test equipment for calibration or service.

- The AEC exposures produce light films, and are not consistent.
  i. The AEC can be affected by humidity, or by light leaking into a photomultiplier system. This is a test for stability during a long exposure time.
  ii. Set exposure factors for minimum mA and kV. Set the backup exposure time to 5.0 seconds. For those systems that provide adjustment of kV only, set kV to minimum, and select fine focus.
  iii. Set the density control to minimum density. Select the centre chamber.
  iv. Close the collimator, and aim the X-ray tube away from the Bucky under test.
  v. Leave all room lights fully ON.
  vi. Perform a radiographic exposure. The AEC should indicate the maximum exposure time was reached, and the exposure was not terminated by the AEC.
  vii. Repeat this test for each chamber in the Bucky or spot filmer under test. Caution; as these are large test exposures, allow cooling time between exposures.
  viii. Does a chamber fail this test?
  ix. Test again, this time with the room lights turned off. If the AEC now gives a longer test exposure, this is a system using photomultipliers, and external light is leaking into the AEC chamber or photomultiplier assembly.
  x. Examine the front edge of the chamber assembly carefully. Cassettes may have hit it, when they were placed in the Bucky. This can damage the chamber, allowing external light to enter. Cover the damaged area with metal foil, and test the chamber again.
  xi. Many AEC systems use ionization chambers to measure radiation. Older versions were sensitive to humidity. Design changes overcame this problem. Contact the manufacturers service department, in case there is a design modification to upgrade your unit.
  xii. If any chambers fail this test, do not use that chamber until the problem is corrected. Contact the service department; advise them of the problem, and the tests carried out.
- One or more chambers have stopped working.
  i. Check the condition of the connecting cable. This especially applies in fixed installations where the cable passes through ducts etc. Some varieties of rats appear to like chewing on small wires.
  ii. On a mobile system, check inside the connecting plug for possible broken connections.
- The film density has changed. It is necessary to adjust the density control several steps to compensate.
  i. Is a similar change of density setting required for both the wall Bucky and the table Bucky? This may be due to the processor instead of the AEC. Check the processor for possible problems with chemicals or temperature.
  ii. Has the problem occurred after a new batch of film? The new film may have a different sensitivity. The AEC will require recalibration.
  iii. Were the intensifying screens changed? The AEC will require recalibration.
- For further assistance, contact the service department. If an electronics technician is available, the technician can carry out further tests after obtaining advice from the service department. This would be specific to the make and model of the AEC requiring attention.

d. When a request is made for service

- After carrying out the maintenance tests for the AEC, problem areas may be located.
  i. Retain all test films, and document the conditions of test. This includes mA station, kV, FFD,

PART III. FAULT DIAGNOSIS AND REPAIR MODULES

depth of water phantom, and indicated exposure time.

ii. Test films made with anatomical test objects may be very interesting. However, testing with a water phantom produces the most consistent results, and allows direct comparison with previous tests.

- When an AEC system has been in use for a while, it may be found that the density control has to be adjusted for different examinations. While it is a simple matter to have service recalibrate the system to "0" density setting, the following information can indicate if attention is also required to kV tracking, or short time compensation.

    i. The type of patient examination.

    ii. The chamber in use, and the density setting in use for that chamber.

    iii. Does density setting have to be changed on chamber selection?

    iv. Does density have to be changed depending on kV used?

    v. The kV and mA values used. Or kV and focal spot if mA is not selected manually.

    vi. The indicated exposure time obtained after an exposure.

- **Be aware that some apparent AEC problems are due instead to film processor drift. Before requesting service for the AEC, ensure the processor performance has been checked.**

PART IV
Automatic film processor

MODULE 11.0
Automatic film processor

Aim

The aim is to provide routine maintenance procedures for the automatic film processor. This module presents a series of regular maintenance schedules. When used with sensitometry techniques, this module can be used to implement a quality control programme. Repair procedures for the processor are provided in module 11.1 page 199.
(**Note:** Reference module page numbers refer to the title page.)

Objectives

A processor maintenance schedule should be followed regularly. Performing this maintenance will ensure optimum quality of processed films, and allow detection of problems before they become serious. On completion of this module, the student will be familiar with maintenance procedures for the automatic film processor. These procedures should be used together with the maintenance instructions in the operators' manual. A routine maintenance check-sheet is provided in appendix 'D' page 186.

Note: This module is based on the procedures previously presented in the 'Quality assurance workbook'. As reference is made to sensitometry techniques, this section from the WHO Quality assurance workbook is included in appendix 'A' page 163.

Contents

a. General precautions
b. Preparation for maintenance
c. Daily maintenance
d. Weekly maintenance
e. Monthly maintenance
f. Quarterly maintenance
g. Annual inspection and service
h. Replacement parts schedule

Equipment required

- An accurate thermometer (alcohol or electronic).
- Hydrometer.
- Sensitometer. (Or pre-exposed sensitometry film). *
- Densitometer. (Or processed reference film for comparison purposes). **
- Chemical stirring rods. These may be stainless steel or PVC. Important; the rods should be labelled 'developer' and 'fixer' to prevent cross contamination of chemicals.
- Measuring cylinder. Graduated 100 ml glass or plastic container.
- Sodium hypochlorite bleach. (For monthly maintenance).
- Tank cleaning brushes, one each for developer and fixer tanks.
- Scouring pads. (Plastic or nylon type).
- Clean disposable cloths.
- Clean hand towels.
- Plastic bucket.
- Mop.

* A packet of pre-exposed sensitometry films may be obtained from the film supplier.
** A previous processed sensitometry film may be used as a reference.

a. General precautions

- **Before removing any panels, ensure the processor power is switched off. The processor power isolation switch should also be turned off.**
- All adjustable settings of the processor should be recorded. This especially applies to microprocessor-controlled systems. These have a large number of settings, or options, and may develop an error. For example, after a power failure, or due to incorrect adjustment.
- The following items should be available for personal protection.
  i. Plastic apron.
  ii. Coat to protect clothing from chemical splashes.
  iii. Rubber gloves.
  iv. Protective glasses and mask, to protect the face from chemical splashes.
  v. An emergency eye kit should be available in the darkroom.
- Do not wear long loose clothing; this may become caught in the rollers.
- Ensure that the darkroom is adequately ventilated.
- Clean up any spills or splashes.

b. Daily maintenance

- **Before start up.**
  This assumes shutdown procedure was not performed, or the processor has been idle for some time.
  i. Remove processor lid.
  ii. Remove crossovers, and wash in warm water, with a sponge or plastic cleaning pad. **(Always do developer first, then fixer, to avoid contamination of developer.)**
  iii. Wash tank covers and splash guards.
  iv. Wipe over all rack rollers that are above solution levels.
  v. Clean interior exposed surfaces.
  vi. Check replenishment tanks/bottles levels. Check for unusual colour or smell.
  vii. Check replenishment hoses for possible leaks or kinks.
  viii. Replace the water drain standpipe, if appropriate.
  ix. Ensure the wash water drain valve is closed. (Some processors may be fitted with an automatic drain valve)
  x. Turn on water, and check that wash tank is filling. Time water flow if necessary.
  xi. **Note.** Depending on make and model, water flow will not commence until the unit is powered up.
  xii. Replace crossovers, and tank lids.

- **On start up.**
  i. With the top cover removed, switch on the processor.
  ii. **Note.** Some processors have sensors, or microswitches, to ensure the cover is correctly fitted. With the lid off, you will need to activate these switches manually.
  iii. Listen for any unusual noise or vibration.
  iv. Check film transport system. Ensure all rollers are operating normally.
  v. If not previously filled, check that wash water is now filling correctly.
  vi. Check replenishment system is working.
  vii. Replace processor lid.
  viii. Feed in one unprocessed 35 × 43 cm film as a clean-up film.
  Note. Do not use processed film, as these are harder, and contain fixer.
  ix. Inspect processed 'clean up film'. Feed in a second film if necessary.
  x. Clean exterior surfaces, including feed tray and receiving bin. **Pay extra attention to the feed tray.**
  xi. Wipe over all darkroom surfaces.
  xii. When the processor has reached normal operating conditions, a routine sensitometry test may be carried out. See **appendix 'A', page 163**.

- **Normal working.**
  i. **Follow manufacturers operating instructions. (Read the manual.)**
  ii. Be aware of any changes in operation, noises, leaks, or deterioration of processed films.
  iii. Do not pull processed films out till they are clear of the rollers.
  iv. Always wait for the 'ready' signal or light before feeding the next film.
  v. When feeding films, insert the wide side as the leading edge. The film should be lined up against one side of the tray, not in the centre.
  vi. Do not allow anyone to stand next to, or lean on, the processor.
  vii. Ensure the darkroom ventilation is correct, and there is no build up of humidity or fumes. This especially applies where a bench top processor is used.

- **On shut down.**
  i. Remove processor lid.
  ii. **Note.** Some processors have sensors, or microswitches, to ensure the cover is correctly fitted. With the lid off, you will need to activate these switches manually. Some processors have several lid safety switches. Please refer to the operating or service manual.

iii. Observe transport system.
iv. Listen for any abnormal noise or vibration.
v. Observe level of solutions and wash water.
vi. Switch off.
vii. Look for any leaks.
vii. Remove and wash all crossovers, splashguards and tank lids.
ix. Wipe over all rack rollers above solution level. (Always do developer first, then fixer, to avoid possible contamination of developer.)
x. Inspect and wash roller drive-cogs and drive mechanism where appropriate.
xi. Replace tank lids. (Do not install crossovers.)
xii. Turn off wash water, if appropriate.
xiii. Remove water drain standpipe, if appropriate.
xiv. Wash off all chemical splashes on interior exposed surfaces.
xv. Wipe any splashes from exterior surfaces.
xvi. Replace processor lid. Leave it slightly raised at one end, to avoid build up of fumes and condensation.
xvii. The darkroom door should be left open, with the ventilation fan operating. (Depending on power constraints.)
xviii. Place crossovers on top of processor, with drain standpipe if appropriate, and cover with a cloth; or store in a cupboard set aside for that purpose.
xix. Observe levels of replenishment tanks. If required prepare a fresh solution.
xx. Observe stocks of films, chemicals, or other depleted supplies. Restock or order as required.
xxi. Record all restocking.
xxii. Report any problem or fault areas. See **module 11.1 page 151**.
xxiii. Update the logbook

c. Weekly maintenance

- **Follow manufacturers' recommendations.**
- Perform a sensitometry test. (For best control, this should be performed as a daily routine.)
- Check solution temperatures, in particular developer temperatures. This is usually around 34~36 degrees Celsius
  i. **Note.** Allow time for the temperature to fully stabilize first.
  ii. Compare with any readout on the processor panel, and manufacturers' recommendations.
  iii. If outside the specified temperature limits, compare to results previously recorded in the logbook. Adjust if necessary.
  iv. In case a drift of temperature is observed, investigate further. Use the manufacturers maintenance manual as a guide. See **module 11.1 page 151**.
- Check replenishment rates.
  i. Remove processor lid.
  ii. Locate lid safety switches, if fitted. Place a small weight, or else a small packing piece held with tape, to keep these switches operated.
  iii. Switch processor on.
  iv. Divert the developer inlet to the tank, into a 100 ml measuring-cylinder.
  v. On some bench top processors, this may not be possible. In which case divert the flow of used developer from the tank, which would otherwise go to the waste tank or the drain. However, pass a least one film in first, to ensure excess developer has commenced to flow. Discard this initial measurement.
  vi. Pass five 35×43 cm fresh films through the processor. Do not use previously processed films, as these are harder, and contain fixer.
  vii. Divide the measuring cylinder contents by five, to find the replenishment rate.
  viii. **Note.** The above procedure is required for some processors, which may not add individual replenishment for each film inserted. This especially applies for microprocessor-controlled units, which calculate several other factors besides film size.
  ix. Repeat the above for the fixer tank.
  x. Record the results.
  xi. Check with previous results for any significant variation, or drift.
  xii. Adjust if necessary.
- Remove and wash all deep rack rollers in warm water
  i. Particularly for the developer section, the rollers may develop a layer of chemical crystals. A nylon or plastic cleaning pad will assist in the removal of these crystals, or 'encrustation'.
  ii. Inspect for correct function, wear or damage.
  iii. Rinse and install.
- Check main drive shaft and chains or drive belt.
- Carry out any other maintenance recommended by the manufacturer.
- Report any problem or fault areas. See **module 11.1 page 151**.
- Update the logbook

d. Monthly maintenance

- **Follow manufacturers' recommendations.**
- Perform weekly maintenance.
- Inspect all racks and component parts during cleaning.

- Clean filters.
- Drain all and clean all tanks.
  i. Pay special attention to the wash water tank.
  ii. The tanks may be filled with a dilute concentration of 0.5% hypochlorite solution. An alternative is a system-cleaner chemical kit. This is a two-part mix, plus a neutralizer, which is added to the water when flushing out the cleaner.
  iii. Let the solution sit in the system for no longer than 30 minutes.
  iv. Rinse the solution from the system, and dislodge 'bio-growth' or algae. Use a clean stiff brush or other recommended tools to clean the surface.
  v. Rinse the system thoroughly.
  vi. Caution. Do not allow concentrated sodium hypochlorite to come in contact with fixer or developer. Dangerous fumes can result.
- Do not forget to add starter to the developer tank.
- Manufacturers recommend replacement of all chemicals on a monthly basis. This will especially apply to developer, where oxidation continues even when not in use. If recharging is not economic, then inspect the condition of the solutions in the replenishment tanks, and change as felt necessary.
- Carry out any other maintenance recommended by the manufacturer, or felt necessary.
- Report any problem or fault areas. See **module 11.1 page 151**.
- Update the logbook.

e. Quarterly maintenance

- **Follow manufacturers' recommendations.**
- Perform weekly and monthly maintenance.
- Discard remaining chemicals in replenishment tanks.
- **Note.** This especially applies to developer, which may be oxidised. The developer will start to turn brown.
- Dispose of chemicals as required by local regulations. *Do not flush down the drain. Especially do not discard so that seepage may end in a well, or in the irrigation water.*
- Wash out replenishment tanks, and flush hoses.
- Mix a fresh solution of developer and fixer.
- **Note.** Only mix sufficient developer to suit short to medium term requirements. This will reduce deterioration due to oxidation.
- Do not forget to add starter to the developer tank.
- Remove processor panels. Inspect carefully for any leaks around pumps.
- Check overall condition of processor.
- Report any problem or fault areas. See **module 11.1 page 151**.
- Update the logbook.

f. Annual inspection and service

Even if you do not have a maintenance contract with a service company, it is advisable to have the unit fully inspected and serviced at least once each year. This service should ensure that the processor is performing to full specification. In addition, wear items such as pump valves, especially for the fixer pump, can be replaced. This service will also enable a new set of reference sensitometry films to be obtained. Prior to such service, it is recommended to have a fresh supply of chemicals available. If film has been stored in suspect conditions, a fresh pack of film should also be on hand, to allow accurate calibration.

g. Replacement parts schedule

Manufacturers usually recommend replacement of items subject to wear or deterioration. This should be carried out at regular intervals. A typical example of these items is provided in table 11–a.

Table 11–a. A typical replacement parts schedule

Replace	Each month	Three months	Six months	Each year
Chemicals, developer and fixer.	●			
Replace fixer rack roller-springs.			●	
Replace entry, developer, wash and drying rack roller springs.				●
Replace developer rollers.				●
Replace 'poppet' valves in replenishment pumps.				●
Developer filters.				●
'E' rings.				●

MODULE 11.1

Automatic film processor

Aim

Many problems with the automatic film processor are avoided by routine maintenance. Unfortunately, routine maintenance may not be performed properly, or not at all. This module provides a list of common problems that may occur, and their solutions. Routine maintenance procedures for the processor are provided in module 11.0 page 147.
(**Note:** Reference module page numbers refer to the title page.)

Objectives

On completion of this module, the student will be familiar with film processor problems, and what to look for when correcting these problems. In the event of a problem, look also in the operation manual, for the manufacturers advice.

(Task 16.'Films appear too dark', and task 17 'Films exhibit symptoms of low fixer', should be attempted on completion of this module).

Note: Detailed instructions for sensitometry are provided in the 'WHO Quality assurance workbook'. If sensitometry is regularly performed during maintenance, then deviations from the recorded characteristic curve will aid diagnosis of film problems. The sensitometry section from the WHO Quality assurance workbook is included in appendix 'A' page 210.

Contents

a. Electrical precautions
b. Plumbing precautions
c. Suggestions for processor service or repair
d. The processed film appears dirty
e. Pressure marks on the film
f. Film is scratched or jammed
g. Film appears under developed
h. Uneven developing across the film
i. Film has high base fog and excessive contrast
j. Films appear poorly fixed
k. Films are discoloured. May appear 'sticky'
l. Insufficient or uneven drying
m. Bands across the film, perpendicular to the film transport direction
n. Film 'fogging'
o. Static electricity marks

a. Electrical precautions

- **Before removing any panels, or performing any internal repair, ensure the processor power is switched off. The processor power isolation switch should also be turned off.**
- **An electrician or electronics technician should perform any electrical tests or adjustments.**
- If testing or replacing a fuse, see **module 5.0 page 65**.
- To make adjustments, it may be is necessary to remove a module or printed circuit board (PCB), and reconnect it with an extension board. This should only be attempted on advice from the service department.
    i. **Take care that power is switched off, before proceeding.**
    ii. Before removing a module or PCB, touch the processor frame. This is to discharge any static electricity.
    iii. Take note of plugs or sockets that may need to be removed or reconnected. Do not rely on memory. Make a diagram of the connections. If connections or wires are not marked, attach a temporary label.
    iv. When a PCB is fitted to an extender card, take care not to bump or dislodge it once power is restored. Damage can result.

b. Plumbing precautions

Many plumbing problems in a processor may be attended to, providing due care is taken. This can include:

- Attention to plumbing or piping leaks.
- Replacement of replenishment-pump valves.
- Replacement of replenishment or recirculation pumps.

Before attempting any repairs where the internal piping or plumbing may be disconnected, take the following precautions.

- Ensure the relevant processor tank has been drained of any solution.
- Flush the system to remove any residual solution.
- **Ensure the power is turned off,** *also at the power isolation switch.*
- Turn off the water supply to the processor.
- Make a diagram of piping connections before removing. Attach labels for identification.
- Take care when disconnecting piping, not to lose small 'O' rings. These can be hidden inside the connection, and fall out later.
- When piping is disconnected, residual flushing water will drain out. Be prepared, and place cloth, or a towel, under the pipe before disconnecting.
- Have a bucket, or container, available for any unexpected problem.
- Wear suitable protection clothing and gloves. See **module 11.0 page 147**.

c. Suggestions for processor service or repair

- When diagnosing a problem, refer also to the operators or service manual for the processor. If in doubt of the cause of a problem, request advice from the manufacturers service division.
- **Hint.** When trying to locate a part in the processor, refer to the diagrams in the parts manual.
- When replacement of a part is required, include any auxiliary components that may be required. For example, if replacing a faulty recirculation pump, include replacement 'O' rings for the piping connections.
- Place any small screws or parts in a container, to avoid loss.
- All adjustable settings of the processor should be recorded. This especially applies to microprocessor-controlled systems. These have a large number of settings, or options, and may develop an error. For example, after a power failure, or due to incorrect adjustment.

d. The processed film appears dirty

- The processing tank rollers are dirty.
    i. Carry out the recommended weekly maintenance.
- Dirt or algae contamination of the wash water.
    i. Replace wash water.
    ii. Ensure wash water trough is clean.
    iii. Examine the water supply filter, and either 'back flush', or exchange the filter element.
    iv. Check the water flow rate.
    v. Check operation of the automatic drain valve. (This is not fitted to all processors)
- Dirt or contamination of the processor solutions.
    i. Carry out a complete cleaning procedure. See **module 11.0 page 147**.
    ii. Replace processor solutions. Make up a complete fresh batch. Ensure filtered water is used. Do not forget to add starter.
    iii. Developer and fixer recirculation filters may be fitted on some processors. These should be cleaned weekly as part of routine maintenance.
    iv. Some processors have a separate developer filter, not installed in the tank. Depending on processor make or model, this filter should be changed each year. If suspect, change immediately.
- A cleaning-film procedure has not been carried out.
    i. This should be carried out each morning.
    ii. Use a full size unprocessed film.
- The feed tray is dirty.

e. Pressure marks on the film

- Clean the film rollers.
- Pay special attention to developer rollers.
- Replace any rollers that do not have a smooth surface, after cleaning.
- A pair of developer or fixer rollers may have developed a flat, or uneven, area.
    i. Test by slowly rolling along a flat surface. Feel for any 'bumps' as the roller is rotated.
    ii. Place a light behind the roller. Move the roller along a flat surface. Look for any gaps as the roller is rotated.

f. Film scratched or jammed

- With the top cover removed, feed a test film through the processor.
    i. Listen carefully for any unusual noise.
    ii. Does the film jerk, or not move smoothly in any area?
    iii. Does the film exit partly rotated?
- Racks incorrectly installed.
    i. Check the position and seating of the racks. Pay careful attention to guide marks or grooves.
    ii. Check that racks are not distorted, or bent out of shape.

- Loose or damaged roller pressure springs.
    i. These are coiled springs shaped in the form of a loop. They pull the rollers together, and provide the correct pressure on the film.
    ii. If springs are damaged, or have uneven tension, then the rollers can feed the film at an angle.
    iii. Compare the suspect spring to other springs. Replace with a new pair, one on each side of the rollers.
- Film crossover guides not properly installed or faulty.
    i. Examine the area around the guides for any sharp edges, or scratched sections.
    ii. Check the rollers. Ensure free movement of the rollers.
    iii. Check crossover alignment. *Ask the service department for advice, before making any adjustment.*
    iv. Check crossover guides are correctly seated, no distortion or cracks.
- Damaged gears.
    i. A previous jammed film may result in broken or damaged gear teeth. This can cause erratic or stopped rotation of the rollers.
    ii. A gear is not sitting in the correct position on the shaft. Check for a missing retaining clip. (Circlip). This fits in a groove of the shaft, to keep the gear in position. Some gears have a plastic retaining clip as part of the gear moulding. If broken, the gear must be replaced.
- Incorrectly set drive shaft.
- Timing belt or chain incorrectly installed or broken.
- Sharp or damaged edges in the film entrance table.
- Incorrectly adjusted film entrance table.
- On systems with a micro-switch for film size sensing, the actuation lever may be damaged.
- Films are fed too close together.
    i. Does a warning light operate, until ready for the next film?
    ii. Does a chime sound when the processor is ready for the next film?

g. Film appears under developed

- Operator error.
    i. Wrong X-ray exposure setting.
    ii. Incorrect cassette. Detail instead of normal screens.
    iii. Excessive starter was added after service.
- Insufficient developer replenishment. Check the replenishment flow rate.
    i. Replenishment pump not working.
    ii. A leaky valve in the replenishment pump.
    iii. The replenishment feed line is blocked. (Or twisted and 'kinked')
    iv. Faulty film size detection.
- The developer supply is oxidized or depleted.
    i. Replace the developer supply, if more than one month old.
    ii. Test specific gravity. Use the temperature correction chart, Fig C-1 page 177.
- Incorrect developer temperature.
    i. Compare the temperature to the previous recorded value, when the processor was last serviced.
    ii. Monitor developer temperature during the day. Look for excessive temperature drift.
- Film transport speed has increased.
    i. Check for incorrect settings in the processor computer.
    ii. Measure film transport time.
    iii. **Note.** If transport speed is incorrect, this will also affect fixing and drying.

h. Uneven developing across the film

- Recirculation pump not working.
- Partially blocked developer filter. Clean as part of weekly maintenance.
- Damaged or blocked recirculation pipe lines.

i. Film has high base fog and excessive contrast

- Operator error
    i. Starter was not added after service. (Or insufficient starter.)
    ii. Add starter.
- Incorrect developer temperature.
    i. Compare the temperature to the previous recorded value, when the processor was last serviced. Reset if required.
    ii. Monitor developer temperature during the day. Look for excessive temperature drift.
- Developer over concentrated.
    i. Check supply specific gravity. Use the temperature correction chart, Fig C-1 page 177.
    ii. Check replenishment rate.
    iii. Add starter.
- Film transport speed has decreased.
    i. Operator error. The speed adjustment was left on low speed, after processing single emulsion films.
    ii. Check for incorrect settings in the processor computer.
    iii. Measure the film transport time.

iv. Motor speed may be reduced due to incorrectly fitted racks.
v. Or motor speed may be reduced due to stiff bearings. Lubricate the bearings. (The bearings may be noisy).

j. Films appear poorly fixed

- Insufficient fixer replenishment.
  i. Check the replenishment flow rate.
  ii. Adjust the flow rate or pump operation time.
  iii. Replenishment pump not working. Some processors have two fixer replenishment pumps working in parallel. One may be faulty.
  iv. Faulty 'Poppet' valves on the replenishment pump. Replace.
  v. Replenishment feed line blocked. (Or twisted and 'kinked')
  vi. Faulty film size detection.
- Fixer supply incorrectly mixed. Check specific gravity.
- Fixer is contaminated, replace with a fresh solution.
- Fixer temperature too low. This may apply where the processor has separate heaters for fixer and developer.
- Poor 'squeegee' action of rollers as film exits the developer tank. This leaves excessive developer on the film, preventing proper contact with the fixer.
  i. Clean the rollers.
  ii. Examine the roller compression springs; ensure correct fit and tension.
  ii. Some processors have a 'mini wash area', with the crossover rollers, where the film is transported between tanks. Ensure water level is correct, and is circulated.

k. Films are discoloured. May appear 'sticky'

- Fixer temperature too low. This may occur if the processor has separate heaters for fixer and developer.
- Fixer is depleted. See 'Films appear poorly fixed'
- Wash water temperature too low.
- Film transport speed has increased.
  i. Check for incorrect settings in the processor computer.
  ii. Measure the film transport time.
  iii. **Note.** If transport speed is incorrect, this will also affect developing and drying.

l. Insufficient or uneven drying

- Incorrect temperature setting. Temperature may need to be increased if humidity level is high.
- The drying heater is faulty.
  i. If more than one element, an element may be burnt out.
  ii. Faulty operation of the over-temperature safety thermostat.
  iii. Power fuse to the heater is open circuit.
- The drying fans are faulty. Possible failure of one fan only.
- The drying thermostat is faulty.
- Fixer may be depleted, or at too low a temperature.
- Wash water temperature low.

m. Bands across the film, perpendicular to the film transport direction

- Dirty rollers
  i. Clean the rollers. The rollers may develop a layer of chemical crystals. A nylon or plastic cleaning pad will assist in the removal of these 'crystals', or encrustation.
  ii. Replace any rollers that do not have a smooth surface after cleaning.
  iii. Check for damage or flat areas on the rollers.
  iv. Test by slowly rolling along a flat surface. Feel for any bumps as the roller is rotated.
  v. Position a light behind the roller. Move the roller along a flat surface. Look for any gaps as the roller is rotated.
- Rollers do not rotate smoothly. They stop and start.
  i. The drive belt or chain may be loose. Adjust according to the maintenance manual.
  ii. Incorrect positioning of rack or rollers.
  iii. A bearing may require cleaning, or lubrication.
  iv. Damage to a gear tooth.
- Film is slipping in the rollers.
  i. Examine roller compression springs; ensure correct fit and tension.
  ii. Look for missing springs.

The following film problems may not be due to the processor.

n. Film fogging

- Darkroom safe light is faulty.
  i. Test by leaving film on bench for a short time, then processing. Next, place film directly into processor, but keep safe light off.
  ii. Aim the light upwards, away from the workbench or processor film table.
  iii. Has the globe been replaced with a wrong type?
  iv. Has the film been changed to orthochromatic film? Contact the film supplier for advice. Obtain correct filters for the safe light.

- Damaged cassette, allowing light to enter.
    i. Test by inserting a film in the suspect cassette, with the safelight switched off. Then place the cassette in different positions, in normal room lighting.
    ii. Process the film.
- All films appear fogged.
    i. The film has been stored under excessive temperature, or humidity conditions.
    ii. The film has passed its expiration date.
- Film has intermittent fogging. Artefacts can also be observed.
    i. Scatter radiation is entering the cassette storage area.
    ii. Possible fault with radiation shield. Test by placing a test cassette for a while in the suspect area.
- Film exhibits fogging towards one edge only. All films of the same size have a similar problem.
    i. The film storage bin has been opened under full lighting conditions.
    ii. Possible light leak into the film storage bin. Check for proper closing and operation of the film bin.
    iii. Improper light shielding of films, due to torn packaging etc.

o. Static electricity marks

- These appear as 'branched', or 'dotted' areas on the film.
    i. This is due to a static discharge, as the film is handled.
    ii. A common cause is dry, or low humidity conditions. Some floor coverings, and type of shoes, can also cause this problem.
    iii. Before handling the film, discharge yourself by touching the metal tray of the processor.
    iv. Use anti-static cleaners for the cassette intensifier screens.

TASK 16

Films appear too dark

You have just returned from holidays. On using your normal exposure techniques, the films appear too dark. Your assistant informs you she has also been having a problem, to obtain the correct exposures.

You suspect a problem with the processor. However, list possible reasons, not caused by the processor, which might cause dark films.

_____
_____
_____
_____
_____

Make a list of possible processor problems, which could cause a dark film. Indicate on this list the order in which you would check these items.

_____
_____
_____
_____
_____

Carry out suitable tests. Describe these tests and their results.

_____
_____
_____
_____
_____
_____

What action is needed to correct the problem?

_____
_____
_____
_____

Tutor's comments

Satisfactory/Unsatisfactory

Signed _____ Date _____
Tutor

PART IV. AUTOMATIC FILM PROCESSOR

TASK 17
Films exhibit symptoms of low fixer

After carrying out initial tests, you replaced the fixer, and adjusted the fixer pump. However, as several days go by the problem repeats itself.
You come to the decision that the fixer pump is faulty and requires attention.

What were the original symptoms? _____
_____
_____

Describe the tests carried out, and action taken to correct the problem. _____
_____
_____
_____
_____

The problem has now repeated itself. You have contacted the processor agents, and discussed the problem. They recommend you replace the pump valves, suspected leaking. You now have the replacement valves and are about to affect a replacement.
Describe some important precautions before attempted disassembly _____
_____
_____
_____
_____

Reassembly has been successful. At the beginning you made some adjustments in an attempt to correct this problem. Now with the processor powered up, and charged with fresh chemicals, what adjustment should again be checked? _____
_____
_____
_____

Tutor's comments

_____

Satisfactory/Unsatisfactory

Signed _____   Date _____
Tutor

MODULE 11.2
The film ID printer

Aim

Film ID printers range from basic versions where the film is first removed from the cassette, and then placed in the printer, to motorized versions; which print the film through a 'window' in the cassette. The suggestions made here are for the basic version only.

Objectives

After completion of this module, simple repairs to a basic film printer may be achieved. Assistance from an electrician is recommended.

Contents

a. Operation of a film printer
b. Precautions for replacing the lamp
c. Failure to print
d. The printing is too light

a. Operation of a film printer

The basic film printer consists of a lamp, which delivers a brief burst of light through the paper ID strip onto the film. In its simplest form, a capacitor is charged to a preset voltage. On closing the printer lid, a microswitch connects the capacitor to the lamp, producing a brief flash of light. A potentiometer controls the voltage level on the capacitor, which in turn controls the lamp output. Additions may include preheating the lamp filament, or providing a flash timer. Later versions replaced the lamp with a xenon flash tube, similar to those employed in a camera. Due to the simplicity of the design, very little can go wrong. However, some problems may still occur.

b. Precautions for replacing the lamp

Before opening the cover ensure the printer is disconnected from the power point. Take care not to touch any of the internal wiring, as there may be significant voltage stored in a capacitor.

The replacement lamp should have a similar power rating. Depending on the actual mode of operation, changing to a higher rated lamp could produce a lower flash intensity.

c. Failure to print

Before investigating, ensure the printer is unplugged from the power point.

- Is the globe faulty? Try a replacement globe.
- Can you hear a small 'click' as the lid is closed? If not the expose switch may need adjustment. Otherwise the switch may be faulty.
- Does the power cord have a broken connection? Check both at the plug end and where the cord enters the printer. **Repairs to the power cord or plug should be performed by an electrician.**
- Is the power point faulty? Check the printer in a known good outlet.
- Has the print density control developed a bad connection? Try adjusting to a different position.

PART IV. AUTOMATIC FILM PROCESSOR

d. The printing is too light

- Did this occur after changing the globe? Check to ensure the correct type was fitted.
- Has the type of paper used for the patient ID been changed to a different version?
- If the first printing attempt is light, but an attempt at printing shortly after produces better results, then a pre-heat resistor or adjustment may be faulty. **Have an electrician check for this possibility. Do not attempt this by yourself; there may be a high voltage charge on a capacitor.**

PART V
Appendices

APPENDIX A
Sensitometry

Note. This section on sensitometry is an extract from the WHO Quality assurance workbook, *by Peter J Lloyd.*

Sensitometry is the study and measurement of the relationship between exposures, films, screens, and processing.

Principal use

- Our interest lies more in its use in checking film processor performance, in particular automatic processors.
- By standardizing exposure, film and screen types, conditions under which films are exposed, handled and stored, leaves only one variable, that of film processing.
- Any variation in film image must then be due to film processing.
- (See 'module four, manual processing', page 78, in *WHO Quality assurance workbook* for an elementary form of evaluating film processing.)
- Sensitometry is a more comprehensive form of monitoring processing performance.

When to do processor control sensitometry

- First thing every morning. This especially applies if used for mammography films.
- After the processor has reached the correct operating temperature.
- After feeding cleanup films through.
- After cleaning or servicing the processor.
- Before processing any patient radiographs.

Outline of procedure

- A standard **step-wedge** image must be produced.
- This image consists of a range of clearly defined images.
- Production of this image must be consistent. Methods for its production are described following.
- The film is processed in the film processor to be monitored.
- The resultant image densities are determined using a **densitometer**. The results are recorded, graphed, and the graphs evaluated.
- The first graph produced is known as a **characteristic curve**.
- To draw a characteristic curve, plot **test film densities** against **test film density step numbers**. Step one must be the lightest density.
- From this characteristic curve, several other graphs may be plotted, each giving additional information.
- This combined information will give a comprehensive picture of processor performance.

The test film

The image on the test film must be a standard series of clearly defined densities, ranging from barely visible to black.

- These densities are usually over a range of 21 steps.
- A smaller number of steps may be used.
- Number the steps from 1. (Lightest step)

Producing the test film

There are several ways of preparing the test film. Four examples are provided here.

METHOD 1

The best and most reliable method is to use a **sensitometer**, which produces a standard range of densities.

Using the sensitometer
- Use a dedicated box of film, of the type commonly used in your department.
- Use the sensitometer in the darkroom where the processor is to be monitored.
- Select the light colour relevant to your films colour sensitivity, (blue or green)
- Under safelight conditions, insert a sheet of film into the sensitometer until it reaches the backstop.
- Press the cover down until the indicating signal, (audio or light), has stopped.
- Raise the cover and remove the film.
- Process the film in the processor to be monitored.

- The film should always be placed in the same position on the film tray, with the step-image parallel to the rollers.

METHOD 2

Make an X-ray image of an **aluminium step-wedge** under standard conditions.

Making the step wedge image
- Place an 18 × 24 cm cassette, loaded with your standard film, face up on the X-ray table.
- Place the step wedge on the face of the cassette.
- Using a 100 cm FFD (SID), centre and collimate to the step wedge.
- To make more than one image on the same film, strips of lead rubber may be used to divide the cassette.
- Set an exposure that will produce a full range of step wedge densities, and make an exposure. You may need to experiment first in order to determine the correct exposure for your particular step wedge/film/screen combination.
- Process the film, under safelight conditions, in the processor to be monitored.
- The film should always be placed in the same position on the feed tray, with the step-image parallel to the rollers.
- Standard conditions must be used each time a step wedge image is made.

METHOD 3

Producing a standard range of densities, using X-ray, **without** the step wedge

Making the image
- Place an 18 × 24 cm cassette, loaded with your standard film, face up on the X-ray table.
- Divide the face of the cassette into 11 strips.
- Cover all but the end strip with lead rubber.
- Set a 100 cm FFD (SID), centre and collimate to the uncovered area. **(Strip '1')**.
- Expose using a low exposure (Enough to produce a barely visible image).
- Move the lead rubber so that **strips '1' and '2'** are uncovered.
- Using the same exposure, expose both strips. (Strip '1' has now been exposed twice).
- Move the lead rubber so that **strips '1', '2', and '3'** are uncovered and expose all three strips, using the same exposure. (Strip '1' has now been exposed three times, and strip '2' twice).
- Repeat this process until all strips have been exposed.
- Process the film, under safelight conditions, in the processor to be monitored.
- Feed the film into the processor in the same way as that described in method 2.
- The film may be cut down the middle lengthways, when removed from the cassette, in order to create a second strip. The unused strip should be placed in a light tight box for further use.

METHOD 4

- Purchase **pre-exposed sensitometry film** produced by a film manufacturer.

Terminology To understand the process of sensitometry, it is necessary to have an understanding of some basic terminology.

Sensitometer A consistent light source, which produces a standard range of densities, when exposed on film.

Densitometer A consistent light source, combined with a light measuring sensor, used for accurately measuring film density.

Film density The degree of film blackening. You will see from the characteristic curve graph, and your own experience, that density increases as exposure increases.

Contrast The difference between two or more densities on a film. The straight line portion and shape of the characteristic curve gives us information about contrast. A high contrast film curve will lie toward the left, whilst a lower contrast film curve will lie more towards the right. (See Fig 5–3).

Gradient The contrast of a film at a given density. When a straight line is drawn tangent to the characteristic curve at a given density, this line forms the slope which is the gradient of that density.

Average gradient A line drawn between the **0.25 and 2.00 density levels** on the characteristic curve.

Toe gradient A line drawn between the 0.25 and 1.00 density levels on the characteristic curve.

Mid gradient A line drawn between the 1.00 and 2.00 density levels on the characteristic curve.

Upper gradient A line drawn between the 2.00 and 3.00 density levels on the characteristic curve.

Base plus fog The density of processed film, without the effects of light or radiation; the density at which the characteristic curve begins.

APPENDIX A. SENSITOMETRY

Exposure Intensity of radiation × time (mAs).

Speed Indicated by the location of the curve along the step (exposure) axis. A faster film curve will lie more toward the left, whilst a slower film will lie more toward the right. To calculate film speed, use a density level of 1.0. (This is considered to be the average of the useful density range of 0.25 to 2.0).

Exposure latitude The range of exposure factors, within which the resultant radiograph is considered to be acceptable. A film that is said to have 'wide latitude' has the ability to accept large changes in exposure, without excessive density changes.

Carrying out the sensitometric test

Frequency of test
- Daily

Equipment required
- Sensitometer, to produce image by Method 1 described previously, *or*
- Step wedge, to produce image by Method 2 described previously, *or*
- Lead rubber sheet and cassette, to produce image by Method 3 described previously.
- Unexposed film or manufacturers pre-exposed sensitometry film.
- Densitometer.
- Specialised graph paper supplied by a film manufacturer, *or*
- Simple small grid, graph paper, available from stationary shops, drawn up in the same way.
- Any scale graph paper may be used, but it is common to use a ratio, X axis to Y axis of 0.15 : 1.0. **The important thing is that you do not vary the ratio once your quality control programme is under way.**

Method
- Allow the processor to stabilize.
- Process the test film in the processor to be checked.
- Tests must be carried out at the same time each day under the same conditions.
- Developer temperature must be taken at the time of test.
- Using the densitometer, read the density of each step on the test film image.
- Record the densities against their step numbers.
- Plot a graph of step numbers on the horizontal axis against densities on the vertical axis and join up the dots. (See Fig A–1).

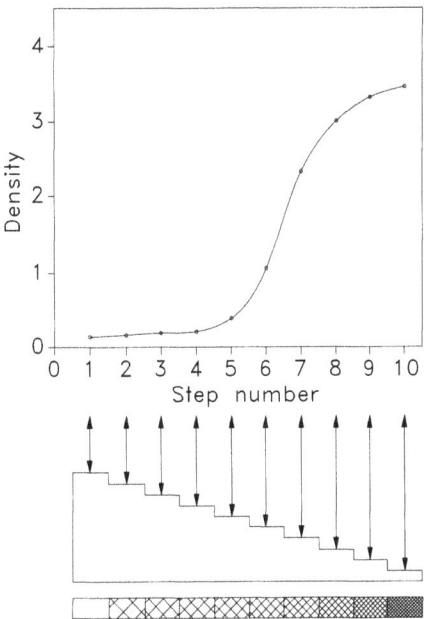

Fig A–1. Typical characteristic graph, showing relationship of density to step number

Using the densitometer
- Switch on the densitometer and allow the unit to stabilize.
- Set the densitometer reading to zero.
- On 'bench top' units and 'clamshell' types, this is performed by making a test reading without film.
- Place the centre of the density to be read directly over the light aperture and under the reading arm.
- Lower the reading arm to the film, press the 'read' switch and hold for a few seconds, until the readout stabilizes.
- Lift the arm
- Record the reading against the step number.
- Repeat for all density steps on the image.
- In case a hand held unit is used, place the densitometer against the film viewer, and then adjust the readout to zero. Important, do not shift the position on the film viewer during subsequent readings.

Plotting the characteristic curve
- The graph paper must have **'Density'** on the **Y (vertical) axis**, and **'Step Number'** on the **X (horizontal) axis**.
- Plot density of step 1 (the lightest density) at the appropriate level on the Y-axis, and directly above Step 1 on the X-axis.
- Repeat for all other step densities.
- Join up all the plots to form a free flowing curve. (See fig A–2).
- Draw in the toe, mid and average gradients as described previously.

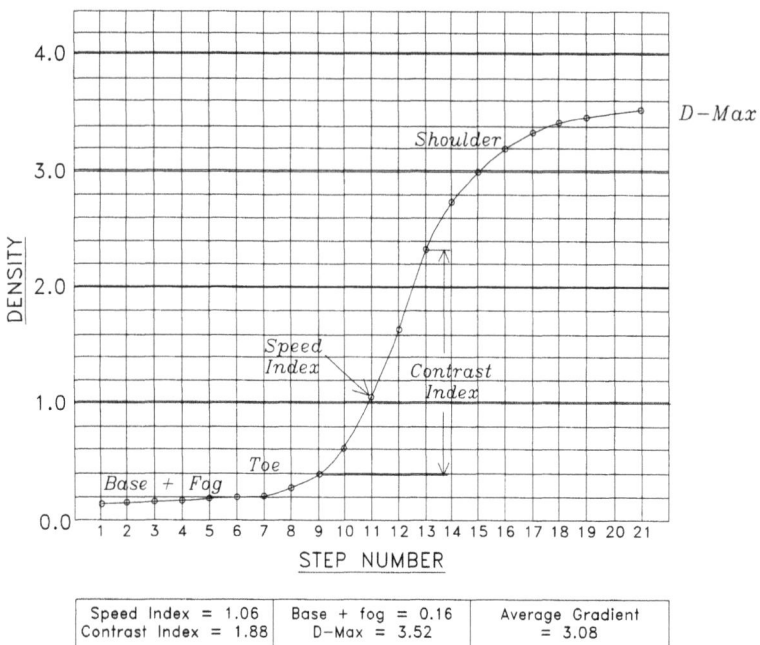

Fig A–2. Example of a characteristic curve

Evaluation
- Compare your characteristic curve with the control characteristic curve that was plotted when the chemistry was first mixed and the processor set up.
- If these curves vary markedly there is reason to believe that the processor is not functioning correctly.
- Calculate the 'Speed', 'Contrast', 'Base plus Fog (D-Min)', 'Temperature' and 'D-Max' and enter on the appropriate graphs. (See Fig A–2 and Fig A–3)

Speed
- Select the step number that has a density within a range of 1.0 to 1.3 on the step wedge image. This step number becomes the speed step.
- Record the step number on the **speed chart** in the space provided (Speed step No. ●●). Do this only on day one, after the chemistry has been freshly mixed.
- Plot the density reading on the zero line under day one. Do this only on day one after the chemistry has been freshly mixed.
- Measure and plot the density of the speed steps daily. (See Fig A–2, Fig A–3, and Appendix C page 177.)
- Observe how much the speed plot varies from the zero line.
- An acceptable variation is **plus or minus 0.15**.
- **Variations above this may need corrective action.**

Contrast
- Select densities of the steps, two steps above, and two steps below, the speed step.
- Record these steps on the contrast chart, in the space provided (Step Below ●● and Step Above ●●)
- Subtract the smaller from the greater of these two densities. This difference is the contrast of your standard reference.
- Record this standard reference against the zero line on the **Contrast Chart**. Do this only on day one, after the chemistry has been freshly mixed.
- Repeat this process daily, plotting the contrast indicator under appropriate dates. (See Fig A–2, Fig A–3, and Appendix C page 177.)
- Observe how much the contrast indicator varies from the zero line.
- An acceptable variation is **plus or minus 0.15**.

Base plus fog (D-Min)
- Using the densitometer, measure the density of the film that has received no exposure.
- Record this reading against the zero line on the **Base plus Fog Chart**. Do this only on day one, after the chemistry has been freshly mixed.
- Repeat this procedure daily, plotting the base plus fog density under appropriate dates. (See Fig A–2, Fig A–3, and Appendix C page 177.)
- Observe how much the base plus fog indicator varies from the zero line.
- Ideally it should not go above **0.02**. **Action should be considered if it exceeds 0.023.**

APPENDIX A. SENSITOMETRY

Temperature
- Take the temperature of the developer
- Record against the zero line of the **Temperature Chart**. Do this only on day one, after the chemistry has been freshly mixed.
- Repeat this procedure daily, plotting Temperature under appropriate dates, (See Fig A–2, Fig A–3, and Appendix C page 177.)
- Observe how much the temperature varies.
- **Action should be taken if the temperature varies more than a few degrees.**

D-Max
- Using the densitometer, measure the maximum density step on the step wedge image.
- Record against the zero line of the **D-Max Chart**. Do this only on day one, after the chemistry has been freshly mixed. (See Fig A–2, Fig A–3, and Appendix C page 177.)
- Repeat this daily, recording the D-Max on the chart under appropriate dates.
- **Noticeable variations give advanced warnings of chemistry problems.**

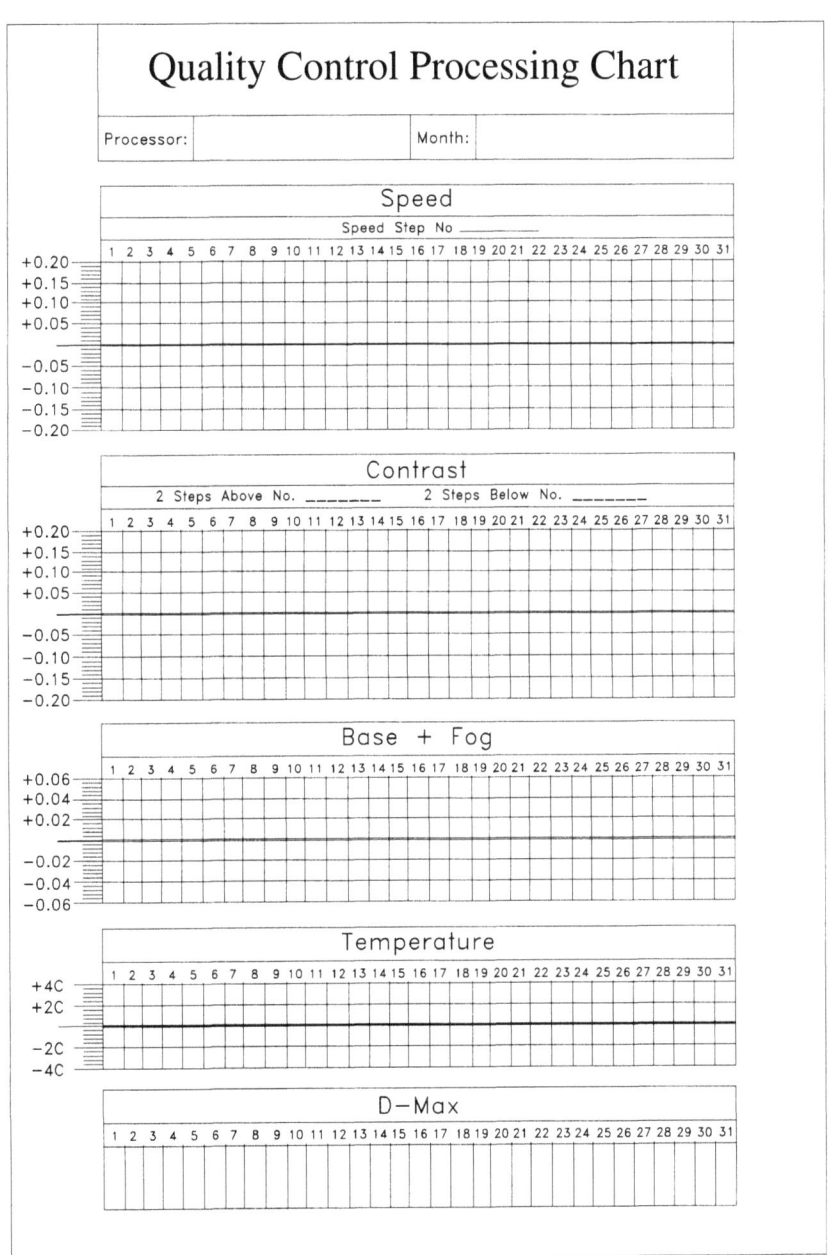

Fig A–3. Sensitometry charts for recording speed, contrast, base + fog, temperature and D-Max data

How to use the sensitometry graphs

All processors will show some variation in results from day to day. However, action should be taken if the variations are sudden or continue to increase or decrease over a period of time and pass beyond the acceptable limits. If the change is sudden, then all possible influencing factors should be checked, or a mistake may have been made. The test should be repeated.

In control

When results are within acceptable limits the processor is said to be 'in control' and no action is required.

Out of control

If one or more of the charts (speed, contrast or fog) is showing results that are outside the acceptable limits, especially if the change is sudden or continues to increase or decrease, then the processor is said to be 'out of control' and immediate action should be taken.

Action to be taken if the processor is out of control

- Immediately stop using the processor
- Inform other users
- Start problem solving process
- File a report

Table A–1. Example of the possible interpretation of sensitometric charts, and recommended action

Chart changes	Possible cause	Action
• Speed and contrast increase • D-min acceptable • ——— • Speed and contrast up • D-min increases	*First stage of over development* 1 Developer temp' high 2 Excessive replenishment 3 Developer too concentrated 4 Check processing time	1 Adjust temp' 2 Adjust replenishment 3 Change developer 4 Developing time too long, add starter
• Sudden increase in speed, D-min & D-max after service.	*Excessive over development.* Starter omitted or insufficient starter in developer.	Add starter to developer.
• Speed decreases • Loss of image contrast • D-min normal • ——— • Image density low over whole image. Speed and contrast low. • D-min also low	*Under development* 1 Developer temp' low 2 Developer exhausted 3 Insufficient replenishment 4 Replenisher used up 5 Developer too dilute 6 Developing time too short	1 Check and adjust temp' 2 Check and adjust replenishment 3 Check and refill replenisher tank 4 Replace replenisher 5 Check processing time
• Sudden decrease in speed and D-max after service • Small decrease in D-min	Excessive amount of starter in developer	Replace developer, add correct amount of starter
• Increase in speed • Shoulder decreases • Loss of contrast • Increase in fog	1 Aerial oxidation 2 Contaminated developer	Replace developer after washing out tank Add correct amount of starter

APPENDIX B

Recommended tools and test equipment

Suitable tools and test equipment are required when carrying out routine maintenance, or repairs to X-ray equipment. These should be suitable for the required work, as unsuitable tools may cause damage or injury.

The tools and test equipment indicated in this appendix will provide a basic tool kit. Some items of test equipment can be constructed, and are described in 'Making simple test tools'.

Contents

a. Basic tools for service and maintenance.
b. Service aids.
c. Test equipment.
d. Making simple test tools.
  i. Aluminium stepwedge.
  ii. Resolution test piece for fluoroscopy TV.
  iii. X-ray alignment template.
  iv. X-ray spinning top.
  v. Tomography resolution test.

a. Basic tools for service and maintenance

- Set of Phillips screwdrivers, Nos 1, 2, & 3.
- Set of flat blade screwdrivers, blade widths from 3mm to 10mm.
- Set of mini or 'Jewellers' screwdrivers. (Economy pack.)
- Set of metric Allan keys.
- Set of imperial Allan keys.
- Retractable blade, utility knife.
- Tape measure. (Minimum length of 3.0 metres)
- Ruler.
- Spirit level.
- Torch.
- An angled inspection mirror. (Dental style.)
- Combination pliers.
- Long nose pliers.
- Tweezers.
- Medium size side cutting pliers.
- Medium (6–8 inch) adjustable wrench.
- Set of metric spanners, up to 15mm size.
- Hammer.
- Protective glasses. (For processor maintenance)
- Processor tank cleaning brushes.
- Compartmented carry case.

b. Service aids

- Rolls of PVC insulating tape. (Several colours.)
- Packet of assorted plastic cable ties.
- Silicon grease. (Dow Corning 'DC-4').
- Small container of light household oil.
- Spray can of multipurpose lubricant spray. (RP-7, WD-40, CRC-2.26 etc).
- Suitable solvent for cleaning sticking plaster etc. from equipment. (For example, eucalyptus oil, methylated spirits. etc).
- Clean disposable cloths.
- Scouring pads, plastic or nylon type.
- Sodium hypochlorite bleach.

c. Test equipment

- A basic multimeter.
  i. A simple 'analogue' meter, capable of measuring up to 500V AC and DC, plus resistance, is sufficient.
  ii. If selecting a digital display meter, avoid 'auto-range' versions. Instead select one that has individual switch selections of the required measurement range.
  iii. When not in use, remove the battery from the meter.
- Accurate thermometer. Alcohol or electronic. To suit processor requirements.
- Hydrometer, to suit processor requirements.
- Measuring cylinder. Graduated 100ml glass or plastic container
- Densitometer. Or, a previous processed sensitometry film. This may be used as a density reference.
- Sensitometer, or a packet of pre-exposed reference film.
- Film screen contact test tool. (See page 134 of the 'WHO Quality assurance workbook'.)

- Optional step-wedge, 21 steps, for simulation of a sensitometer. (See pages 133 and 134 of the 'WHO Quality assurance workbook'.)
- *Standard aluminium step-wedge.
- *X-Ray alignment template
- *Tomography phantom.
- *Resolution test piece, for fluoroscopy TV.
- *X-ray spinning top, for use with single-phase or portable X-ray generators.

* See 'Making simple test tools'.

d. Making simple test tools

Aluminium step-wedge

A step-wedge is one of the most useful tools for measuring relative radiation output from an X-ray generator. This tool may be used to measure radiation consistency between selected mA stations of an X-ray control, or make a comparison between other X-ray controls in a department.

The step-wedge is sensitive to both changes of kV and mAs. In use, a series of test exposures are made for each ma station, using the same kV and mAs. If one of the mA stations shows a lighter or darker strip, then make another test of that mA station; together with a small change of either kV or mAs. The calibration error is the amount of kV or mAs required to achieve identical results.

Making a step-wedge

If a commercial step-wedge is not available, one may be made from several strips of 2.0 mm thick aluminium glued together.

- Obtain a 2100 mm length of 25 mm by 2.0 mm aluminium strip from the local building hardware store.
- Cut to lengths of 30, 50, 70, 90, 110, 130, 150, 170, 170, and 170 mm. When glued together as shown, they will form a 20 mm high stepwedge. **See Fig B–1**
- The glue used may be 'contact' adhesive or epoxy resin. (Recommended) If using epoxy, then roughen the surface of the aluminium strips before gluing together.
- As an aid to assembly, obtain a small cardboard box. (A shoebox is suggested.) Place the strips against an inside corner during assembly, to form a neat stack.
- An optional five 170 mm strips may be glued together to form a separate 10 mm filter. This is placed under the stepwedge when required. This allows the stepwedge to be used at higher kV or mAs settings.

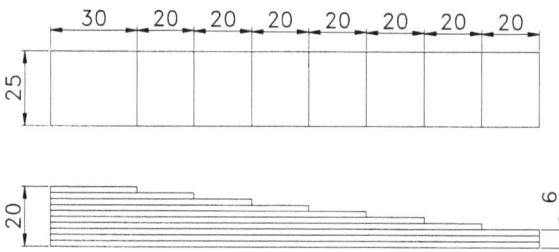

Fig B–1. Formation of a step-wedge from 2.0 mm aluminium strips glued together

The illustration in fig B–2 illustrates two simulated test exposures. The film is cut into individual strips with a pair of scissors. The strips are then placed alongside each other, and moved beside each other, until the density steps are matched. If a densitometer is available, then a direct comparison can be made of each step. This avoids cutting the film into strips.

Using the step-wedge

General considerations

- Besides showing a change in generator output, the stepwedge density will be changed by the processor calibration. Before testing a generator, check the processor, and make sure its performance is inside the required limits.

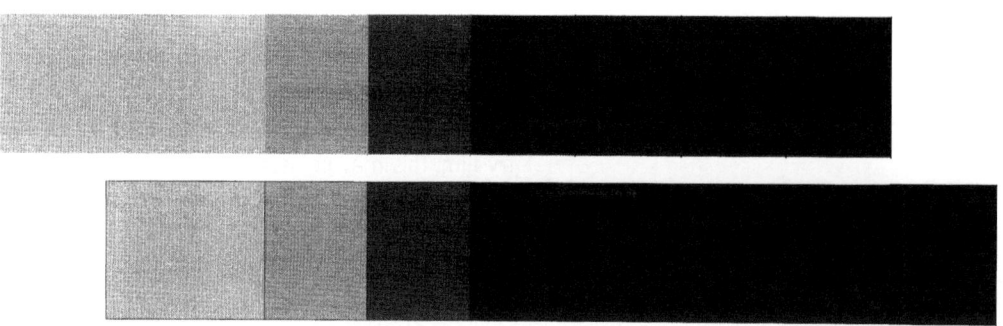

Fig B–2. Simulation of two test exposures obtained with a stepwedge

- For consistent results, use the same cassette for all tests, and the same type of film.
- Do not use the Bucky; instead place the cassette on the tabletop.
- Use a standard 100 cm focal spot to film distance. (FFD)
- Several exposures may be made on the one piece of film. Place two pieces of lead rubber on top of the cassette, positioned against each side of the stepwedge. As the stepwedge is repositioned, the lead rubber is also moved, to shield the film from unwanted radiation.
- If testing at high kV levels, additional filtration may be needed. This can be an extra 10 mm of aluminium placed under the stepwedge, or else 1~2 mm of copper placed in front of the collimator.
- If the test exposure is over exposed, and extra filtration is not wanted, then insert a sheet of dark paper between the film, and one of the intensifying screens of the cassette.
- Depending on generator design, exposure times shorter than 0.02 seconds may not be accurate. This can apply to older single or three-phase generators. Current high-frequency systems have greater accuracy, however, the HT cable still has a small effect on very short exposure times. This is due to energy stored in the HT cable capacitance. For this reason, some designs have a minimum mAs limit of 0.5 mAs.
- The capacitor discharge (CD) mobile has a non-linear output. For example, the kV will drop by 10 kV during a 10 mAs exposure. However, it is still possible to make a test of radiation consistency using the stepwedge. Also, if the same FFD, kV, and mAs values are used, then relative outputs between CD mobiles can be compared.

Step-wedge calibration

This could be called 'Getting to know your stepwedge'.

At first, the optimum exposure factors for a particular stepwedge will require some experimentation. Two factors are involved, kV and mAs.

- Select a value of mAs that will not require too short an exposure time, when used with the highest mA position for your X-ray generator.
- Select a nominal value of about 65~70 kV and make a test exposure.
- If the exposure is too light, then increase the exposure time, or the kV. If too dark, and the exposure times are short, then reduce kV.
- The test exposure should show a range of density values towards the centre of the stepwedge. Record the FFD, mAs, and kV values for later use.
- When an optimum exposure has been obtained, repeat this test with small increases of kV. Record the increase of kV to change the density by one step.
- Increase the value of mAs in small steps until you again obtain a density change of one step. Record this as the percent increase of mAs.
- If extra filtration (Such as the added 10 mm of aluminium) is available, then make another series of test exposures. This time, change kV only, until a density match is obtained. Record this value for later use.

Typical stepwedge techniques

There are two important tests the stepwedge can perform. These are radiation reproducibility and X-ray output linearity. These tests are described in **module 1.1 page 19**.

Resolution test for fluoroscopy

The overall resolution of an imaging system is quoted in 'Line pairs/mm'. This test requires the use of a highly accurate line-pair gauge.

Two versions are shown. Fig B–3 has individual groups, of line-pairs. These are arranged in ten steps, 0.5 LP/mm to 5.0 LP/mm. Fig B–4 has lines converging to a point. Calibration marks are provided to indicate the equivalent line-pair resolution.

Many other versions are available. The type of pattern and material used depends on the specific test required. Common types are similar to fig B–3, or B–4. The line-pair pattern is made from 0.1 mm lead for normal use.

Fig B–3. A typical line pair gauge, 0.5 to 5.0 LP/mm

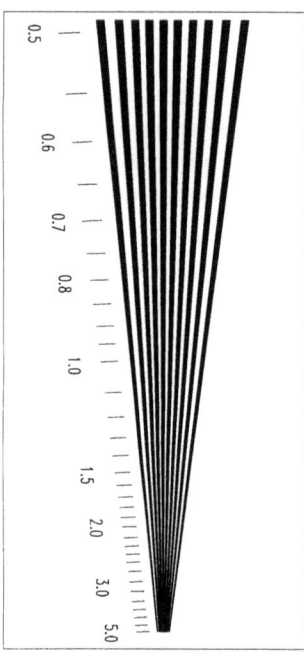

Fig B–4. Convergence style line pair gauge, 0.5 to 5.0 LP/mm

Making a convergence style focus aid

If a commercial line-pair gauge is not available, a convergence style focus aid can be constructed, using sewing needles. See Fig B–5.

Material required

- Seven thin sewing needles, about 60~70 mm in length.
- Five small pins.
- Photocopy of Fig B–5
- A piece of 3 mm thick medium density fibreboard (MDF), or a piece of thin acrylic plastic, cut to the size of Fig B–5.
- Adhesive. Epoxy-resin is preferred.
- Pair of tweezers.

Construction

- Using adhesive, attach a photocopy of Fig B–5 to piece of 3.0 mm MDF, or thin acrylic, cut to the size of Fig B–5. Use this diagram for positioning the needles.
- Place a thin film of epoxy resin over the top the diagram. This is to help hold the needles in place during positioning.
- Using Fig B–5 as a template, place the needles on top of the outlines shown in the drawing. Start first from the two outside positions, working from each side towards the centre. A pair of tweezers will help in positioning. The 'rotation' position of the needle eye is not important.
- (The eye end of the needles should have a gap between each needle of about 2 mm)
- Position small sewing pins to one side as indicated. These are to indicate the resolution reference position.
- Once all needles and pins are in position, wait for the epoxy resin to harden. Now mix another lot of epoxy resin, and cover the pins and needles, so they remain firmly attached.

Using the line pair gauge

- Tape the gauge onto the centre of the input face of the image intensifier. If access is difficult, then tape the gauge to underneath the serial changer.
- To avoid interaction with grid lines, attach the gauge (Fig B–3) so it is rotated approximately 25~45 degrees. (On some CCD TV cameras, this also avoids interaction between pixels.)
- Raise the serial changer to maximum height above the tabletop.
- If the system has automatic kV control, this should be turned off. Set manual fluoroscopy kV to 50~55 kV.
- With fluoroscopy 'on', adjust kV or mA to obtain a normal brightness and contrast image on the monitor.
- Carefully observe the line-pair patterns. The limiting definition is the line-pair group that is reasonably visible, while the next group is completely blurred out.
- If using the focus aid (Fig B–5), observe the distance from the apex before blurring occurs, which indicates focus quality.
- **Note.** This test is relative only. A direct comparison in terms of line-pairs is possible, only if previously compared with a commercial line-pair gauge.
- Record the line-pair resolution, or 'V' pattern distance obtained, and compare with any earlier tests.
- When a multi field image intensifier is installed, repeat this test for all other field sizes.
- The combined TV and II focus should be better than 12 LP/mm for a 9" image intensifier. (A typical

Fig B–5. A focus aid, made with sewing needles

result might be 14 LP/mm with a CCD camera, and 16 LP/mm with a 'vidicon' camera.)

X-ray alignment template

The template is designed to provide evaluation of the X-ray collimator light beam accuracy. In use the template is placed on top of the cassette, and the collimator light beam is positioned to the template markers. After making an X-ray exposure, the markers, shown on the film, indicate the radiation position.

Making a template
Material
- A piece of 3 mm thick medium density fibreboard (MDF), cut to the required size. Stiff cardboard may also be used.
- Packet of large paper clips
- A length of single strand wire, about 1 mm diameter, and 1.5 metres long
- Adhesive. (Epoxy resin is recommended)
- Small angle cutting pliers
- Pair of tweezers
- Drawing instruments

Drawing the template
Refer to the diagram of Fig B–6. This indicates the template design before placing markers on the template.

- Draw a rectangle, with sides 24 by 30 cm. This is indicated by the dashed line in the diagram.
- Draw a horizontal and vertical line in the centre of the rectangle.
- Draw a short line every 1 cm along the horizontal and vertical lines, starting from the centre.
- Using the 1 cm marks as a guide, now draw three rectangles. 10 by 10 cm, 16 by 16 cm, and 20 by 26 cm.
- The circle indicates a nominal placement point for an orientation marker.
- Now trim the MDF, or cardboard, to the edge of the 24 by 30 cm rectangle. (Shown as a dashed line in the diagram.)

Making the markers
The markers are made from short pieces of wire. Prepare the required markers before attempting to glue them in position.

- The markers need to be about 3.0 cm in length to mark the position of the rectangles, and about 1.0~1.5 cm to mark the 1 cm divisions.
- Straighten a paper clip to the shape required for the corners, cutting off the remainder with the pliers.
- The length off wire will need to be straightened. Fasten one end to a fixed object, and the other end to the pliers handle. Pull firmly; stretching the wire will cause it to become straight.
- Prepare two long pieces of wire, 240 mm and 300 mm in length.
- Use the remainder for the short 1.0~1.5 cm markers.

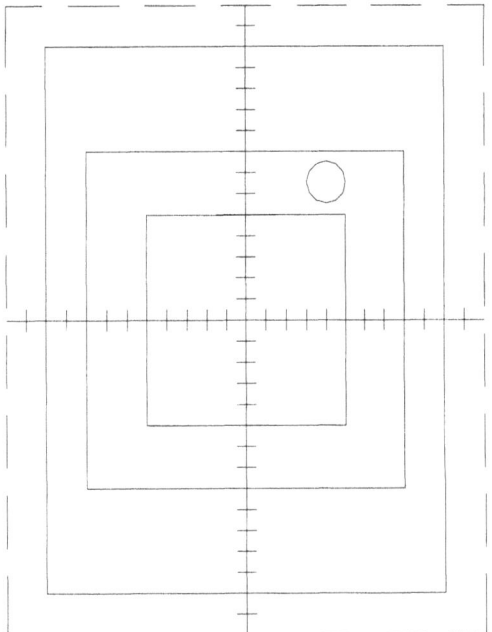

Fig B–6. The template outline, before placing the markers

Placing the markers
- Refer to the diagram B–7. This indicates the markers in position, and the outline of the template as it could appear on a test film.

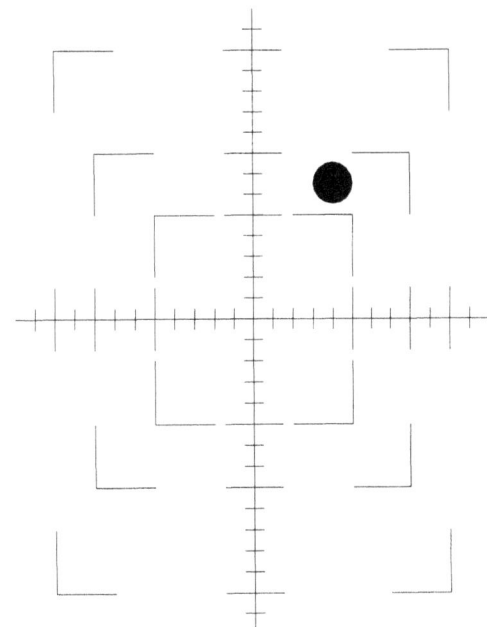

Fig B–7. The markers in position on the template (The outline of Fig B–6 is not shown.)

- Mix a suitable quantity of epoxy resin. Use this to fasten the markers in position to the previously drawn outlines. (A pair of tweezers will aid in positioning)
- After all the short markers are in position, the long pieces of wire are placed on top, and attached with epoxy resin.
- Finally, attach a small coin, or other suitable marker, to indicate position of the template on the test films.

Using the template
- Place the template in the centre of a 24/30 cm cassette.
- Collimate the light beam to the outer 20 by 26 cm rectangle.
- Make an exposure, using a low mAs and kV output.
- Evaluate the accuracy of the collimator from the resultant film.

The spinning-top
The spinning-top is a tool for determining the exposure time of single phase, or portable X-ray generators. With these controls, power is applied in the form of high voltage pulses, every half cycle, or once every cycle, of the input power frequency. This produces individual pulses of X-rays. The spinning top is a means of spreading these pulses over an area of film, enabling individual pulses to be counted as a function of exposure time. The spinning-top cannot be used with three-phase, high frequency, or CD mobiles. Commercial spinning tops are available, but may be expensive. They are now supplanted by electronic detectors that require an oscilloscope, or as part of a radiation measurement device.

Fig B–8 illustrates a spinning-top, which may be constructed locally. You may attempt this yourself, or have it made at a local engineering shop. Alternately, please refer to page 134 of the 'WHO Quality assurance workbook' for additional information.

Material
- An 80 mm diameter mild steel disc, 2.0 or 3.0 mm thick.
- A 50 mm diameter mild steel disc, 5.0 mm thick.
- A 10 mm 'cap head' or 'hex head' bolt, with an overall length of 35~40 mm. Note, this screw or bolt should be mild steel, NOT high tensile. This would be very difficult to drill.
- Two 10 mm nuts.
- A 3.0 mm countersunk metal thread, 35 mm long.
- Two 3.0 mm nuts.

Tools
- Drill. (Preferably a drill-stand and vice)
- Drill bits of 2.0 mm, 3.0 mm and 10 mm
- Countersink bit
- Bench vice
- Flat file
- Spanner
- Centre punch
- Hammer
- Pencil
- Rule

Construction
- See Fig B–8.
- Very carefully determine the centre of the 80 mm disc. Mark the centre with the centre punch. This will prevent the bit moving off-centre when drilling.
- Drill a 3.0 mm pilot hole in the centre of the disc. Then drill the required 10.0 mm hole.
- Drill a 2.0 mm hole about 7.5 mm from the edge of the 80 mm disc.
- Mark the centre of the 50 mm disc with the centre punch, and drill a 3.0 mm hole. Use the countersink bit to provide a chamfer for the countersunk screw.
- Prepare the 3.0 mm countersunk screw by filing the end to a sharp point.
- Insert the screw into the plate. Check that the head does not protrude below the plate. If necessary, apply the countersink bit again.
- Fix the screw firmly in position, using the second nut as a 'locknut'.
- Carefully mark the centre of the 10 mm bolt, using the centre punch.
- Drill a 3.0 mm hole through the centre of the bolt, to depth of about 10 mm.

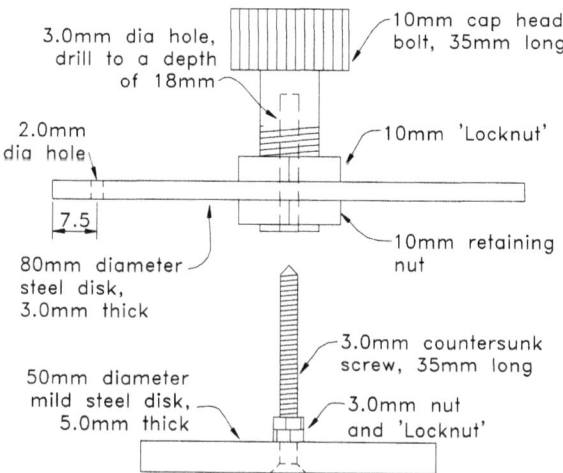

Fig B–8. Spinning-top construction details

- Make a trial fitting of the 3 mm metal thread into the bolt. If it is a tight fit, then carefully file a small amount from around the sides of the metal thread.
- Continue drilling the 3 mm hole into the bolt, checking to ensure the hole does not become too deep.
- Install a 10 mm nut to the bolt, then the disc, then the other nut. Fasten firmly.
- Place the large disc and bolt down over the 3.0 mm 'spindle' and check that it rotates freely.
- Your spinning-top is now ready for use.

Using the spinning-top
- Place the top on one corner of a 24/30 cassette, and cover the rest with lead rubber.
- Set the X-ray tube at its normal operating height.
- Select a low to medium mA station, and low kV.
- Select the exposure time for testing. Take care to record this setting.
- Note. The top can normally record exposure times of 0.01 to 0.3 seconds. Longer times can lead to overlap of the dots.
- Give the top a firm spin, and place the generator into preparation.
- Watch the top rotation. As it slows down to a low speed, make your exposure.
- Place the top on the next position of the cassette, and shield the rest of the cassette with lead rubber.
- Select another test exposure time, and once more make an exposure.
- The top described in fig B–8, can perform a total of six tests on one 24/30 film
- Develop the film, and analyse the results.

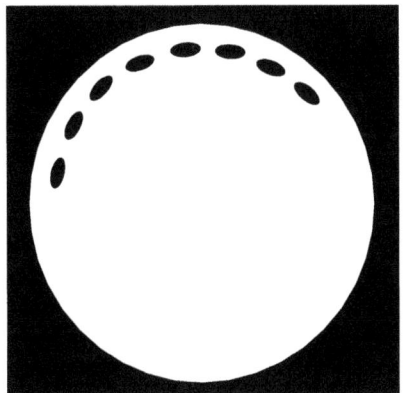

Fig B–9. A typical test result

Test results
- The number of dots obtained depends on the time selection, and the type of rectification.
- Fig B–9 shows 8 dots. This would indicate an exposure time of 0.08 sec for a 50 Hz full wave rectified generator.

Table B–1. Spinning-top dots and exposure times

Dots	60 Hz self rectified	50 Hz self rectified	60 Hz full wave rectified	50 Hz full wave rectified
1	0.008*	0.01*	0.008	0.01
2	0.033	0.04	0.017	0.02
3	0.05	0.06	0.025	0.03
4	0.067	0.08	0.033	0.04
5	0.083	0.1	0.042	0.05
6	0.1	0.12	0.05	0.06
7	0.12	0.14	0.58	0.07
8	0.133	0.16	0.067	0.08
9	0.15	0.18	0.075	0.09
10	0.167	0.2	0.083	0.1

* A single dot with self-rectified generators may be considered similar to the full wave rectified generator time.

- Table B–1 indicates the number of dots, and the equivalent exposure times. (The 60 Hz times have been 'rounded off')

Test result problems
- The dots overlap around the full circumference.
  i. The exposure was made with the top spinning too quickly.
- The dots merge together.
  i. The spin speed is too slow.
  ii. X-ray output is too high. Try a lower kV or mA setting.
- If the test is required at high mA or kV, then place a sheet of paper between the film and the intensifying screens in the cassette, or use non screened film.
- A 'half dot' indicates a problem with the exposure contactor.
  i. If this occurs at the start of the exposure, especially with exposure times of 0.03 or 0.05, this could be due to incorrect contactor phase adjustment, or slow pull-in of the contactor.
  ii. If it occurs at the end of the exposure, and is an extra half dot, this is due to the contactor sticking or slow release. Most X-ray controls have an adjustment for both the contactor pull-in and release times.

Tomography resolution test tool

This tool is designed for two requirements.

1. To test tomography performance. The centre test objects 'A' should show a clear image under all modes of tomography operation.
2. Test the tomography height calibration. When set at 60mm the centre test 'A' should be clear, with the lower 'B' and upper 'C' test objects equally blurred.

Material

You will need the following.

- Two pieces of 5mm thick acrylic plastic; cut to 100 by 90mm.
- Or, 6.0mm medium density fibreboard (MDF). The platform legs then become 90 by 54mm.
- Two pieces of 5mm acrylic; cut to 90 by 55mm.
- Seven small paper clips.
- A small coin.
- Epoxy resin.

Note: While it is important that the platform is supported parallel to the tabletop, the dimensions supplied are as a guide only. For example, depending on materials available, you may wish to construct this test to a different size, or to a different height above the tabletop.

Construction

- Refer to the drawing, Fig B–10.
- Use epoxy resin to attach the two 90 by 55mm legs to the 90 by 100mm platform. (This should result in a 60mm height of the platform top.)
- Prepare the paper clips by spreading the sides apart slightly, to leave a small air gap between the sides.
- On the top centre of the platform, glue three paper clips in the pattern shown as 'A'. Allow time for the epoxy to harden.
- Mix another small quantity of epoxy. Glue two paper clips underneath the platform to form the pattern 'B' Allow time for the epoxy to harden.
- Finally, with a fresh epoxy mix, attach the second platform over the top of the first platform (and the clips). Glue two paper clips and a small coin to the top of the second platform, to form the pattern 'C'
- There should now be three different pattern groups of paper clips, each separated in height by 5mm. (Or 6.0mm if using 6.0mm MDF)

Operation

- Place the tool in position on the tabletop.
- Set the fulcrum height to 6.0cm. Set the tomography angle to maximum.
- Ensure the X-ray tube is set to the correct FFD. (Normally 100cm)
- Manually check the tomograph operation, ensuring the required tube-stand locks are set either 'on' or 'off' as required.
- Select a low mA position, and a low kV setting.
- Make a test exposure.
  i. If the film is too dark, reduce the kV, or, if possible, select a lower mA station.
  ii. If the test results are still overexposed, then place a piece of paper between one side of the film, and the intensifying screens of the cassette.
  iii. If necessary, adjust the fulcrum height so the patterns 'B' and 'C' show equal blurring, and the centre pattern 'A' is clear.
- Repeat this test at different tomography operation speeds, and in all modes.

Evaluating the results

- Horizontal shaking or jerking will cause poor resolution of the paper clip perpendicular to the direction of travel.
- Lateral shaking will cause poor definition of the centre paper clip.
- If a particular operation mode provides a poor result, provide a notice warning against use of that mode until the problem is corrected. See **module 8.1 page 127**.
- If the fulcrum height requires recalibration, contact the service department for advice.
- Record the results in the logbook. Retain the films as part of the service test records.

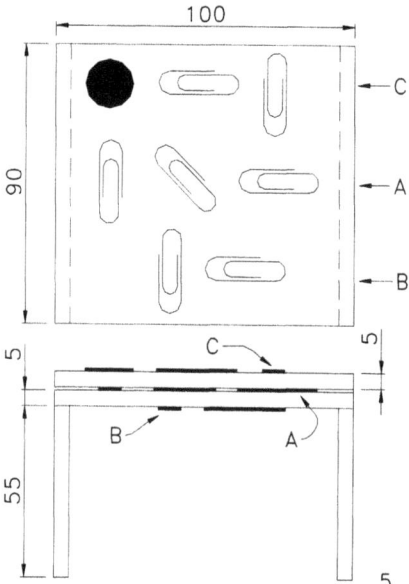

Fig B–10. Tomographic resolution test tool

APPENDIX C
Graphs, check sheets, and record sheets

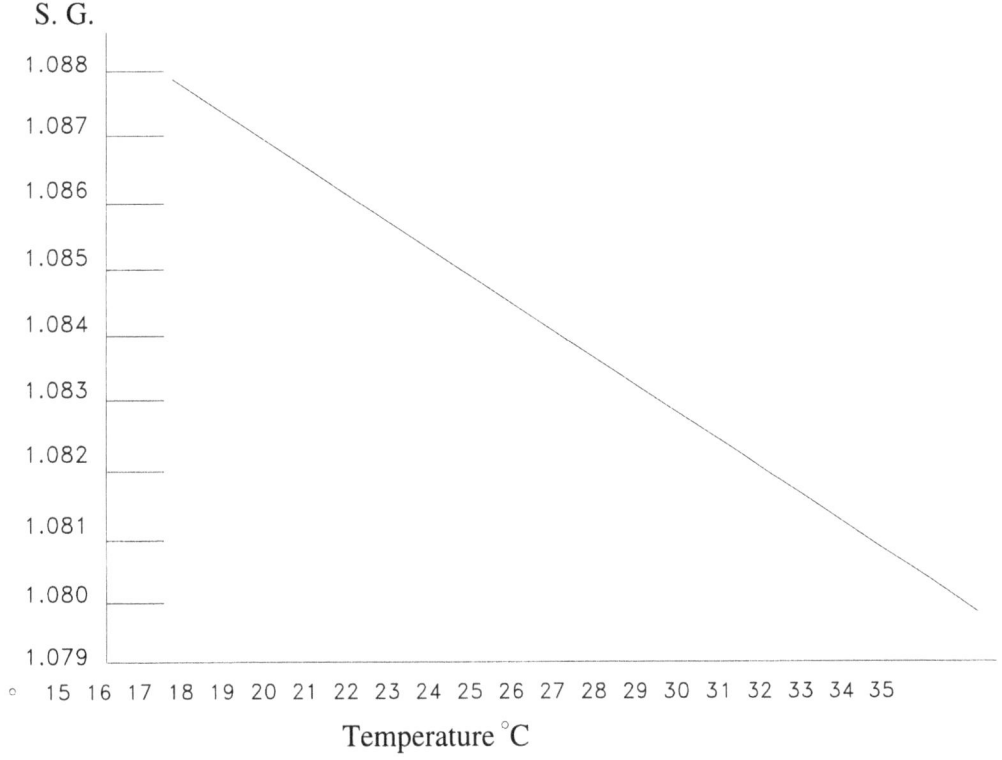

Fig C-1. Specific gravity / temperature graph

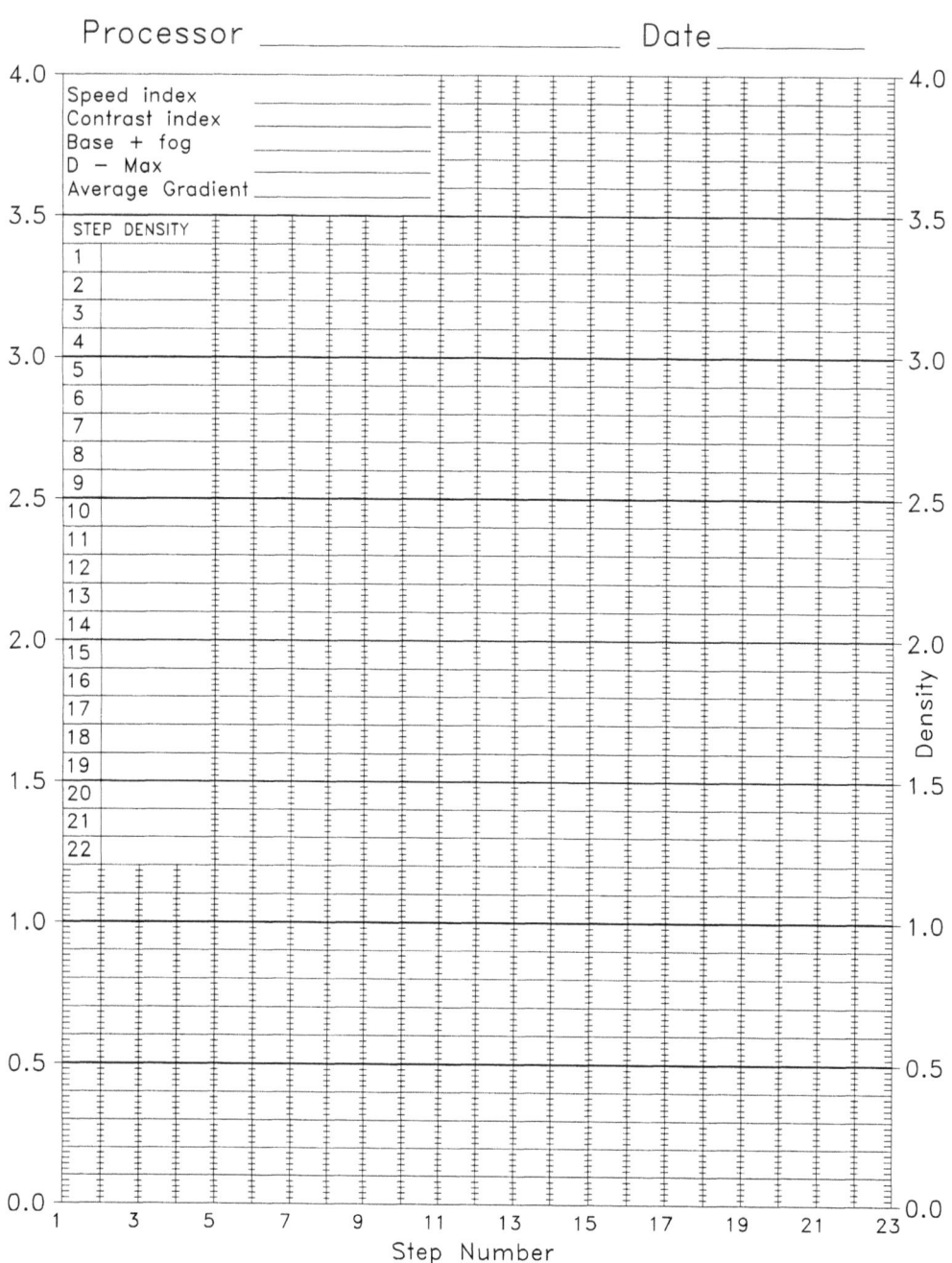

Fig C-2. Characteristic curve chart

APPENDIX C. GRAPHS, CHECK SHEETS, AND RECORD SHEETS

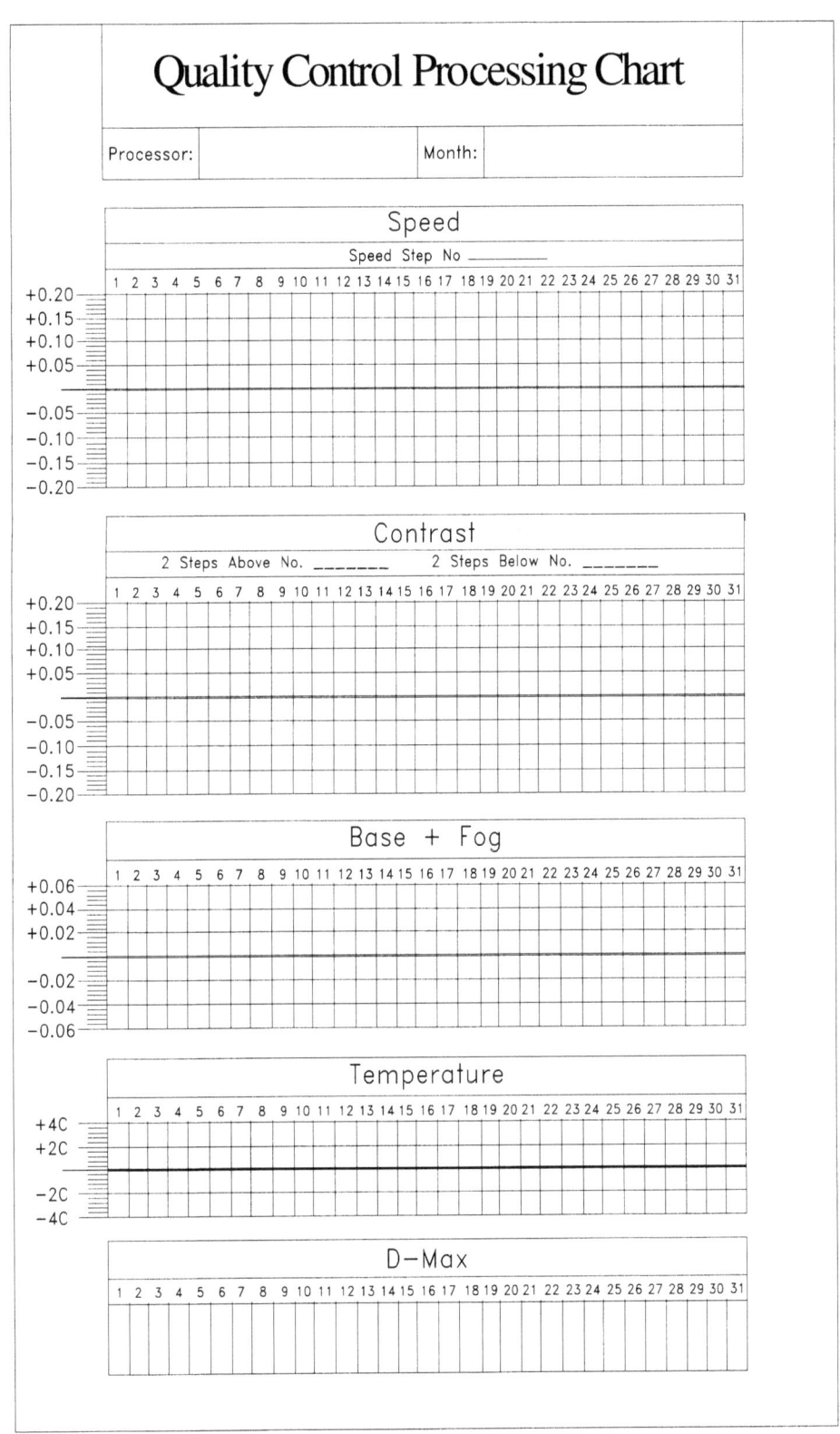

Fig C-3. Quality control processing chart

X-ray equipment records

Room function _____   Room No _____

X-ray generator

Manufacturer _____   Model No _____

Serial No _____   Supplier _____

Date installed _____   Warranty expires? _____

Maximum kVp _____   Maximum mA _____

Generator type: Single phase _____ Three phase _____ Mobile _____

Inverter/high frequency _____ Portable _____ AEC included? _____

Manuals supplied. (Include publication number)

Operation	Installation	Service	Circuits/connection diagrams	Parts list

X-ray tube

Manufacturer _____   Model No _____

Serial No _____   Supplier _____

Date installed _____   Warranty expires? _____

Maximum kVp _____   KW rating _____

Focus size: Broad focus _____   Fine focus _____

High or low speed operation? _____

Manuals supplied. (Include publication number)

Operation	Installation	Rating charts

APPENDIX C. GRAPHS, CHECK SHEETS, AND RECORD SHEETS

X-ray tube suspension

Manufacturer _____   Model No _____

Serial No _____   Supplier _____

Date installed _____   Warranty expires? _____

Mounting _____ Floor/Ceiling _____   Tomography attachment? _____

Manuals supplied. (Include publication number)

Operation	Installation	Service	Circuits/connection diagrams	Parts list

Collimator

Manufacturer _____   Model No _____

Serial No _____   Supplier _____

Date installed _____   Warranty expires? _____

Rotating/fixed _____   Light switch, Clockwork/Electronic _____

Globe type/part No _____   Spare globe supplied? _____

Manuals supplied. (Include publication number)

Operation	Installation	Parts list

Table

Manufacturer _____   Model No _____

Serial No _____   Supplier _____

Date installed _____   Warranty expires? _____

Table movements. Fixed _____   Longitudinal _____ Lateral _____

Elevating _____ Tilting _____   Integral with tube stand _____

Potter-Bucky, YES/NO _____   Oscillating or fixed grid _____

Grid ratio _____ Lines _____   Focal range _____

Manuals supplied. (Include publication number)

Operation	Installation	Service	Circuits/connection diagrams	Parts list

Upright Potter-Bucky

Manufacturer _____   Model No _____

Serial No _____   Supplier _____

Date installed _____   Warranty expires? _____

Grid ratio _____ Lines _____ Focal range _____ Fixed/oscillating _____

Manuals supplied. (Include publication number)

Operation	Installation	Circuits/connection diagrams	Parts list

Fluoroscopy table

Manufacturer _____   Model No _____

Serial No _____   Supplier _____

Date installed _____   Warranty expires? _____

Table angulation: _____ 90/15 _____ 90/30 _____ 90/60 _____ 90/90 _____

Fluorescent screen? _____ Image intensifier? _____ TV? _____

Potter-Bucky, YES/NO _____   Oscillating or fixed grid _____

Grid ratio _____ Lines _____   Focal range _____

Manuals supplied. (Include publication number)

Operation	Installation	Service	Circuits/connection diagrams	Parts list

Stationary grids

Manufacturer	Supplier	Model	Serial No	Grid ratio	Grid Lines	Date Supplied

Film processor

Manufacturer _____   Model No _____

Serial No _____   Supplier _____

Date installed _____   Warranty expires? _____

Processor type: Bench top? _____   Floor mounted? _____

Daylight loading? _____   Processing time _____

Manuals supplied. (Include publication number)

Operation	Installation	Service	Circuits/connection diagrams	Parts list

Request for service

Equipment requiring service. _____ Date _____

Make _____ Model _____ Serial No _____

Location. _____ Reported by _____

Service required: Routine maintenance ☐   Calibration ☐   Repair ☐

In case of an equipment fault or problem, provide a full description of the symptoms.

```
_____
_____
_____
_____
_____
_____
```

Describe any tests or adjustments carried out, and the results obtained.

```
_____
_____
_____
_____
_____
_____
```

Service request made by _____ Department _____
Phone No _____ Date _____ Request order No _____
Service Company _____ Representative _____
Address _____
_____

Phone No _____ Fax No _____
Service Company job No _____ Attendance date _____
Are any parts or additional service required?
_____
_____
_____ Completion date _____

Equipment maintenance and repair log

Room or equipment No _____

Date	Requirement	Action	Performed by	Reference No

APPENDIX D

Routine maintenance checklist
X-ray generator: Fixed installation

Room No _____ Inspection date _____ Performed by _____

Area		✓ = Pass X = Fail NA = Does not apply	Attention required
Power 'OFF'	Knobs and switches		
	Function labels		
	Meter 'Zero'		
	Cleanliness.		
Power 'ON'	Indicator lamps.		
	Digital readouts.		
	Mains voltage compensation		
X-ray tube overload protection	Maximum anode load	(Broad Focus)	
		(Fine Focus)	
	Maximum kV		
	Minimum kV		
MA Calibration	By mA meter		
	By mAs meter		
	By other technique.		
	Filament pre-heat test.		
Radiation consistency	With added preparation time.		
	With minimum preparation time.		
Radiation linearity	At 60~70 kV		
	At 90~100 kV		
Other areas for attention			
Problems not corrected since last routine maintenance check			
Faults or problems requiring further attention.			

APPENDIX D. ROUTINE MAINTENANCE CHECKLIST

X-ray generator: Mobile unit

Unit/Serial No _____ Inspection date _____ Performed by _____

Area		✓ = Pass ✗ = Fail NA = Does not apply	Attention required.
Power 'OFF'	Knobs and switches.		
	Function labels.		
	Meter 'Zero'		
	Cleanliness.		
	Loose panels or screws.		
	Vertical suspension wire rope.		
	Bearings.		
	Lubrication.		
	Manual locks.		
	Mobile brakes.		
	HT cables.		
	Electrical cables.		
	Plugs, sockets.		
	Battery water.		
Power 'ON'	Indicator lamps.		
	Digital readouts.		
	Mains voltage compensation.		
	Electromagnetic locks.		
	Battery charge indication.		
	Power assist drive		
	Anti-crash bumper.		
X-ray tube	Overload protection.		(Broad Focus)
			(Fine Focus)
	Oil leaks.		
	Bearing noise.		

Collimator	Collimator timer.		
	Collimator scale / knob alignment.		
	X-ray to light beam alignment.		
	As above, but +/− 90 deg' rotation.		
MA Calibration	By mA meter.		
	By mAs meter.		
	By other technique.		(Refer installation/service manual)
	Filament pre-heat test.		
Radiation consistency	With added preparation time.		
	With minimum preparation time.		
Other areas for attention			

Problems not corrected since last routine maintenance check.	
Faults or problems requiring further attention.	

X-ray generator: Capacitor discharge mobile

Unit/Serial No _____ Inspection date _____ Performed by _____

Area		✓ = Pass ✗ = Fail NA = Does not apply	Attention required.
Power 'OFF'	Knobs and switches		
	Function labels.		
	Meter 'Zero'.		
	Cleanliness.		
	Loose panels or screws.		
	Vertical suspension wire rope.		
	Bearings.		
	Lubrication.		
	Manual locks.		
	Mobile brakes.		
	HT cables and cable ends.		
	Electrical cables.		
	Plugs, sockets.		
	Battery water.		
Power 'ON'	Indicator lamps.		
	Digital readouts.		
	Mains voltage adjustment.		(Only some units)
	60 kV charge.		
	kV 'Top-up'.		
	90 kV charge.		
	Reset, 90 kV to 60 kV.		
	Electromagnetic locks.		
	Battery charge indication.		
	Power assist drive		
	'Anti-crash' bumper.		

X-ray tube	Oil leaks.		
	Bearing noise.		
Collimator	Collimator timer.		
	Collimator scale alignment.		
	X-ray to light beam alignment.		
	As above, but +/– 90 deg' rotation.		
	Collimator 'Dark Current' shutter.		
mAs Calibration	By kV fall.		
	By other technique.		(Refer installation/service manual)
Radiation consistency	Using step-wedge.		
	Using water phantom.		
Other areas for attention			
Problems not corrected since last routine maintenance check.			
Faults or problems requiring further attention.			

X-ray generator: Portable

Unit/Serial No _____ Inspection date _____ Performed by _____

Area		✓ = Pass ✗ = Fail NA = Does not apply	Attention required.
Power 'OFF'	Knobs and switches.		
	Function labels.		
	Meter 'Zero'		
	Cleanliness.		
	Loose panels or screws.		
	Height adjustment system.		
	Bearings.		
	Lubrication.		
	Manual locks.		
	Electrical cables.		
Power 'ON'	Indicator lamps.		
	Line voltage adj.		
X-ray tube head	Oil leaks.		
Collimator	Collimator timer.		
	Scale accuracy.		
	X-ray to light beam alignment.		
	As above, but +/− 90 deg' rotation.		
MA Calibration	By mA meter.		
	By other technique.		(Refer installation/service manual)
Radiation consistency	Using step-wedge.		
	Using water phantom.		
Other areas for attention			
Problems not corrected since last routine maintenance check.			
Faults or problems requiring further attention.			

X-ray tube stand

Room No _____ Inspection date _____ Performed by _____

Area		✓ = Pass ✗ = Fail NA = Does not apply	Attention required
Mechanical and electrical inspection. Floor ceiling tube stand version	Suspension wire rope.		
	Vertical movement bearings.		
	Vertical guide rails.		
	Clean floor track & check bearings.		
	Lubrication.		
	Ceiling rail bearings.		
Mechanical and electrical inspection. All versions of tube stands.	Manual locks.		
	Electromagnetic locks.		
	Lateral centre; table Bucky.		
	Lateral centre; wall Bucky.		
	Electrical cables.		
	HT cables.		
Command Arm	Trunnion ring rotation lock.		
	Indicator lamps and switches.		
	Bucky centre light.		
	Function labels.		
	Focal spot to Bucky distance.		
	Angulation indicator.		
	Cleanliness.		
Other areas for attention.			
Problems not corrected since last routine maintenance check.			
Faults or problems requiring further attention.			

APPENDIX D. ROUTINE MAINTENANCE CHECKLIST

X-ray tube

Room No _____ Inspection date _____ Performed by _____
Tube model No _____ Serial No _____ Tube position, UT _____ OT _____

Area		✓ = Pass ✗ = Fail NA = Does not apply	Attention required.
X-ray tube.	Rotation in trunnion rings.		
	Loose attachments.		
	Electrical cables.		
	HT cable earth shield at cable end.		
	Loose HT cable ends.		
	HT cables not stretched as tube rotated.		
	Oil leaks.		
	Bearing noise.		
X-ray tube 'ageing'	Carried out?		
	Any instability?		
Other areas for attention			
Problems not corrected since last routine maintenance check.			
Faults or problems requiring further attention.			

Over-table collimator

Room No _____ Inspection date _____ Performed by _____
Collimator globe type No _____

Area		✓ = Pass ✗ = Fail NA = Does not apply	Attention required
General maintenance	Electrical cables.		
	Lamp timer.		
	Light beam intensity.		
	Spare globe in stock?		
	Collimator blades do not slip when adj' knob released.		
Alignment tests	Crosshair centre.		
	Bucky centre light.		
	Collimator field size scale & knob alignment.		
	Alignment of light field to X-ray beam.		
	As above, collimator rotated 90 degrees.		
	As above, collimator rotated minus 90 degrees.		
Other areas for attention.			
Problems not corrected since last routine maintenance check.			
Faults or problems requiring further attention.			

APPENDIX D. ROUTINE MAINTENANCE CHECKLIST

The Bucky table and vertical Bucky

Room No _____ Inspection date _____ Performed by _____

Area		✓ = Pass ✗ = Fail NA = Does not apply	Attention required
Bucky Table	Tabletop movement.		
	Magnetic locks.		
	Switches and indicator lamps.		
	Lateral centre stop.		
	Profile rail screws.		
	Lubrication.		
	Clean & polish.		
Table Bucky	Bucky movement.		
	Bucky lock.		
	Electrical cables.		
	Cable support arm.		
	Lubricate track.		
	Lost markers?		
	Tray handle not loose.		
	Tray cleanliness.		
	Grid oscillation.		
Vertical Bucky	Lost markers? (Remove front cover to check).		
	Electrical cable.		
	Tray handle not loose.		
	Tray cleanliness.		
	Grid oscillation.		
	Vertical lock.		
	Clean & lubricate vertical track.		
	Rotation or tilt lock.		
Other areas for attention			
Problems not corrected since last routine maintenance check			
Faults or problems requiring further attention			

Tomography attachment

Room No _____ Inspection date _____ Performed by _____

Area		✓ = Pass ✗ = Fail NA = Does not apply	Attention required
Mechanical and electrical inspection	Loose or missing screws.		
	Connecting cables, plugs and sockets.		
	Tube-stand rotation lock 'off'.		
	Bucky lock 'off'.		
	Bucky movement.		
	Coupling arm bearings & pivot points.		
	Coupling arm clamp.		
	Fulcrum height adjustment.		
	Compression band		
	Cleanliness.		
Operation test. Tube stand travel	No slipping.		
	Stop position ok.		
	No jerking.		
Performance test	Layer height calibration.		
	Image sharpness.		
Other areas for attention			
Problems not corrected since last routine maintenance check			
Faults or problems requiring further attention			

APPENDIX D. ROUTINE MAINTENANCE CHECKLIST

Fluoroscopy table

Room No _____ Inspection date _____ Performed by _____

Area		✓ = Pass ✗ = Fail NA = Does not apply	Attention required
Mechanical and electrical inspection.	Loose panels or fittings.		
	Suspension cables or chains.		
	Electrical cables.		
	Spot filmer movement from 'Park' to 'Operate'		
	Image intensifier balance.		(Demountable units)
	Image intensifier mounting clamp.		(Demountable units)
	Spot filmer radiation shield.		
	Lost film markers?		
	Undertable Bucky.		(See 5.1 Bucky Table & Wall Bucky)
	Footrest.		
Operation test, table body.	Locks.		
	Switches and indicator lamps.		
	Function labels.		
	Tabletop movement.		
	Tabletop centre stop.		
	Table tilt vertical.		
	Table tilt Trendelenburg.		
	Electrical cables not pulled when table tilts or spot filmer moves.		
	Power assistance.		
	Anti-crash device.		
	Compression lock safety release.		
	Spot filmer vertical balance.		
	As above, table vertical		

Operation test, spot filmer.	Compression cone movement		
	Manual 'close to film' shutters. (Basic tables).		
	Manual cassette movement. (Basic tables).		
	Manual cassette film format. (Basic tables).		
	Motorized cassette.		
	'In' 'Out' movement.		
	Motorized cassette film format.		
X-ray beam alignment	Table horizontal. (Test in 'four spot' mode).		
	As above, table vertical.		
Other areas for attention			

Problems not corrected since last routine maintenance check.	
Faults or problems requiring further attention.	

APPENDIX D. ROUTINE MAINTENANCE CHECKLIST

Fluoroscopy TV systems

Room No _____ Inspection date _____ Performed by _____

Area		✓ = Pass ✗ = Fail NA = Does not apply	Attention required
Mechanical and electrical inspection, Image Intensifier	II balance when dismounted.		
	Vertical movement.		
	Ceiling suspension travel.		
	Electrical cables.		
	Knobs and switches		
	Function labels.		
Mechanical and electrical inspection, TV system	Electrical cables.		
	Coaxial cable and connectors.		
	Is 75 ohm switch in correct position?		
	Monitor fastened to monitor trolley.		
Image sharpness			Record the value of resolution obtained.
	For large field.		
	For medium field.		
	For small field.		
Automatic brightness control.	Auto kV or mA, with 3.0 cm water.		
	Auto kV or mA, with 18.0 cm water.		
	Camera control only with 5.0 cm water.		
	Camera control only. kV towards 100 kV with 5.0 cm water.		
Other areas for attention			
Problems not corrected since last routine maintenance check			
Faults or problems requiring further attention			

Automatic exposure control

Room No _____ Inspection date _____ Performed by _____

Area		✓ = Pass ✗ = Fail NA = Does not apply		Attention required	
Drift test, for long exposure times.	Centre chamber				
	Left chamber				
	Right chamber				
Table Bucky, film density.	Set ~90 kV and 18 cm water.		Record film Density.	Record AEC exposure time.	
	Centre chamber				
	Left chamber				
	Right chamber				
	Record exposure factors.				
	kV	mA	Backup time	Cassette used	
Table Bucky, kV tracking.	Set ~60 kV and 10 cm water.		Record film Density.	Record AEC exposure time.	
	Centre chamber				
	Left chamber				
	Right chamber				
	Record exposure factors used.				
	kV	mA	Backup time	Cassette used	
Table Bucky, short time compensation.	Low mA exposure				
	High mA exposure				
	Record exposure factors used.				
	kV	Low mA	High mA	AEC exposure time for high mA	

APPENDIX D. ROUTINE MAINTENANCE CHECKLIST

Wall Bucky, film density	Set ~90 kV and ~18 cm of water		Record film Density.	Record AEC exposure time.
	Centre chamber			
	Left chamber			
	Right chamber			
	Record exposure factors used.			
	kV	mA	Backup time	Cassette used	Water Phantom

Fluoroscopy Table. Spot Filmer AEC	Set ~90 kV and 18 cm water.		Record film Density.	Record AEC exposure time.
	Centre chamber			
	Dual chambers			
Other areas for attention				

Problems not corrected since last routine maintenance check	
Faults or problems requiring further attention.	

The X-ray room

Room No _____ Inspection date _____ Performed by _____

Area		✓ = Pass ✗ = Fail NA = Does not apply	Attention required
Accessories.	Foam wedges		
	Sandbags		
	Patient markers		
	Callipers		
Radiation protection.	Lead rubber gowns		
	Lead rubber aprons / patient protection		
	'Radiation' door warning light.		(Should operate on prep' as well as expose)
	Radiation leakage in operator area.		
The room.	Lighting.		
	Ventilation.		
	Air conditioning.		
	Sufficient shelves or storage area.		
	Patient cubicles.		
	Tidiness		
	Cleanliness		
	General condition of wall paint, floor covering etc.		(Is this a pleasant area for a patient?)
Other areas for attention			
Problems not corrected since last routine maintenance check			
Faults or problems requiring further attention			

APPENDIX D. ROUTINE MAINTENANCE CHECKLIST

Automatic film processor

Processor No _____ Inspection date _____ Performed by _____
Developer type _____ Developer temperature _____ Cycle time _____
Developer replenishment rate _____ Fixer replenishment rate _____
Water temperature _____

Area		✓ = Pass ✗ = Fail NA = Does not apply	Attention required
Chemistry	Developer temp'		
	Water temp'		
	Developer replenish' rate		
	Fixer replenishment rate.		
	Water flow		
Chemistry supply tank	Developer specific gravity		
	Fixer specific gravity		
	Chemistry appearance		
Cleaning	Feed tray		
	Cross over rollers		
	Deep racks		
	Foreign matter (algae)		
Filters replaced	Water (each month)		
	Developer (6~12 months)		
Rack sections, drive systems.	Looseness & geometry		
	Shaft retainers, gear wear.		
	Helical gear looseness		
	Drive motor chain tension		
	Abnormal noise		
Replenishment system	P/E film detection system		
	Replenisher solenoid valve function		
	Replenisher pump function		
	Circulation pump function		

Dryer	Temperature		
	System check		
Plumbing	Chemical supply hose leaks		
	Water supply		
General items	Microswitch (lid in place)		
	Panel fit, light leaks.		
	Audible warning signal		
Problems not corrected since last routine maintenance check			
Faults or problems requiring further attention			

APPENDIX E
X-ray equipment operation
Introduction to X-ray equipment operation

Aim

The aim is to provide an overall view of current X-ray equipment design and operation. This information is intended to enhance the maintenance and repairs sections of this workbook, by providing a detailed examination of equipment operation requirements. In addition, to provide some of the technical knowledge required by an electrician, or electronics technician, assisting in repairing the equipment.

Object

When carrying out routine maintenance, and in particular, diagnosing incorrect equipment operation, a good knowledge of how equipment operates is required.

The material in this appendix is intended both as a revision of equipment operation, and to provide specific information of equipment internal operation. This includes operational sequence of events, and the internal tests and checks carried out by the equipment. This is also an introduction to X-ray systems for an electrician or electronics technician, who may be asked to assist in the event of a problem. The first three parts have been provided as the background for this introduction.

Contents

Part 1. Production of X-rays	205
Part 2. The X-ray tube	208
Part 3. High voltage generation	213
Part 4. The X-ray generator control unit	218
Part 5. The high-tension cable	232
Part 6. X-ray collimator	233
Part 7. X-ray tube suspension	235
Part 8. The grid and Potter Bucky	236
Part 9. Tomography	239
Part 10. The fluoroscopy table	240
Part 11. The automatic film processor	243

PART 1 THE PRODUCTION OF X-RAYS

Contents

a. The X-ray tube
b. Bremsstrahlung radiation
c. Characteristic radiation
d. X-ray properties
e. Filters
f. Specification of minimum filtration
g. The inverse square law

a. The X-ray tube

The X-ray tube consists of an anode and cathode inside an evacuated glass envelope. The cathode is a filament, which when made very hot, emits electrons. When a high voltage supply is placed between the cathode and anode, the electrons from the cathode strike the anode, releasing X-rays. See Fig E-1. There are two main types of X-ray radiation generated: Bremsstrahlung (braking radiation) and characteristic radiation.

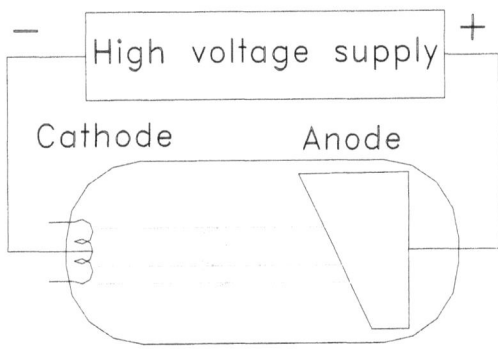

Fig E-1. The X-ray tube

b. Bremsstrahlung radiation

When an electron passes close to the nucleus of an anode atom, it is deflected, and its speed or energy reduced. At the same time, an X-ray photon is produced, which has an energy level equal to that lost by the electron. See Fig E-2. Peak X-ray energy, expressed in 'electron-volts' or 'keV', occurs only when an electron strikes the nucleus, giving up all its energy immediately. The electron will continue to pass through the anode atoms, and produce further X-ray photons. However, about 99.5% of the electron energy is lost in generating heat.

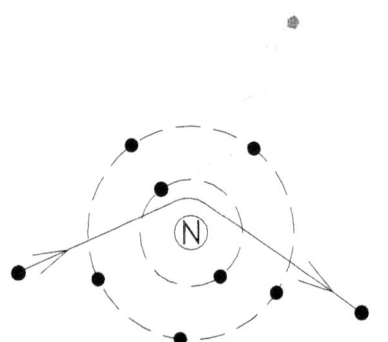

Fig E-2. Bremsstrahlung radiation

c. Characteristic radiation

This occurs when an incoming electron collides with an electron in the inner 'K' shell. To replace the missing electron, an electron moves from the 'L' shell to the K shell, giving up its energy as an X-ray photon. This has a predominant energy of 59 keV. See Fig E-3.

There are other transitions, notably from the 'M' shell to the 'K' shell (67.2 keV) and 'N' shell to the 'K' shell (69 keV). The above energy levels are specific for tungsten, and are known as 'Characteristic radiation'.

Note. To eject an electron from the K shell, the incoming electron requires energy greater than 70 kV, which is the binding energy of the K shell electron to the nucleus of a tungsten atom. Below 70 kV, radiation is entirely due to Bremsstrahlung. At 80 kV, characteristic radiation is about 10%, and at 150 kV is about 28% of the total usable X-ray beam.

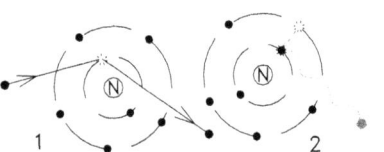

Fig E-3. Characteristic radiation

d. X-ray properties

X-ray beam quality and quantity depends on three main factors.

—The kV applied between anode and cathode
—Filtration to remove low energy X-rays.
—The amount of electron emission from the cathode, which affects quantity only.
—The film focus distance (FFD). Radiation is reduced by the inverse square law.

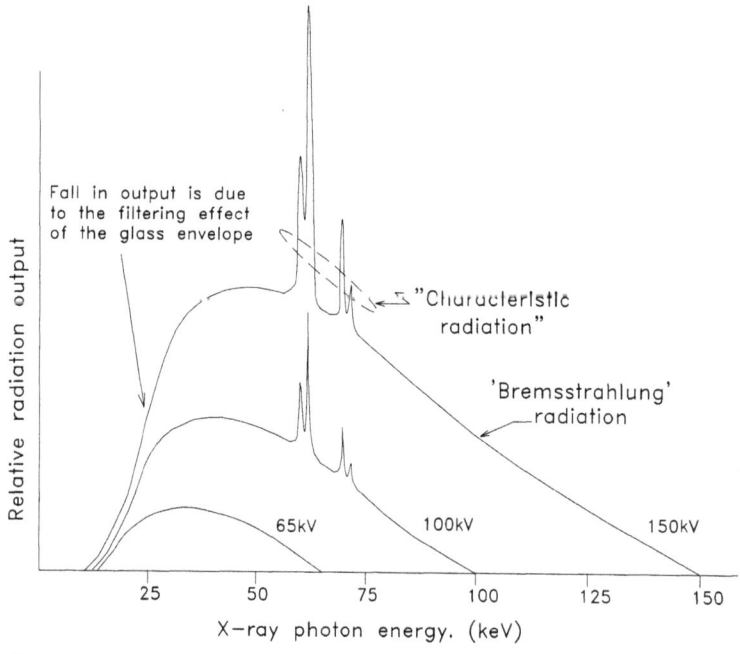

Fig E-4. Illustration of relative kV output, for three values of kV

e. Filters

X-ray photons below ~40 keV have little penetrating power in standard diagnostic X-ray procedures, and only contribute to unwanted radiation of the patient. To remove these lower energy X-rays, added filters are placed in the X-ray beam. The filter material is normally made of pure aluminium. For special applications filters made of different materials may be used. These are called 'K Edge' filters. An example of this is an X-ray tube used in mammography, which may have a molybdenum filter.

Where it is desired to make most use of low keV radiation, some collimators have a removable filter. This has a safety switch, so that if the filter is removed, X-ray generation is not permitted above a specified kV level.

f. Specification of minimum filtration

Most countries specify a minimum filtration that will be used for diagnostic X-ray. The total filtration is the combination of the X-ray tube glass, the mirror in a collimator, plus the added filter in the X-ray beam. To ensure the minimum required filtration is obtained, tables are provided for measurement purposes.

Typical half value layers are provided in table E–1. The actual specification may differ in some countries.

How to measure the half-value layer

- At a specified kV, a radiation meter measures the radiation from the X-ray tube. Added aluminium filtration is placed in the beam. The amount of aluminium to reduce the beam by 50% is called the half-value layer.
- Referring to table E–1, at 100 kV, this should require at least 2.7 mm of aluminium.
- If the specified value of aluminium reduces radiation by more than 50%, total filtration is insufficient, so more permanent aluminium must be placed in the X-ray beam.

Table E–1. Minimum half value layer, at different kV levels

X-ray tube voltage (kV)	Minimum permissible first HALF-VALUE LAYER (mm Al)
50	1.5
60	1.8
70	2.1
80	2.3
90	2.5
100	2.7
110	3.0
120	3.2
130	3.5
140	3.8
150	4.1

g. The inverse square law

The quantity of X-rays available for a given area depends on the distance from the X-ray tube. For a given distance, the X-ray beam may cover an area of 10 × 10 cm. If we double the distance the same beam will now cover an area of 20 × 20 cm, in other words, four times the previous area. However, the radiation available for each 10 × 10 cm section is now only one quarter its previous value. See Fig E–5.

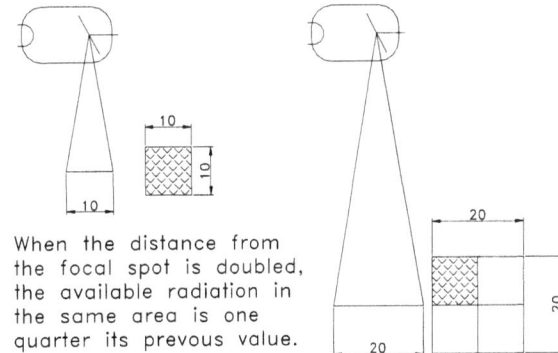

When the distance from the focal spot is doubled, the available radiation in the same area is one quarter its prevous value.

Fig E–5. Illustration of the inverse square law

PART 2 THE X-RAY TUBE

Contents

a. The stationary anode X-ray tube
b. The rotating anode X-ray tube
c. The X-ray tube housing
d. The X-ray tube focal spot
e. Anode angle
f. Maximum anode heat input
g. Anode rotation speeds
h. Effect of rotation speed on output
i. Anode heat and cooling time
j. The X-ray tube filament
k. Filament focus
l. Grid controlled X-ray tube

a. The stationary anode X-ray tube

This is usually found in portable X-ray generators, or in dental units. The anode is a small insert of tungsten, inside a large copper support. The copper is to help adsorb the heat produced. As a general rule focal spots are larger than for the rotating anode type, as the heat produced is in a very small area.

b. The rotating anode X-ray tube

By rotating the anode, the heat produced is spread around a wide area. This allows time for heat to be absorbed into the body of the anode. As a result, much smaller focal spots may be used, together with an increase in output.

Rotation is achieved by attaching a copper cylinder to the anode. This forms the 'rotor' of an induction motor. Special ball bearings are required, designed to withstand the heat from the anode. A stator winding is placed over the anode end of the X-ray tube, to form the energising section of the motor. See Fig E–7.

c. The X-ray tube housing

The housing is lead lined, so that radiation only exits via the port in front of the focal spot. This port is usually a truncated plastic cone, extending from the surface of the housing close to the X-ray tube glass. This reduces the absorption of X-rays due to the oil. Oil provides the required high voltage insulation, and serves to conduct the heat from the anode and stator winding to the outside surface. A bellows is provided to allow the oil to expand as it becomes hot. A thermal safety switch is fitted to ensure protection against excessive housing heat. In some cases, this may be a micro switch, operated when the bellows expands beyond its operating limit. See Fig E–9.

d. The X-ray tube focal spot

By focussing a vertical beam of electrons, onto the anode, which has a specific angle, an effective small area of X-rays results. This is known as the 'focal spot', and the method of generation as the 'line focus principle'.

As indicated in Fig E–10, this effective focal spot becomes enlarged as the useful beam is projected towards the cathode end of the X-ray tube. While the spot will become smaller towards the anode side, a point is reached where X-ray generation rapidly becomes less. This is known as the 'heel effect'. See Fig E–11.

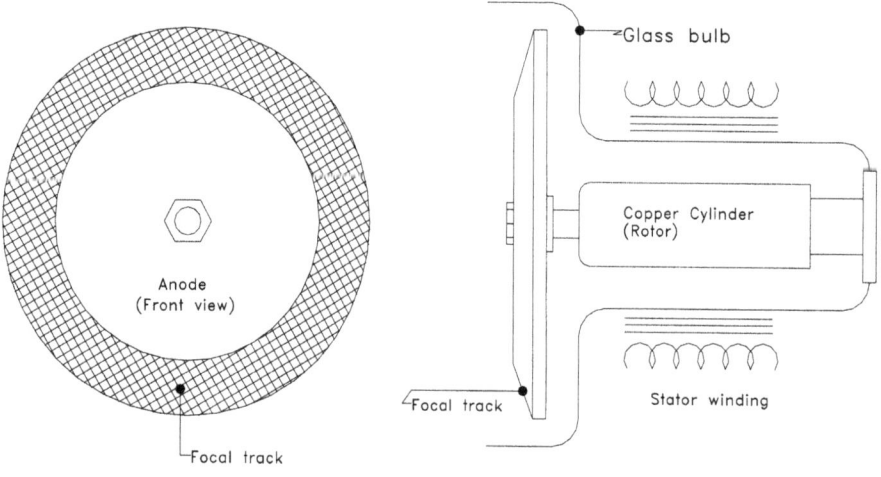

Rotating anode X-ray tube

Fig E–7. Anode and motor for a rotating anode X-ray tube

APPENDIX E. X-RAY EQUIPMENT OPERATION

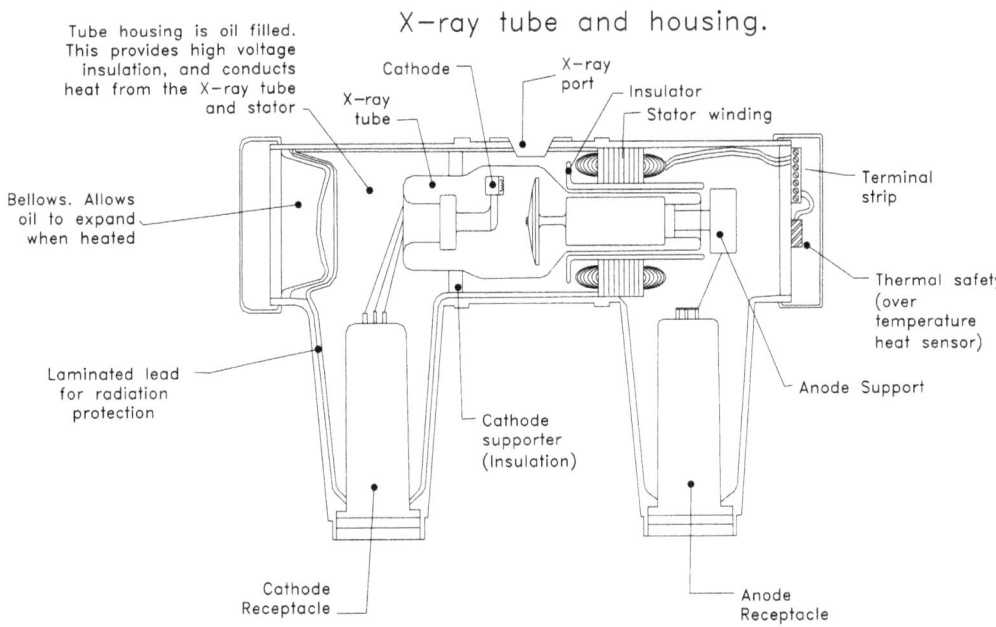

Fig E–9. The X-ray tube and housing

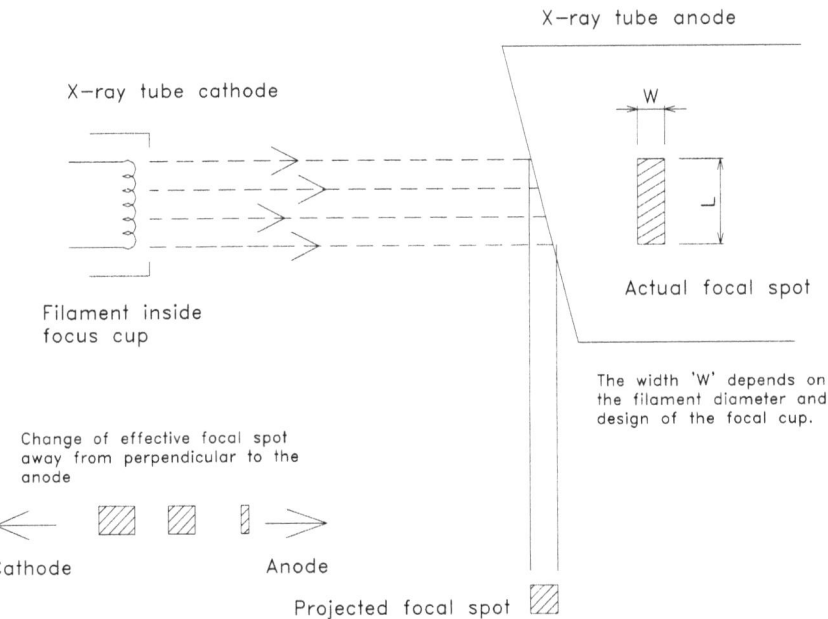

Fig E–10. Formation of the focal spot

e. Anode angle

The wider the anode angle, the greater will be the film coverage at a specific distance. However, to maintain the same focal spot size, the length 'L' of the electron beam must be reduced. This results in a smaller area to dissipate the immediate heat, so the maximum output of the tube has to be reduced. See Fig E–10.

A common angle for an over-table tube is 128. An under-table tube in a fluoroscopy table may have an angle of 168. With a 128 angle, radiation may cover a 35 × 35 cm film at a FFD of 100 cm, while a 168 angle

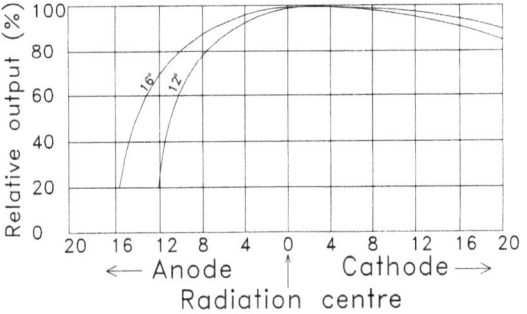

Fig E–11. Relative radiation output for two anode angles

would cover the same film at a distance of 65 cm. Fig E–11 indicates the relative radiation output for two common anode angles. The rapid fall off to the anode side is due to heel affect.

f. Maximum anode heat input

The maximum heat input for the X-ray tube anode is determined by:

- The anode material.
- Anode rotation speed.
- Anode diameter.
- Focal spot size.
- The kV waveform. (Single-phase, or three-phase)

An X-ray tube anode load capacity is rated as the number of kilowatts for an exposure time of 0.1 second. This is calculated from the rating chart for a specified mode of operation. For example, In Fig E–11a, the product of mA and kV at 0.1 second is 38 kW.

g. Anode rotation speeds

There are two anode rotation speeds in use, low speed and high speed. These depend on the power main supply frequency. High speed was originally obtained from static frequency-triplers, which generate the third harmonic of the mains frequency. Later high-speed systems use solid-state inverters, so high speed is now usually at the higher 10800 frequency, even with a 50 Hz supply.

With the simple form of induction motor used to rotate the anode, there will always be some slip, so the anode does not reach the full possible speed. The nominal speed that may be reached is indicated in brackets in table E–2.

Table E–2. Common anode speeds. The speed shown in brackets is the actual obtained speed, versus the theoretical maximum speed

Frequency	Low speed	(Low speed)	High speed	(High speed)
50 Hz	3000	(~2850)	9000	(~8700)
60 Hz	3600	(~3450)	10800	(~10500)

h. Effect of rotation speed on output

High-speed operation is of maximum benefit for short exposure times. (The generator should also sufficient output, to take advantage of high-speed anode rotation.) In Fig E–12b two load lines are indicated, one for high-speed, and one for low-speed operation. While this example is for 100 Kv operation, a similar result is obtained for other load factors.

i. Anode heat and cooling time

A stationary anode X-ray tube can have the copper section of the anode extended outside the glass container, and into the oil. This allows direct conduction of anode heat. This is not possible for a rotating anode, and heat is dissipated by direct radiation from the anode disk. Depending on anode diameter and thickness, this can take a long time time.

A typical cooling chart is provided in Fig E–13, and the formulas for calculation of the heat unit provided in table E–3.

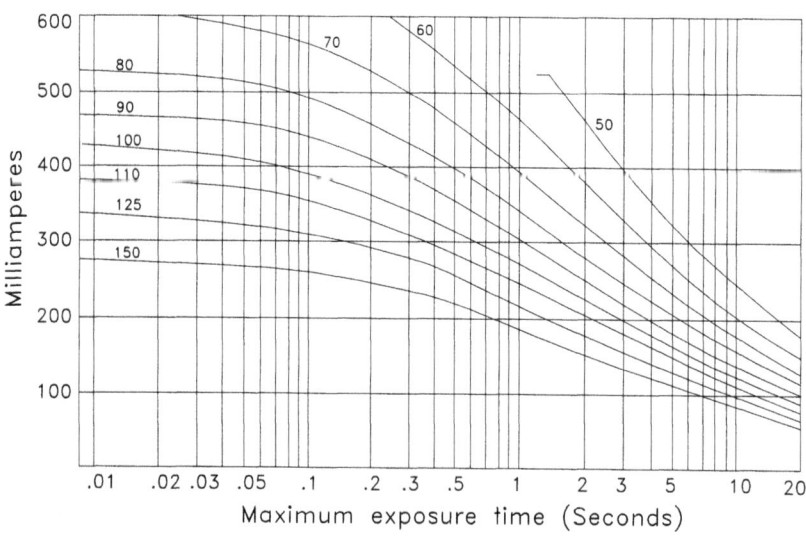

Fig E–12a. A typical anode-rating chart

APPENDIX E. X-RAY EQUIPMENT OPERATION

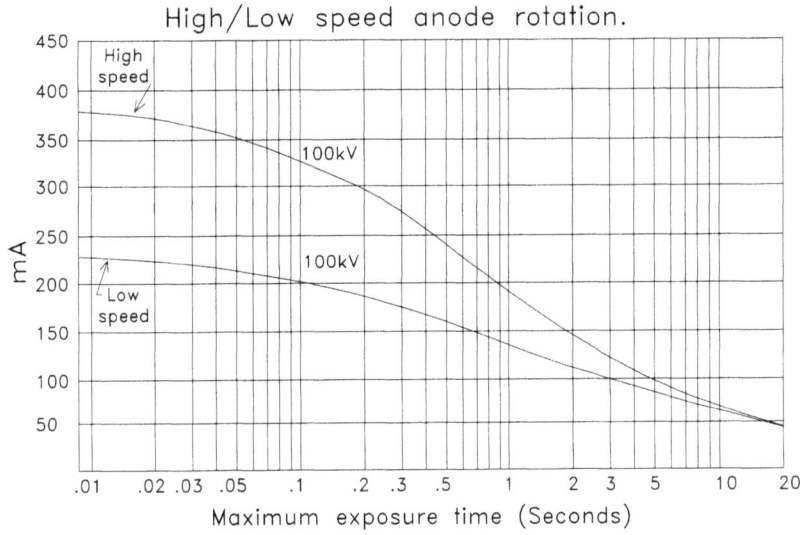

Fig E–12b. High-speed operation allows an increased anode load

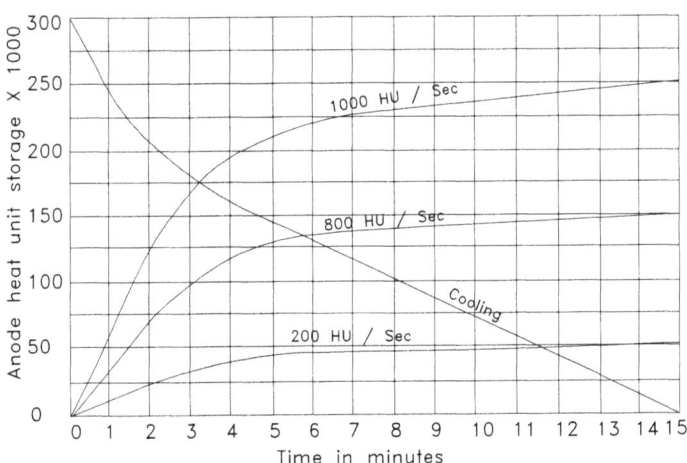

Fig E–13. A typical chart to indicate the rise in anode heat versus the cooling time

Table E–3. Formulas used for anode heat-unit calculation

kV waveform	Per exposure	Continuous
Single phase, full wave operation.	HU = kV × mA × s	HU/s = kV × mA
Three phase, full wave operation.	HU = kV × mA × s × 1.35	HU/s = kV × mA × 1.35
Medium or high frequency inverter.	HU = kV × mA × s × 1.35	HU/s = kV × mA × 1.35

j. The X-ray tube filament

To emit electrons, the filament must be brought to a white heat temperature. As the temperature increases, a point is reached where, despite further increases in temperature, only a small increase in emission results. In this area tungsten evaporation also increases, greatly reducing the filament life. This determines the maximum usable emission from the filament. Fig E–14 indicates the non-linear characteristic of the filament.

When the kV is increased, electron emission from the filament to the anode also increases. This is commonly known as the 'Space charge' effect. As an example, Fig–14 shows the change of mA that can take place as kV is increased. In this example, with a

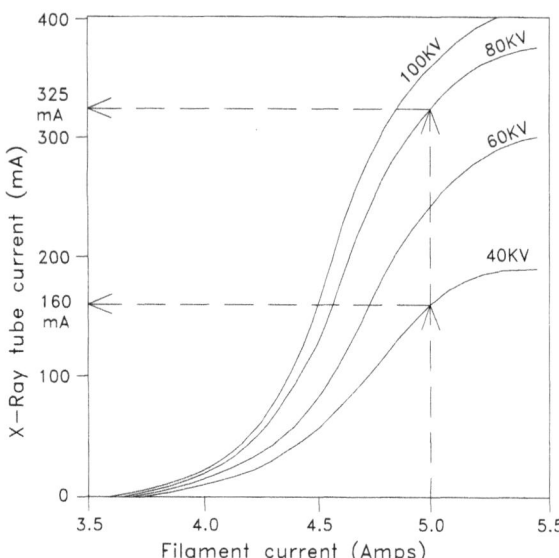

Fig E-14. A typical filament emission chart

filament current of 5.0A, at 40kV the emission is 160mA, and increases to 325mA at 80kV. To keep mA constant, as kV is changed, the generator control must change the filament current. This is called 'space charge compensation'.

k. Filament focus

To enable a tight beam of electrons to the anode, the filament is placed inside a 'focus cup'. The focus cup is connected directly to the common centre point of the cathode.

Normally the two filaments are placed side by side, and angled, so to strike the same anode position. Some designs instead have the filaments placed end. This allows formation of two separate tracks on the anode. These tracks can have separate angles to suit the required application. There is, however, a problem with two separate tracks, as exact alignment of the collimator to both tracks is not possible.

l. Grid controlled X-ray tube

In this design, the focus cup is brought out to a separate connection. By applying a strong negative voltage between the focus cup and the filament, electron emission is suppressed.

With this change of connection, the cathode cup is now referred to as a 'grid'. Grid control allows control of the X-ray exposure, while high voltage is continuously applied between anode and cathode. In operation, the grid is kept negative with respect to the filament, until an exposure is required. During an exposure, the negative voltage is removed, permitting emission from the filament. To terminate the exposure the grid is again made negative in respect to the filament.

Grid control may be used where rapid precise exposures are required, such as in special procedure rooms. However the most common use of grid control is in capacitor discharge mobiles.

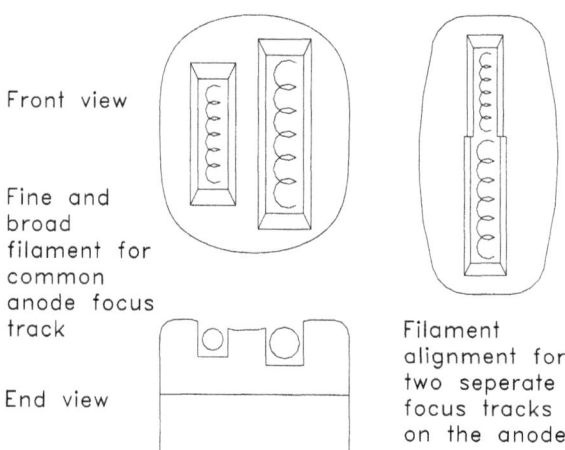

Fig E-15. Two versions of filament design for the cathode

APPENDIX E. X-RAY EQUIPMENT OPERATION

PART 3 HIGH VOLTAGE GENERATION

Contents

a. Single-phase, self rectified
b. Single-phase, full-wave rectified
c. Three-phase generators
d. Three-phase 'Six Pulse' generator
e. Three-phase 'Twelve Pulse' generator
f. The 'Constant potential' generator
g. High-frequency generators
h. The capacitor discharge (CD) mobile

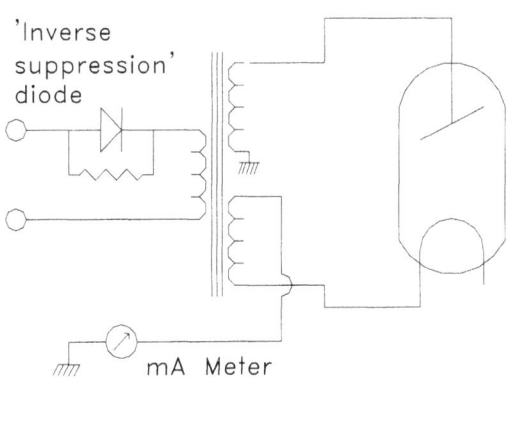

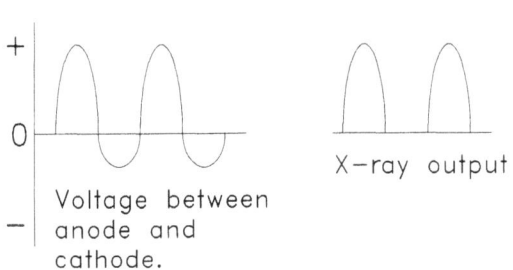

Fig E–17. Single-phase self-rectified generator

a. Single-phase, self rectified

The X-ray tube can also be considered as a rectifier, in that electrons emitted from the cathode filament travel to the positive anode. If the anode is negative in respect to the cathode, no electron flow occurs.

However, in case the anode is very hot, electron emission can also occur from the anode, in which case electron flow can exist from the anode to the cathode. This is called 'back-fire', and would damage the filament. To prevent this, an external diode and resistor is fitted to the primary of the HT transformer. The effect is to greatly reduce the available high voltage on the negative half cycle. This is called 'inverse suppression'.

The high-tension winding is 'centre tapped', so that both anode and cathode have equal voltage applied above ground potential.

Single phase self rectified systems are normally found in small portable X-ray generators, or may be used in dental units. Efficiency is low, and long exposure times will be required.

b. Single-phase, full-wave rectified

Full wave rectification results in both half cycles of the ac voltage used for X-ray production. There is no danger of back-fire, as no negative voltage is applied to the anode. Much higher output is now available. Full wave rectification is used on systems ranging from portable, dental, mobile, and up to heavy duty fixed installations. While self rectified generators may have a maximum output of 10~15mA, full wave rectified units have been produced with up to 800 mA output.

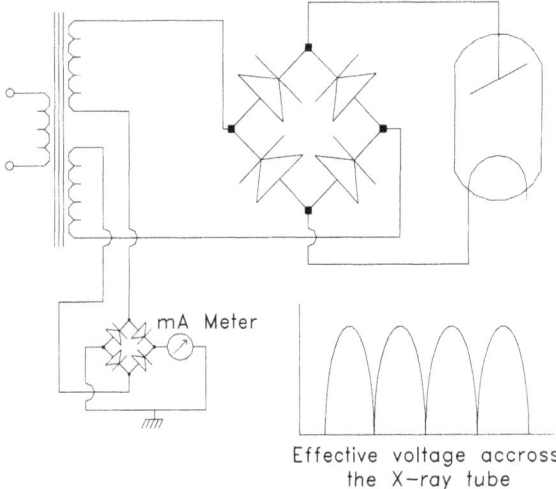

Fig E–18. Single-phase full-wave generator

The high-tension winding is centre tapped, with the centre position connected to ground. This ensures anode and cathode voltages are equally balanced above ground potential.

As the current in the transformer winding is AC, an additional rectifier is required for the mA meter (normally mounted on the control front panel). Exposure times are in multiples of the power main supply frequency. For a 50 Hz supply, exposure time calculation is simple. See table E–4.

With a 60 Hz supply, each pulse is 8.3 milliseconds wide. So some generators may indicate exposure times below 0.1 second as a number of pulses, rather than a set time.

Table E–4. Indication of exposure time for a single-phase, 50 Hz generator

50 Hz supply	10 milliseconds for each 'pulse'	0.01 second exposure = 1 pulse.	0.05 second exposure = 5 pulses	0.2 second exposure = 20 pulses

c. Three-phase generators

By operating with three-phase power supply, several advantages occur:

- The peak power demand per phase is reduced, with the input power equally shared between all three phases.
- Rather than pulsed high voltage, the X-ray tube now has continuous voltage supplied, so radiation for a given kV and mA is considerably greater. This results in shorter exposure times for a given setting, while the radiation absorbed by the patient is also reduced.
- Shorter exposure times, down to 0.003 seconds, are available. Exposure time calculation for 60 HZ is more accurate.
- The X-ray tube has higher anode load capacity for short exposure times, although for long exposure times this will be less.
- Three phase generators have typical outputs of 500 mA up to ~1200 mA.

d. Three-phase 'Six Pulse' generator

This system uses an identical style of winding for both the anode and cathode side. The windings may be configured 'star' or 'delta'. The system obtains its name due to the six joined together pulses that are generated each cycle. The 'ripple factor' for six-pulse is ~13%.

In the example shown below, the secondary windings are both delta configuration. The two isolated sets of windings and rectifier systems allow independent voltage supply to both anode and cathode. By connecting the common centre point to ground, both anode and cathode are equally balanced above ground.

See Fig E–19.

e. Three-phase 'Twelve Pulse' generator

With the twelve-pulse generator, one winding is configured delta, and the other star. The voltage peaks between these two windings have a 30 degree phase-shift, so that a peak of the rectified output from the delta winding will coincide with a trough from the star rectified output. This result in twelve joined together pulses for each cycle.

The overall ripple-factor is considerably improved, to a possible 3.5%. This improved ripple factor allows higher effective radiation output for a set kV, compared to six-pulse generators. With special exposure contactor systems, exposure times down to 0.001 seconds have been achieved. Conventional exposure contactors however, have the same exposure time limitations of the six-pulse systems.

See Fig E–20.

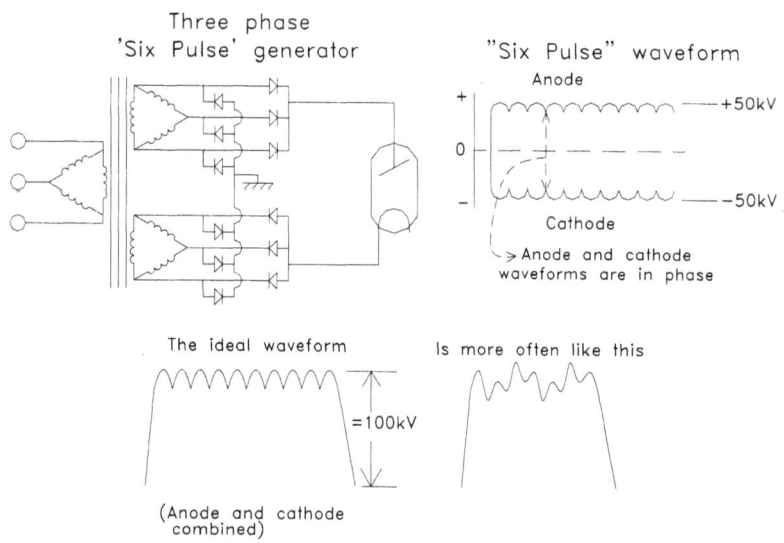

Fig E–19. Three-phase, six-pulse generator

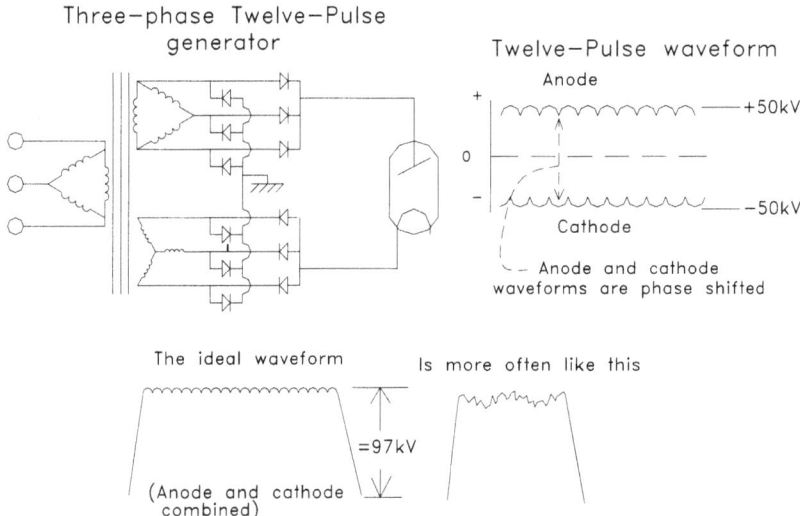

Fig E–20. Three-phase twelve-pulse generator

f. The 'Constant potential' generator

With this generator, there is NO ripple factor, and the voltage applied to the X-ray tube is pure DC. To achieve this, the output of a conventional six-pulse generator is smoothed by high voltage capacitors. The high voltage is then passed through a pair of high voltage tetrode valves. These serve to control the exposure, and regulate the actual high voltage supplied to the X-ray tube.

To achieve good regulation, the high voltage obtained from the generator is set about 50 kV higher than actually required. During the exposure, the tetrodes control the voltage at the required level to the X-ray tube. Constant-voltage generators were used for special procedure rooms, and CT scanners. The construction and maintenance of these systems is expensive. They have been largely replaced by high-frequency inverter systems. However, they are still in use for providing a very accurate X-ray calibration standard.

g. High-frequency generators

These are sometimes known as 'medium frequency' generators, depending on the maximum frequency of the inverter.

Generally, if maximum frequency is below ~20 kHz, the generator is called 'medium frequency'. Current high-frequency generators can operate up to 100 kHz, although most systems will operate below 50 kHz.

Inside the high-frequency generator, the AC mains power is rectified, and smoothed by a large value capacitor, to become a DC voltage supply. The 'inverter' converts the DC voltage back into a high-frequency AC voltage. This in turn is fed into the primary winding, of the high-tension transformer.

High-frequency generators have many advantages over conventional generators, operating at 50 or 60 Hz power main frequency.

- The high-tension transformer now uses ferrite instead of an iron core, with an increase in efficiency.
- The required inductance of the transformer winding is reduced, resulting in a big drop of copper resistive loss, again improving efficiency.
- Transformer manufacturing costs are reduced.
- High-voltage output is tightly regulated, so normal changes in power main voltages have no affect on the exposure.
- The high-voltage waveform is similar to between an ideal six-pulse to twelve-pulse generator for a medium-frequency system. A high-frequency generator waveform has less ripple, in many cases less than 2%. However, final ripple depends on other design considerations.
- High-voltage production is highly consistent, with little variation in residual kV ripple. (Unlike three phase systems, this can suffer distortion of the kV waveform.)
- Used in a mobile system, the inverter may operate directly from storage batteries, or else from large capacitors charged via the power point. In both these cases, kV waveform remains similar to large fixed installations.
- While earlier medium frequency systems had high development costs, present high-frequency systems are more cost effective than conventional generators.

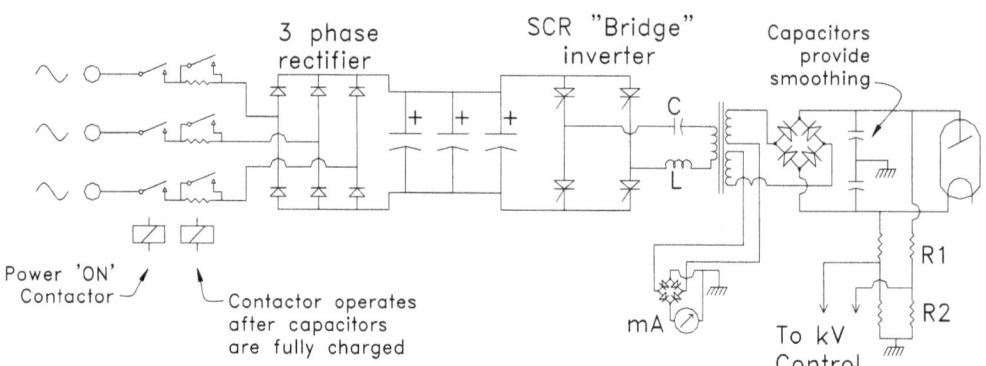

Fig E–21a. Diagram to illustrate the principle of a high-frequency generator

- On initial power up, a resistor limits the charging current of the capacitors. This is necessary, as otherwise with the capacitors discharged; it would be equivalent to placing a short circuit on the output of the rectifiers.
- After the capacitors are charged, another contactor shorts out the resistors. The system is now ready for operation.
- The energy stored in the capacitors supplies the high peak current required by the inverter.
- The inverter illustrated is an SCR 'bridge' inverter. The output of this inverter is coupled via a resonant circuit to the primary of the HT transformer,
- The capacitor 'C', and the inductance 'L', together with the inductance of the transformer winding form a series resonant tuned circuit. The resonant circuit has two functions.
  —As the pulse rate of the inverter increases towards resonance, the energy each pulse produces in the HT transformer secondary also increases. This allows a very wide range of control.
  —The resonant circuit has a 'flywheel affect', so that on the reverse half cycle, the back EMF attempts to reverse the current in the pair of SCRs that produced the initial pulse. This causes that pair to switch off. (The other pair will produce the next pulse, but this time in the opposite direction)
- The high-tension transformer is operated similar to a single-phase generator, with two exceptions.
  —For medium-frequency generators, added capacitors to provide waveform smoothing. For many high-frequency generators however, the inherent capacitance of the HT cables provides the required smoothing, without added capacitors.
  —A built in resistive voltage divider provides measurement of the high voltage during the exposure. This measurement is compared to a reference voltage equivalent to that for the required kV. If there is any difference, the inverter control circuit changes the pulse rate to correct the error. This is called 'closed loop' or 'feedback' regulation.

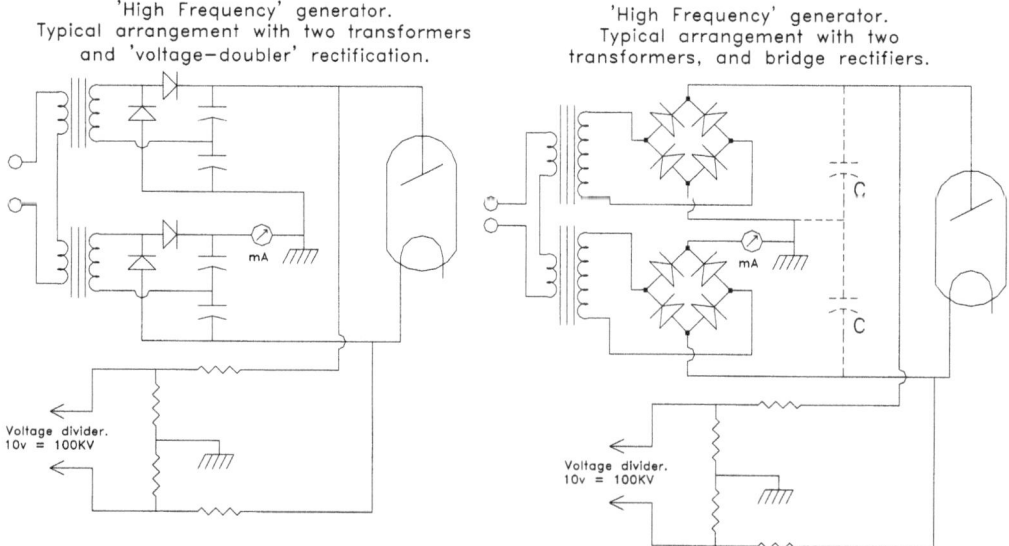

Fig E–21b. Two versions of high voltage generation, used with a high-frequency system

h. The capacitor discharge (CD) mobile

The capacitor-discharge or 'CD' generator obtains high voltage for an exposure directly from a pair of capacitors. These are charged to the required kV before making an exposure. As the kV for an exposure is applied to the X-ray tube prior to an exposure, a 'grid controlled' X-ray tube is fitted. A negative voltage applied between the 'grid', or focus-cup, and the filament. This prevents an exposure until the negative voltage is removed.

Although there is a slow capacitor charging time, the capacitor can rapidly discharge through the X-ray tube, with peak mA currents up to 500 mA. Actual peak mA depends on the X-ray tube used, not the capacitor system. During an exposure, the charge on the capacitor drops by 1 kV per mAs.

Operation

- The high voltage capacitors are charged prior to preparation for an exposure. This may take up to a minute depending on the kV setting required. A resistor in series with the transformer primary limits the charging current, allowing operation from a standard power point.
- The CD mobile has two capacitors, connected in series, with the common point connected to ground. This ensures the high-voltage to anode and cathode of the X-ray tube is equally balanced above ground potential. The capacitors are usually each of two microfarads capacity, and as they are connected in series, make up a total value of one microfarad.
- The transformer secondary and rectifiers are connected to the capacitors to form a 'voltage doubler'. On the positive half cycle D1 conducts, charging C1. On the negative half cycle, conduction is via D2, charging C2. The charge on C1 and C2 add together to produce the total kV available for an exposure. **See Fig E–22.**
- The resistors R1 and R2 provide a voltage measurement for the charging circuit, and for the kV meter.
- At the start of an exposure, the mAs timer operates a high voltage relay. This removes the negative voltage applied between grid and cathode. At end of the exposure, the relay stops operating, and the negative voltage is once more applied to the grid.
- Once the capacitors are charged to the required kV, the charge will slowly drop, partly due to the conduction of the kV measurement resistors, and partly due to a small 'dark current' current of the X-ray tube. (This term is used to describe conduction with a cold cathode filament.) As a result, when the kV drops a small amount, the charging circuit will again operate, 'topping up' the charge on the capacitors. Topping up is disabled during an exposure, and recharging occurs only after the charge button is again pressed.
- Due to dark-current, a very small emission of X-rays will be produced once the capacitors are charged. To prevent external radiation, the collimator is fitted with a motor or solenoid operated lead shutter. This shutter blocks all radiation, and is only opened just prior to a radiographic exposure, or at start of preparation for an exposure. Sometimes after charging the capacitors, a reset to a lower kV may be required. This is performed by a low mA exposure. During this time, the collimator lead shutter remains closed.
- The CD mobile on preparation will operate anode rotation and filament boost as for a standard generator. The filament however does not have pre-heating, as this would increase leakage current through the X-ray tube during standby.
- Control by time and mA selection is not practical, as the starting mA depends on kV selected, and falls during the exposure as kV drops. For this reason direct measurement of mA to operate a mAs timer is required.

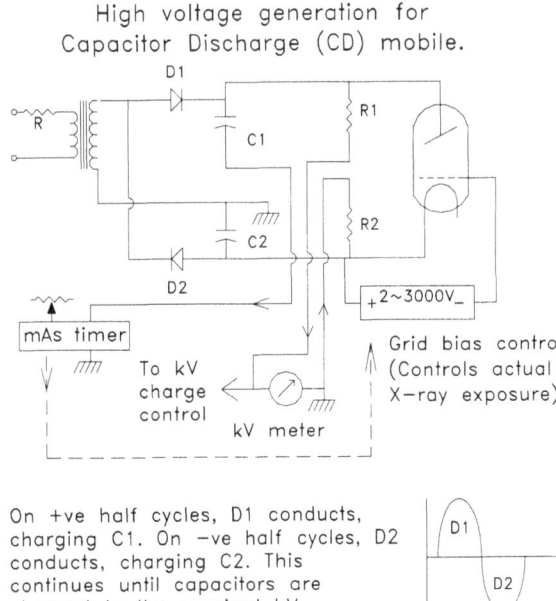

Fig E–22. The capacitor discharge generator

Relation between kV and mAs

As sown in Fig E–23, a 30 mAs exposure will cause the kV to reduce by 30 kV. As the quantity of radiation from an X-ray tube is controlled as much by kV as mAs, large mAs exposures are not practical. For example, if the above example were for 40 mAs, the last 10 mAs of the exposure would be from 60 to 50 kV, and have little effect.

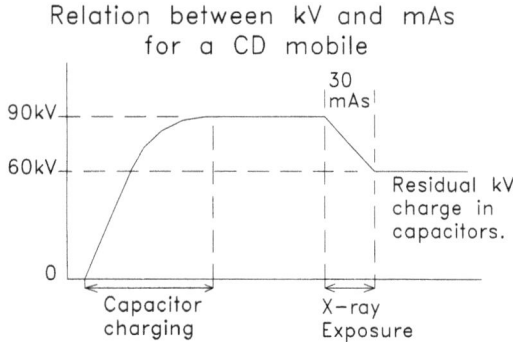

Fig E–23. Relation between kV and mAs, CD mobile

PART 4 THE X-RAY GENERATOR CONTROL UNIT

Contents

a. X-ray control functions
b. High-voltage control and load compensation
c. mA control
d. X-ray radiographic timer
e. Automatic exposure control (AEC)
f. Fluoroscopy timers
g. Exposure contactors
h. X-ray tube anode rotation
i. X-ray tube load calculation
j. Operation sequence control
k. Fault detection and safety systems

a. X-ray control functions

The X-ray control provides the following functions for radiography

- Radiographic kV selection.
- High-voltage load compensation. (For different mA outputs)
- Mains-voltage regulation. (May not be required for most high-frequency generators.)
- X-ray tube filament heating and space-charge compensation for each X-ray tube focal spot, and mA station selection.
- Selection of required mA output.
- Selection of X-ray tube focal spot. In some systems, this is automatically linked to the required mA position.
- Exposure timer. For single and three phase systems, the timer must be synchronized to the mains power supply.
- Exposure contactor, to connect the HT generator to the preselected primary voltage.
- Anode rotation control (or starter). Some systems allow for operator selection of low or high speed.
- X-ray tube safety calculation. Basic requirement is anode load and maximum kV. Calculations may also include maximum filament heating, stored heat in the anode, and a safety factor for multiple exposures.
- Technique selection of external equipment. Eg, table Bucky, vertical Bucky, tomography, etc.
- Operation sequence timing and control. Eg, after the preparation time delay, X-ray exposure request is sent to the Bucky; signal returned from the Bucky starts the exposure.

- Safety provision for operator error, radiation 'ON' warning light etc.
- System fault detection, both prior or during an X-ray exposure.

The following additional functions are provided for fluoroscopy
- Fluoroscopy kV selection
- Automatic fluoroscopy-kV control. (May be an option)
- Fluoroscopy mA control. Depending on the design, this may be not available for the operator. Instead the level of mA may be controlled directly by the fluoroscopy kV selection.
- Fluoroscopy exposure timer. Depending on system design, this may either stop exposures, or just sound an alarm; after a maximum accumulated time (normally five minutes) has expired.

The control may have these optional features
- Automatic exposure control, or 'AEC'. Often known as 'photo timer', and sometimes by the Siemens title of 'Iontomat' or Philips title of 'Amplimat'. The AEC measures the quantity of radiation as it enters a cassette. This measurement is used to control the exposure time.
- Anatomical programmed radiography or 'APR'. APR is a system of preset exposures, depending on the area of the body to be examined. Current systems, with microprocessor controls, allow a high degree of flexibility, and may be treated as a pre-programmed exposure memory system.

b. High-voltage control and load compensation

Adjustment of high voltage for conventional systems is by preselection of the primary voltage. This voltage is sent to the primary of the high-tension transformer, when an exposure is made. This preselection of primary voltage must allow for voltage drop in the generator transformer, as well as the power mains voltage falling when under load. As we change the selection of mA, this also changes the amount of voltage drop that will occur. To compensate, as we increase the mA selection, so we must also increase the primary voltage to keep the previous kV selected correct.

Fig E–24 illustrates the relation between kV, mA, and primary-voltage for a single-phase 400 mA generator. Example: If 80 kV at 200 mA were required, the voltage for this exposure would be preset at 114 V. However, if 400 mA were required instead, then the voltage would be increased to 134 V.

- In the example shown in Fig E–25, a simple method is shown to achieve load compensation. This method may be used for a portable, or mobile, X-ray generator.
- A line voltage adjustment switch allows compensation for different input voltages. The switch is adjusted until the voltmeter is on a calibration mark. If the meter is not set to this mark, than kV will not be correct.

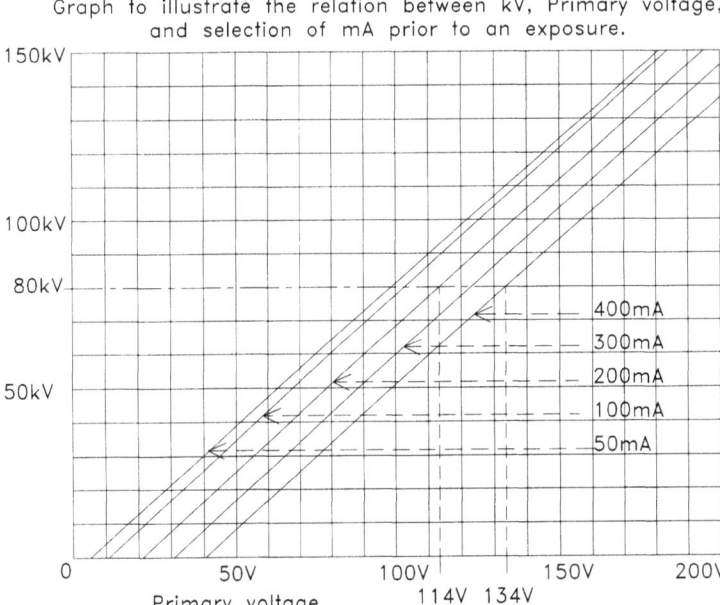

Fig E–24. Load compensation requirement as mA is changed. If 80 kV is required, then primary voltage should be 114 V for 200 mA, or 134 V for 400 mA

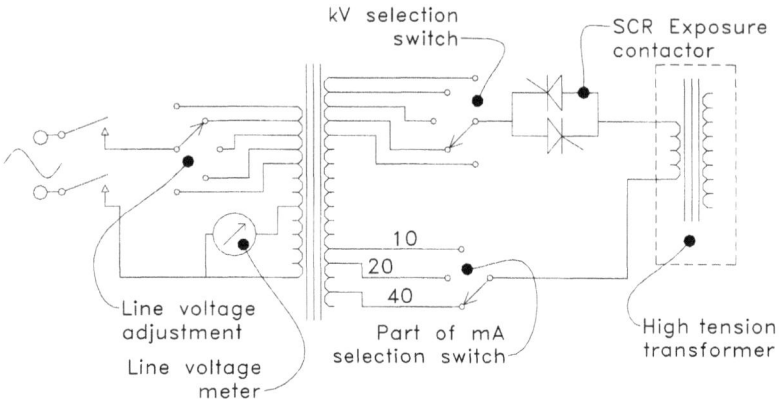

Fig E–25. High-voltage selection, and load compensation, for a portable X-ray generator

- The kV selection switch is set to the required kV.
- On selection of the required mA position, a section of the mA selector switch selects the required load compensation voltage.
- This method may be also used for larger fixed installations. More complex compensation is then applied to allow for mains supply impedance etc.

Another method is to indicate the preselected primary voltage on a voltmeter directly calibrated in kV. These systems often have two kV selection switches, one for coarse settings of about 10 kV, and the other for fine settings of 1 kV. (Although a large multi-step switch may be used instead.)

On selection of a different mA station, the meter will either increase or decrease its indicated kV, depending on the change of mA. By resetting the kV selection switches so the meter again reads the required kV, load compensation is achieved.

A similar method to the above is a scale calibrated in kV, and a pointer moved by the kV selection knob. On selection of each mA station, a different kV scale is brought into view on the control panel. (The mA selection switch also selects the required load compensation, as shown in Fig E–25.)

Many X-ray controls, especially three phase versions, have motorized 'servo' controlled selection of kV, and automatic line voltage compensation. These often use graphite rollers moving along a 'step-less' transformer winding. (Eg, the roller passes along individual turns on the outside of the transformer, making direct contact with each turn as it moves.)

Servo systems measure the voltage as the rollers pass along the transformer, and compare the voltage obtained to a required value. This value is the required kV to be generated, plus an additional voltage for load compensation. For automatic line voltage regulation, the obtained voltage is simply compared to a fixed reference voltage. In both cases, when the required voltage is obtained the servo motor is stopped. On preparation or on making an exposure, these motors are locked out to prevent movement when the mains supply voltage drops.

High frequency generators control kV by comparing directly the actual kV across the X-ray tube with a reference voltage set by the required kV. For example, if the operator selected 80 kV, the reference voltage may be 8V. On start of the X-ray exposure, the voltage at first on the X-ray tube will be 0 kV. Very rapidly, it will approach 80 kV, at which point the measured voltage from the generator will match the reference voltage of 8V. However, as the required kV becomes close to 80 kV, the inverter will reduce its output, so that as 80 kV is actually reached, inverter output is regulated to maintain 80 kV precisely.

High-frequency generators are not affected by mains supply voltage drop during an exposure, due to the self-regulating closed-loop mode of operation. However, some systems do require automatic mains voltage regulation, as well as correct mA calibration, to ensure kV generated at the start of the exposure is correct, and does not 'overshoot'.

Fluoroscopy control by comparison to radiographic control is much simpler, as no load compensation is required. On older systems this may be via a switch selecting 10 kV steps, or by a sliding contact on a circular 'step-less' transformer (sometimes called a 'variac').

Some generators may employ electronic control of kV, using the properties of the SCR radiographic contactor. By this method the effective voltage to the transformer primary is controlled by changing the timing pulses to the SCR contactor. In effect this is a high power version of a lamp dimmer.

With high-frequency systems, control is similar to radiographic output; however in some systems the resonant frequency of the inverter system may be raised, to reduce audible noise.

Automatic fluoroscopic kV may be obtained by having a motor drive or else by direct control of the SCR or inverter as previously mentioned. The control signal may come directly from a TV camera, or else via a photomultiplier sampling the light directed to the TV camera. Automatic fluoroscopic kV control is used to optimise the light level into the camera, as well as avoiding excessive radiation to the patient.

c. mA control

The control of X-ray tube emission, expressed in milliamperes or 'mA' requires consideration of several factors. In particular, the level of filament heating to obtain the required emission, and the affect of generated kV on actual emission.

The following requirements need to be considered.

- Filament current to obtain the required mA emission level.
- Modify the filament current as the set kV, before an exposure, is selected. This is to ensure emission is constant over the range of available kV, and is called 'space charge' compensation. See Fig E–27.
- Provide a level of 'pre heating' so the filament will quickly reach the required temperature during radiographic preparation. The filament may be preset to half the radiographic current in stand-by mode, or in some systems, adjusted to the point where emission would just occur. (~1.0 mA) Additional boost may be applied for quick heating during preparation. This is called 'flash' boost. However, some systems only provide pre-heating for a fluoroscopy tube, and rely on longer preparation time for the over-table tube. See Fig E–28.
- The power supply for filament heating must be well regulated, so that a drop in power mains voltage does not affect heating level. Earlier systems used a 'Constant voltage' or 'Ferro resonant' transformer for this purpose. Later systems use electronic regulation, which precisely measures and controls the current through the filament. This is done by monitoring either the current through the filament

transformer primary winding, (constant current), or the voltage across the same winding, (constant voltage).

- Provide protection so the filament is not overheated. On earlier generators, no protection was provided. Later systems included protection in X-ray tube overload calculation. Present microprocessor controlled systems can have elaborate protection circuits.
- The mA control has a safety system to prevent an exposure, in case filament heating is incorrect. For example, if the filament has become disconnected due to a faulty high-tension cable, or in case the filament is broken. *In this case, there would be no load on the high-tension transformer, and the generated kV could become dangerously high.*
- Compensation for drop in mA output during exposure may be provided. As electrons are attracted away from the filament, the filament temperature falls a small amount. This effect is more noticed as the filament reaches the non-linear section of its operation. Many systems now provide feedback or closed-loop compensation for this effect. By sampling the mA generated during an exposure, comparing it to a reference level set for each mA station, a correction factor is applied to filament heating. Some microprocessor systems use this technique to automatically re-calibrate the filament control, by memorizing the final required value of filament heating.

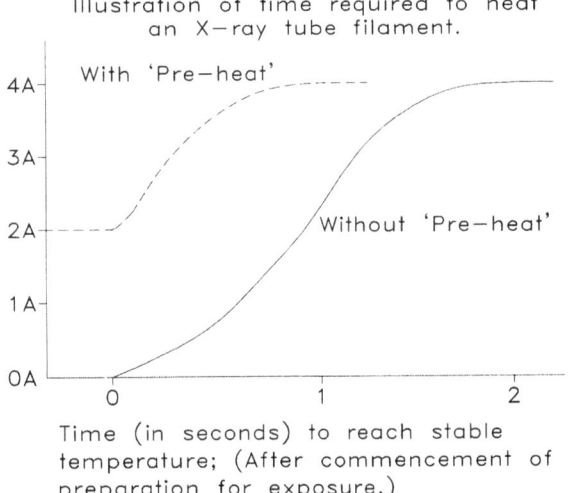

Fig E–26. Comparison of filament heating time. With and without pre-heat

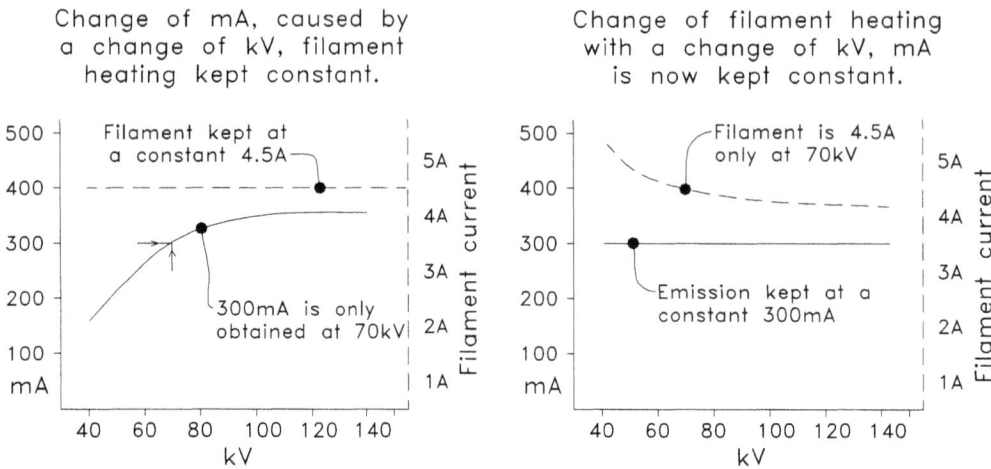

Fig E-27. These two graphs illustrate the need to modify filament heating, as kV is changed

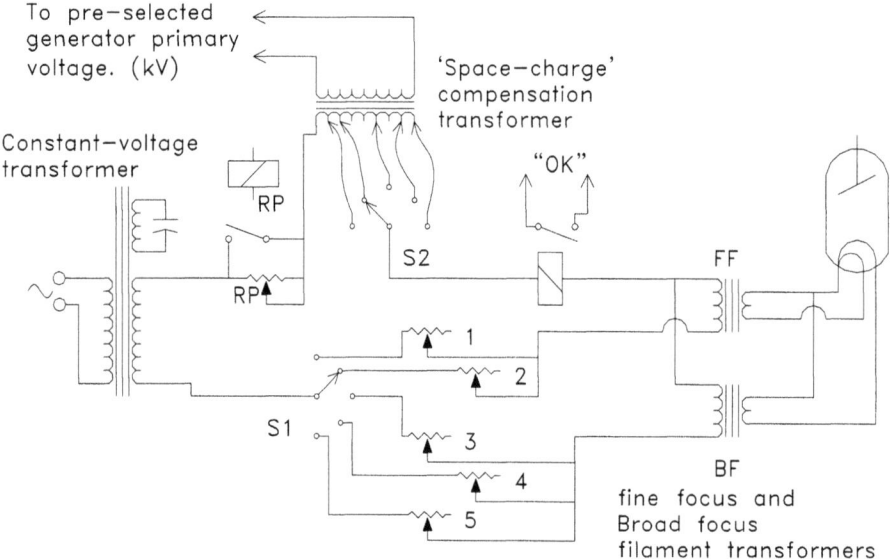

Fig E-28. A basic filament control system

The circuit in Fig E-28 has the following points of interest.

- The switches S1 and S2 are coupled together.
- S1 selects the calibration resistor for filament heating and the change over from fine-focus to broad-focus.
- S2 selects the degree of correction voltage for space-charge compensation. Although not shown, the primary of this transformer is connected so as to reduce filament heating above 70 kV, and increase filament heating below 70 kV. Taps are provided to select the correct amount of compensation for each mA position.
- The resistor RP provides reduced current through the filament transformers for pre-heating. On preparation, the contact of relay RP shorts out resistor RP, providing full radiographic heating.
- The relay and contact marked 'OK' is for filament safety. This relay is a current operated version, and if the filament is broken, or has a bad connection, the relay does not operate. This prevents the control entering the 'ready for exposure' mode.
- The two filament transformers, broad and fine focus, are mounted inside the high-tension generator tank.

Filament control systems have had considerable development since the basic method described in Fig E-28. The heavy-duty resistors were replaced with transistors, and the constant-voltage transformer is no longer required.

- Further development saw the use of highly regulated DC power supplies and a low frequency 'square wave' inverter powering the filament transformers. Such systems monitor either the current through the transformer primary or the voltage across the primary winding. This is compared to a reference voltage for each mA station, and in turn regulates the DC voltage supply to the inverter. Electronic generation of the space-charge compensation allowed an optimum kV relation to be generated. Calibration is usually carried out by small preset potentiometers and in some cases by direct output from a microprocessor. In this case the control voltages are obtained via a digital to analogue (D/A) converter.

- Current systems now tend to use high-frequency inverters, where the operation mode of the inverter itself is controlled to provide the required filament heating. This eliminates the need for a regulated DC supply to the inverter. Control is often via a microprocessor, again using a D/A converter. Filament calibration may be performed manually, by entering calibration settings at designated mA and kV positions. Some microprocessor controls also feature automatic calibration when in a service mode. In this mode the control automatically steps through a series of test exposures, and stores in memory the required data. This feature depends on the design philosophy of the manufacturer.

d. X-ray radiographic timer

The X-ray timer has several functions
- Accurate timed exposures.
- Synchronize start and stop of exposure to mains supply frequency, so the start of the exposure is achieved at 'zero crossing' of the AC power waveform. (Not required for high-frequency systems.)
- For three-phase systems, provide timing signals so each of the three phases will be connected at the correct phase interval. (Not required for inverter systems.)
- For single and three-phase systems, provide a 'phase memory', this is required for exposures that are uneven multiples of the power mains frequency. Eg, as an example 0.03, 0.05, 0.07 seconds. (Other systems may instead use two-stage power switching via a damping resistor, or else pre-magnetize the transformer iron-core by a DC current prior to exposure.)
- Supply the preset exposure time to the X-ray tube load calculation.
- Provide time settings for a safety 'backup' timer. This timer is normally a separate system, set to a little longer time than the exposure timer. If the generator is still producing high voltage after the backup time, the safety system stops the exposure by operating a safety contactor.

Timers with thyratron valves
- These are only on quite old systems, however, many are still in use.
- The thyratron valve may look like an ordinary valve, in that it has a cathode, control grids, and anode. The electrical symbol is also similar, except for a round dot indicating it is gas filled.
- The thyratron may be considered an electron relay, as conduction does not commence until a negative grid voltage is reduced below a specified level. In which case full conduction occurs, and remains conducting until the voltage between anode and cathode is removed or reverses itself. Conduction current is high, and enables the thyratron to directly operate relay coils, and in some cases, large contactors.
- In a thyratron timer, a capacitor is pre-charged to a high negative voltage. The time selector switch, via the exposure start relay, places a selected resistor across this capacitor, which starts to discharge. The time of discharge is controlled by the resistor value. When the negative voltage drops to the required value, the thyratron fires, and operates a relay to end the exposure.
- Thyratron timers have at least one adjustment to control calibration of exposure times. For example, to adjust the timing-capacitor charge voltage. There may be other adjustments, to adjust the pull in or drop out times of the exposure contactors. This ensures start and finish of an exposure is at zero crossing of the primary voltage. (If not correct, severe arcing of the contactor may occur.)

'Solid state' analogue timers
- The thyratron timer is an early version of an analogue timer. In the solid-state version, a capacitor is charged at a pre-determined rate, until the capacitor voltage reaches the level of a comparison voltage.
- The rate of charge of the capacitor is determined by the resistor value selected by the timer switch. Once the capacitor is charged to the comparison voltage, the timer removes the enable signal to the exposure contactor. In this case, this contactor is most often a 'silicon controlled rectifier' (SCR) system, however it might be a mechanical contactor in older systems.
- These timers usually have two adjustments. One adjustment is for the reference voltage, which acts

a calibration for long times. Another adjustment, not always fitted, is in series with the timer switch resistors, and adjusts the short times.

Digital timers
- These use a highly accurate crystal oscillator, which is divided down to provide timing clock pulses.
- The time selector switch loads a binary code into a digital counter. On start of the exposure, clock pulses subtract from the number in this counter. When the counter reaches zero content, the exposure time is finished.
- Although a digital timer does not require a time correction adjustment, there are adjustments for phase synchronization to the power mains supply. (Not required for a high-frequency generator.)
- Digital timers often have a separate analogue timer as a backup unit.
- **Note**. Some X-ray controls, although fitted with a digital display for selected time, may not have a digital timer, and instead use an analogue timer.

Microprocessor timers
- These are usually found with high frequency systems. There are no adjustments to calibrate the exposure time.
- The selected time is entered via a keypad, rather than a switch. The microprocessor downloads the required time to a separate backup timer, which is independent of the microprocessor timer once the exposure starts.
- On many systems, the microprocessor does not start the exposure countdown until the X-ray HT has reached 75–80% of its required value. This allows very accurate timing of exposures, down to 1.0 millisecond or less.

Milliamp second (mAs) timers
There are three versions of the mAs timer.

- **Microprocessor controlled mAs timer**
Some X-ray controls have the option of selecting the exposure by kV, mA, and time. This is called 'three knob' technique. Or the operator may prefer to use a selection of just kV and mAs, called 'two knob' technique. Of this last method, the computer looks at the X-ray tube data stored in memory, and selects the optimum combination of mA and time selection to provide the shortest exposure. Of course, this does mean at times the X-ray tube may be working close to its maximum ratings. To avoid this, some controls have an added selection of ~80% load, instead of maximum tube load.

- **Digital mAs timer**
The digital mAs timer directly operates from the mA produced during the exposure. By passing the generated mA through a resistor, a proportional voltage is produced. This voltage in turn controls a voltage to frequency (V/F) converter, which may produce ten pulses per mAs. When a required mAs is selected, this value is loaded into a counter. The pulses from the V/F converter subtract from this preset value, till the counter reaches zero. This ends the exposure. Digital mAs timers may also be supplied with microprocessor controlled high-frequency generators, for example, a mobile generator. Typically in this case the microprocessor also selects a suitable mA output to match the required kV and mAs.

- **Analogue mAs timers**
With a CD mobile, mA generated during an exposure falls together with the kV. For this reason CD mobiles are fitted with mAs timers, of which the analogue version is the most common type.
With an analogue mAs timer, the mA generated during an exposure produces a voltage across a reference resistor. This voltage in turn is used to charge a timing capacitor in the same fashion as a standard analogue timer. The important difference is that the voltage for timing control is now proportional to the mA generated during the exposure, and not a fixed voltage as in a standard timer.

e. Automatic exposure control (AEC)

Automatic exposure control or photo-timer is usually an option on purchase of an X-ray system. The AEC control is normally mounted inside the X-ray control cabinet. Older systems may instead have an external control unit.

In setting up for an exposure with AEC, the X-ray control timer is set for about 25% to 50% longer exposure time than expected. Some X-ray controls may not require a preset exposure time, instead using a principle called 'falling load'. Falling-load allows the generator to start at a high output, then as the exposure continues, lower the mA generated by reducing filament heating. This allows for extended exposure times, but still within the X-ray tube rating. Compensation is required, to keep kV at the correct level as mA drops.

In operation, the AEC measures the quantity of radiation entering the cassette. When this radiation reaches a predetermined level, an 'end of exposure' signal is sent to the X-ray control timer, terminating the exposure.

APPENDIX E. X-RAY EQUIPMENT OPERATION

Ionization	Sometimes known as 'Iontomat' (Siemens) or 'Amplimat' (Philips). These depend on the minute current generated as gas molecules are ionized by X-ray radiation. Earlier types used atmospheric air as the medium. Later types may use a special gas, such as xenon, to improve sensitivity. Earlier ionization systems were sensitive to high humidity levels. Later systems have a pre-amplifier sealed into the same container incorporating the ion chambers. Adjusting the voltage gain of amplifiers, and the voltage reference for the exposure integrator, controls sensitivity.
Solid state	These depend on the detection material, which when energized by X-rays produces a small electric current. Adjusting the voltage gain of amplifiers, and the voltage reference for the exposure integrator controls sensitivity.
Photomultiplier	The detection area is formed by a thin pocket of luminescent material, similar to that used in a cassette, inside a sheet of translucent acrylic. Light is focussed from the edge of this acrylic into a photomultiplier. The output signal from the photomultiplier is very high compared to the other two methods, and is reasonably immune to humidity problems. The photomultipliers add to the total size of the system, so it can only be installed in a Bucky designed for a particular unit. Adjusting the voltage supply, and / or the last dynode voltage controls the photomultiplier sensitivity. Final adjustment to this type of AEC is otherwise similar to the other two systems.

Measurement of radiation for different sections of the anatomy is required. This is provided by measurement chambers in selected positions. These may be in the centre, or offset to either side for chest exposures. AEC exposure-controls allow for a selection of density, normally in +/− 5% steps, and may also incorporate sensitivity adjustment for different film/cassette combinations.

Provision is sometimes made to obtain an 'average' measurement by adding two or more chambers together prior to an exposure. In other systems, separate controls are incorporated for each chamber. By selecting two or more together, whichever chamber obtains most radiation controls the exposure.

To be successful, the chamber for measuring radiation should be placed between the X-ray grid and the cassette in the Bucky. This reduces sensitivity to scattered radiation. A kV correction signal is required, so the AEC can match the characteristic of the film-screen combination in use.

With conventional generators, especially single-phase versions, special contactor arrangements are required to avoid 'jitter' when approaching short times. This is due to SCR contactors not switching off immediately, but waiting for the next 'zero crossing' point. On a 50HZ system this may cause a variable extension of up to ten milliseconds exposure time. (This problem does not exist with high-frequency systems.)

The film processor must also be accurately maintained, especially if the AEC is being calibrated or tested. There are three varieties of AEC systems for measuring radiation, ionization, solid-state, and photomultiplier systems. The differences are listed above.

f. Fluoroscopy timers

On older generators, small motors, similar to those used in electric clocks, have operated these timers. Later systems used a separate digital timer and display. With current systems, this function is integrated with the microprocessor control.

The timer only operates during a fluoroscopic exposure, and normally has a maximum time setting of five minutes. When the timer approaches the time limit, a warning buzzer may sound. At the end of five minutes, further exposures are prevented unless reset. (Some systems have a switch to bypass this requirement, depending on individual country regulations). Other facilities may be provided with digital fluoroscopy timers, such as total elapsed fluoroscopy time etc. This depends on individual features of a control system.

g. Exposure contactors

Exposure control on old systems was by mechanical contactors. (A contactor is a heavy-duty relay). These contactors require accurate adjustment to ensure the 'make' and 'break' of power to the high-tension generator occurs at the correct phase interval, otherwise severe arcing results.

When silicon controlled rectifiers (SCR) became available, these replaced the mechanical contactors for control of the actual exposure. Mechanical contactors are still required, however, in case of a faulty SCR, as these may develop a short circuit. This is called a 'backup contactor'.

An SCR has the property that once conducting, it remains conducting until the voltage across the device either falls to zero, or changes its polarity. A short positive pulse of voltage is applied to the SCR 'gate', relative to the cathode, to switch on the SCR. The SCR, as it is also a rectifier, will only conduct in one direction. To form a contactor two SCR units are connected in parallel, with one facing the opposite direction.

A basic SCR contactor system is shown in Fig E–29.

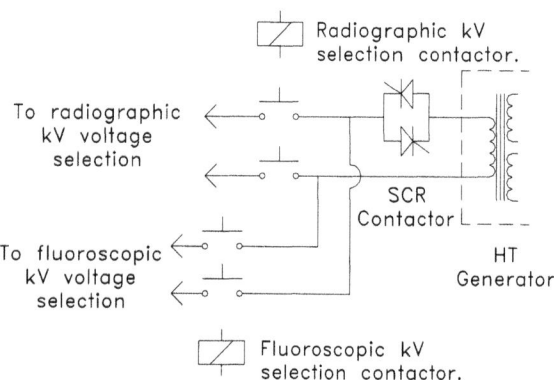

Fig E–29. A basic 'SCR' exposure contactor, for a single-phase generator

h. X-ray tube anode rotation

On preparation for an exposure, two main events happen. The anode is caused to rotate, and the X-ray tube filament is boosted to full operating temperature.

A time delay is required to ensure both anode rotation and filament boost has been completed before permitting an exposure. In many systems, the timer in the anode rotation control allows for both these requirements.

Anode rotation for low speed operation is controlled by the power mains frequency, either fifty or sixty Hertz. (50/60 Hz). **See Fig E–30**.

- On preparation request, relay R4 operates connecting 200 V to the stator winding.
- A phase difference of ninety degrees between the current flowing in the 'start' and 'run' windings of the stator is required. This is supplied by the capacitor 'C'.
- **Note**. This capacitor requires a specific value to match the X-ray tube stator winding. If a different make of X-ray tube is installed, the value of 'C' may need to be changed.
- After the preparation time is completed, the control unit timer operates relay R3, which changes the supply voltage from 200 V to 40 V. This is to ensure the anode keeps rotating at full speed, but without the heat that would be generated if 200 v were continuously applied.
- The two relays, R1 and R2, are current operated relays. These are for safety, in case either the start or run stator windings become disconnected, and prevent an exposure occurring on a stationary anode.
- Relay contacts R1, R2, and R3 form the preparation sequence safety system, and unless all are operated, an exposure is not permitted.
- Present day low speed starters have more elaborate fault detection, and may also have different preparation times, eg for over-table or for fluoroscopy.

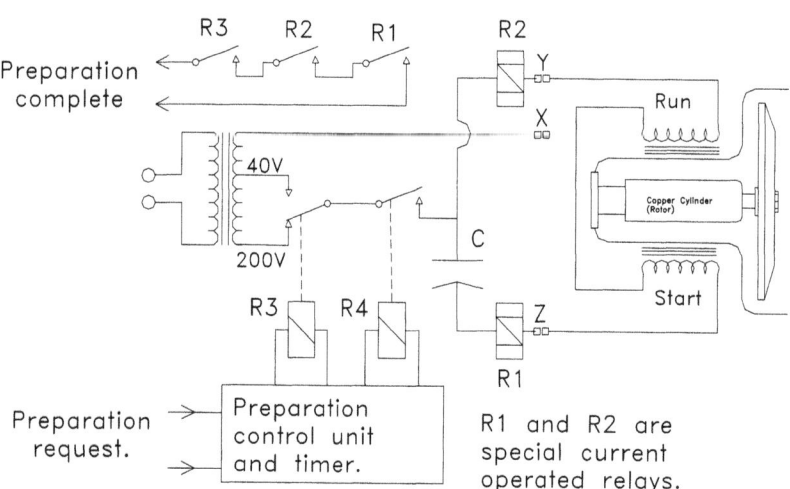

Fig E–30. A basic low-speed anode rotation control

High-speed rotation is provided by either a passive 'frequency tripler', or by an inverter system. The frequency tripler uses special transformers, connected to three-phase power. These transformers are driven in a saturated mode, and produce a highly distorted output, rich in harmonics. The output voltages are connected in series, so that the fundamental 50 or 60 Hz supply frequency is suppressed, and the third harmonic instead is selected. For a 50 Hz input, the output will now be 150 Hz. When applied to the X-ray tube stator, the anode will now rotate three times faster than at low speed.

With an inverter system, incoming single-phase power is converted to DC, and via the inverter back to AC. The inverter may operate at 150 or 180 Hz, depending on make or model. The majority however operate on 180 Hz, as this allows higher anode loads. For special applications, other drive frequencies may also be available.

An X-ray tube operated at high-speed will have greater stress on the bearings, and must not be allowed to coast down to a stop after an exposure. If this happens, there is a strong possibility of severe damage to the bearings, due to resonance affects at some anode speeds. To prevent this occurring, a high-speed starter provides a 'brake' cycle at the end of an exposure. This may be via a DC current through the stator windings, which quickly brings the anode to rest. A more common method is to apply a 50 Hz start signal, which brings the anode from high-speed quickly past the resonant positions to 3000 RPM. The anode then coasts to a stop.

High-speed operation, unless high power starters and special stators are used, may take twice as long to reach full speed. To overcome this problem, especially with a fluoroscopy table, two modes may be used.

- High-speed maintenance or 'hangover'. The starter remains in high-speed operation for up to 20 or 30 seconds from the last exposure.
- Low-speed maintenance or hangover in a fluoroscopy mode. This may last several minutes. As the anode is already rotating just below 3000 RPM, high-speed preparation time is reduced. In some starters, both techniques may be combined.

i. X-ray tube load calculation

An X-ray tube needs to operate within the maximum ratings for that tube, otherwise damage will occur. The manufacturer publishes rating charts, which specify the *maximum* operating conditions for the particular tube. **These parameters are:**

- The maximum rate of heat input for each focal spot, this is calculated from the product of kV and mA, and the exposure time. This will be modified by the speed of the anode, and if operated on single or three-phase. (Inverter systems are treated as three-phase)
- The maximum filament current of each focal spot, this limit occurs at low kV settings. This is required for filament protection.
- The maximum kV that may be applied.
- The maximum amount of heat stored in the anode, and the rate of anode cooling.

Note. These ratings are the *maximum* permitted. Regarding heat input to the anode, this is calculated for a cold anode, so it is unwise to make several exposures close together, **and** close to the maximum permitted input.

All X-ray controls have protection for anode heat input and kV limit. The other parameters depend on the level of design complexity. Present day microprocessor controlled systems may take all calculations into consideration, except tube housing temperature.

A typical rating chart is shown in Fig E–31. A number of load lines are provided for convenience. If we examine the 100 kV load line and compare this to 150 kV and 80 kV lines, we will find that the product of mA and kV is the same. See table E–5.

Table E–5. Maximum anode load at 0.1 second

100 kV and 660 mA	100 × 660 = 66 kW
150 kV and 440 mA	150 × 440 = 66 kW
80 kV and 825 mA	80 × 825 = 66 kW

By using the data shown in Fig E–31, we can obtain the maximum available output for a range of different conditions. These are shown in table E–6.

Table E–6. Maximum anode loads for the rating chart of Fig E-31

0.05 second	125 kV	560 mA
0.05 second	100 kV	700 mA
0.1 second	100 kV	660 mA
0.1 second	80 kV	825 mA
0.3 second	110 kV	500 mA
0.5 second	80 kV	600 mA

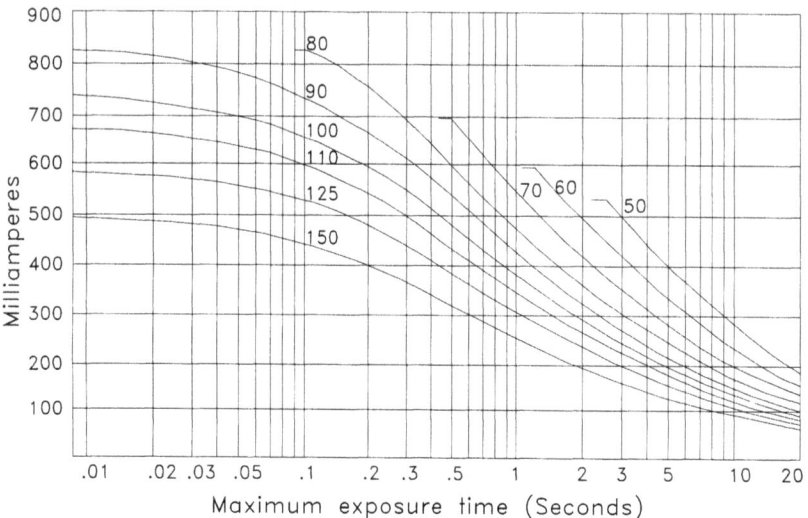

Fig E–31. A typical X-ray tube rating chart

The load lines for 50, 60, 70, and 80 kV stop after reaching a specified mA. This is because the filament will reach the maximum filament current, if required to produce that mA at the specified kV. For example, if at 60 kV an attempt was made to expose at 800 mA, 0.1 second. Although this is well within the anode load limit, the filament current would be excessive, and damage the filament.

In a microprocessor system, the computer carries out all the required calculations. Providing an X-ray tube from the same manufacturer is supplied, a code number is entered for the designated tube. In case the tube to be installed is not included in the list of codes, many controls allow input of full parameters derived from the rating charts, as a 'non standard' tube. With older X-ray controls it is necessary to adjust a series of potentiometers to obtain the correct calculation. These are normally adjusted at specified time positions, with separate adjustments for maximum kV and filament protection (if provided). Still other systems may use a patch-board, with wire jumpers connected between a series of selected pins, but with a similar goal in mind.

j. Operation sequence control

An X-ray control will perform a number of functions besides the actual X-ray exposure. These are concerned with ensuring the system is ready for use, preparation, exposure, fault detection etc.

A digital logic diagram for operation sequence control is shown in Fig. 32.

On power up
- X-ray tube selection in the HT transformer is activated.
- The X-ray tube filament pre-heating is commenced.
- Safety system tests for immediate faults.
- Microprocessor (if fitted) is initialized.
- With high-frequency systems, the inverter power supply capacitors are charged up.

Before radiographic preparation is permitted
- A valid technique selection is required. Hand switch operation should not be possible, if the fluoroscopy table has been selected. Some selections may not be available, eg, accidentally selecting tomography position, if not installed.
- Exposure factors must be within the X-ray tube capacity. (Maximum kV, anode load etc.)
- The X-ray tube housing over temperature switch should not be operated.
- For systems with servo (motor driven) mains voltage correction, or kV selection, preparation should wait till adjustment is finished. (But some designs omit this precaution.)
- On initial power up, or if a tube change is selected, a time delay may be inserted to allow preheating stabilization of the X-ray tube filament. This will also occur if tube change over is by a motor driven switch.
- No fault conditions should exist, eg, faults occurring from a previous exposure or preparation problem.
- In some countries, a door safety switch is required, to prevent exposures if the door is opened. This may also prevent preparation.

On commencement of radiographic preparation
- The preparation hand-switch is operated.
- Preparation request is sent to the tube starter, anode rotation commences.

- The preparation timers commence timing out.
- The X-ray tube filament is boosted to full preset temperature for the required mA.
- Tests are made to ensure no faults with anode rotation or X-ray tube filament.
- Warning light on the X-ray room entrance is illuminated.
- On conventional generators with SCR contactors, a backup safety-contactor connects radiographic power to the SCR contactor. A test is made to ensure there is no short circuit in the SCR contactor.
- Lockout any inputs for fluoroscopy request.
- Change over mA measuring circuits from fluoroscopy mA to radiographic mA.
- Servo-motors for line voltage compensation and kV selection are locked out to prevent operation.
- Peripheral equipment will also go into preparation mode. Eg, on remote controlled systems, the film will move into position.

On completion of preparation, to obtain 'ready for exposure'
- Preparation timer has 'timed out'.
- Anode rotation safety check is satisfied.
- No fault has occurred with filament heating.
- No system fault has occurred.
- Peripheral equipment is ready.
- 'Ready' signal appears on the control panel.

To obtain a radiographic exposure
- Preparation is completed. Exposure hand-switch is operated.
- Operate 'X-ray On' warning signal.
- Send an exposure request to peripheral equipment, eg, Bucky.
- Bucky operates, and returns the exposure signal to the X-ray control.
- On conventional equipment, the timer commences operation. The timer controls correct closing of the exposure contactor. (Mechanical or SCR version).
- On high-frequency systems, both timer and inverter commence operation.
- If fitted, a backup safety timer commences operation.

During a radiographic exposure
- Measure and display mA or mAs during exposure. (Only some systems).
- Measure mA, and operate mAs timer. (If fitted).
- Measure mA, and correct X-ray tube filament heating to ensure correct mA. (Only on some systems).
- Test for kV or mA faults.
- Test for system faults.

At the end of a radiographic exposure
- Send a time-up signal to peripheral equipment, such as a fluoroscopy table.
- Test to ensure high voltage generation stopped. (In case of SCR contactor fault).
- Send a time-up signal to a double-exposure protection circuit. Preparation must be released and started again before another exposure is made. (Not on all systems).

At the end of preparation, following a radiographic exposure
- Filament heating is returned to standby pre-heating.
- A one second 'filament-cooling' timer commences operation. This is required in case the next exposure is for fluoroscopy. For example, if a 500 mA radiograph exposure has just been made, the filament must cool down before allowing a 2.0 mA fluoroscopy exposure.
- Return mA measuring circuits from radiographic mA to fluoroscopic mA.
- If anode rotation was high-speed, the starter now generates a brake cycle.

To obtain a fluoroscopy exposure
- The fluoroscopy timer has not reached the time limit. (Usually five minutes.)
- The X-ray tube filament is not in a cool-down cycle. (After a radiographic exposure).
- The fluoroscopy table safety interlocks are satisfied.
- The room entry door is closed. (Required in some countries).
- A valid technique and X-ray tube has been selected.
- Power to the generator is now selected from the fluoroscopic kV control section.
- In three phase systems, fluoroscopy is most often performed in single-phase mode. The capacitance of the high-tension cables provides the required smoothing of the kV waveform, so very little kV ripple occurs.
- With high-frequency systems, the inverter may select a different mode of operation than for radiography kV. This depends on system design.

During a fluoroscopy exposure
- Operate 'X-ray On' warning light at the entrance door.
- The generated fluoroscopic mA is displayed. (Most systems).

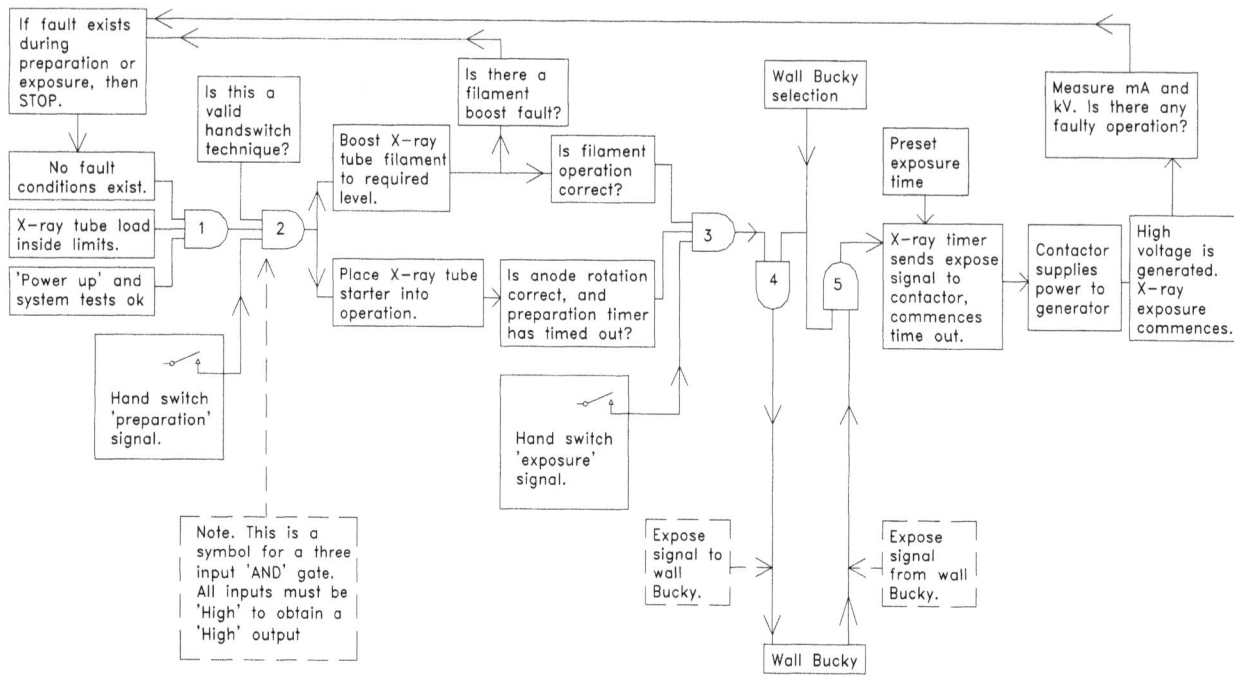

Fig E–32. Operation sequence control

- Fault detection is set for fluoroscopy conditions.
- The fluoroscopic timer is activated.
- A warning audible signal during fluoroscopy may be generated. (Required in some countries, especially if in 'boost fluoroscopy' mode).
- X-ray 'on' signal is sent to the TV system. Digital TV systems with a 'last image hold' feature will now display images in 'real time.' These images will also cycle through a digital memory.

At the end of a fluoroscopic exposure
- With digital TV systems, the TV memory retains the picture obtained just before the exposure finished, and displays that image as 'last image hold'.

k. Fault detection and safety systems

The X-ray control has many systems for detecting faults in operation, and provide for safe operation. Safety systems are provided to ensure correct sequence of operation. These take the form of 'interlocks', which mean that a predetermined series of events must be satisfied before proceeding. Many such interlocks are provided to avoid operator error. For example:

- X-ray tube load protection.
- Selection of an incorrect technique.
- Include which X-ray tube in the technique selection.
- Not permitting exposures that exceed country regulations.
- Exposure time selected is too short for the selected mA. (Some microprocessor systems do not permit exposures less than 0.5 mAs).

Interlocks also ensure that if a fault is detected, further operation is disabled. This may even extend to switching off the complete system.

As X-ray control design has become more complex, there has been an increase of provisions for detecting possible faults, or wrong operation. Later systems often display a code on the front panel, to indicate the type of problem. In many cases the meaning of this code will be found in the operator or installation manuals.

Some models allow clearing of 'non fatal' faults by pressing the 'power on' button. For example, the X-ray tube might have been unstable, and caused the previous exposure to terminate. By pressing the power-on button, another exposure can be attempted. (It is advisable to reduce the exposure settings first, before making a test exposure.). In the case of 'fatal' faults,

APPENDIX E. X-RAY EQUIPMENT OPERATION

these will prevent operation unless the control is switched 'off' then 'on' again. In such a case, considerable caution is required, and any warning signals or codes should be investigated first.

While recent systems may display a fault code, or message, older controls may indicate a fault symbol, or just light up the same indicator used for X-ray tube overload. However, inside the control there are often many indicator LED indicators provided to indicate sequence operation or fault indication. Table E–7 indicates some of the safety interlocks and fault detection requirements that may exist.

Table E–7. Typical safety interlocks and fault detection requirements

On power up and system check.	Interlock test for operation of X-ray tube high-tension selection switch and stator connection relays. Has X-ray tube housing over temperature switch operated? On inverter systems, have the bank of inverter power supply capacitors charged up to the correct voltage?
Before preparation is permitted.	Has a valid technique been selected? Are the exposure factors within the safe operating area of both the X-ray tube and the generator? Entrance door safety switch not activated.
During preparation.	Is the current through the X-ray tube filament transformer above a minimum level? (If below, can indicate open filament connection). Is the filament current inside the maximum limit? Is the current flowing in both the 'start' and 'run' stator windings of the X-ray tube the correct value? Look for illegal voltage on generator transformer primary winding. (In case of an SCR contactor fault in conventional systems). Energize a warning light. 'Do not enter'.
At end of preparation.	Interlock for preparation timers. (Older systems may depend only on the stator control; later systems include a timer for minimum filament heating time.)
On exposure request with peripheral equipment, eg Bucky.	Hand switch exposure request is sent to the required Bucky. The Bucky must move the grid and trigger an interlock to indicate the Bucky is ready for an exposure. This interlock relays the exposure request back to the X-ray control.
To commence actual exposure.	On conventional systems, the expose signal places the timer into operation. The timer waits for a synchronization pulse derived from the mains supply voltage, and at the correct phase interval operates the SCR contactor. With a mechanical contactor, the time for the contactor to operate requires a compensation adjustment. With an inverter system, the signal to the timer and the inverter may occur at the same time. Mains voltage synchronization is not required.
During exposure.	The mA is measured. If mA is higher than a preset detection limit, stop the exposure. With high-frequency systems, if kV is excessive, or, after a short measurement time too low, stop the exposure. Some inverter systems measure the transformer primary current, and if too high stop the exposure.
At the end of a radiographic exposure.	A time-up signal may be sent to peripheral equipment. If in high-speed mode, the X-ray tube starter will now produce a brake cycle. A filament cool-down timer will operate, so a fluoroscopic exposure cannot be made until this timer has finished.

PART 5 THE HIGH-TENSION CABLE

The high-tension cable used in X-ray generators has three main requirements.

- It must be able to withstand more than 75 kV, plus a safety margin. A typical value is 100 kVp.
- The cable requires good flexibility.
- In case of a fault or damage, not to cause danger of electric shock.

With the exception of mammography and some dental units, X-ray generators operate with balanced +/− high-tension above ground. When used with a 150 kV generator, each cable must be able to withstand a minimum of 75 kV.

To provide electrical safety, a woven mesh shield is placed on the outside of the bulk insulation, underneath the protective surface cover. This shield is connected to ground potential at both the X-ray tube and the high-tension transformer. Should a spark occur due to insulation failure, the shield conducts the discharge safely to ground. The cable capacitance plays an important part for high-frequency generators. Typical capacitance for a ten meter length is ~1800 pF.

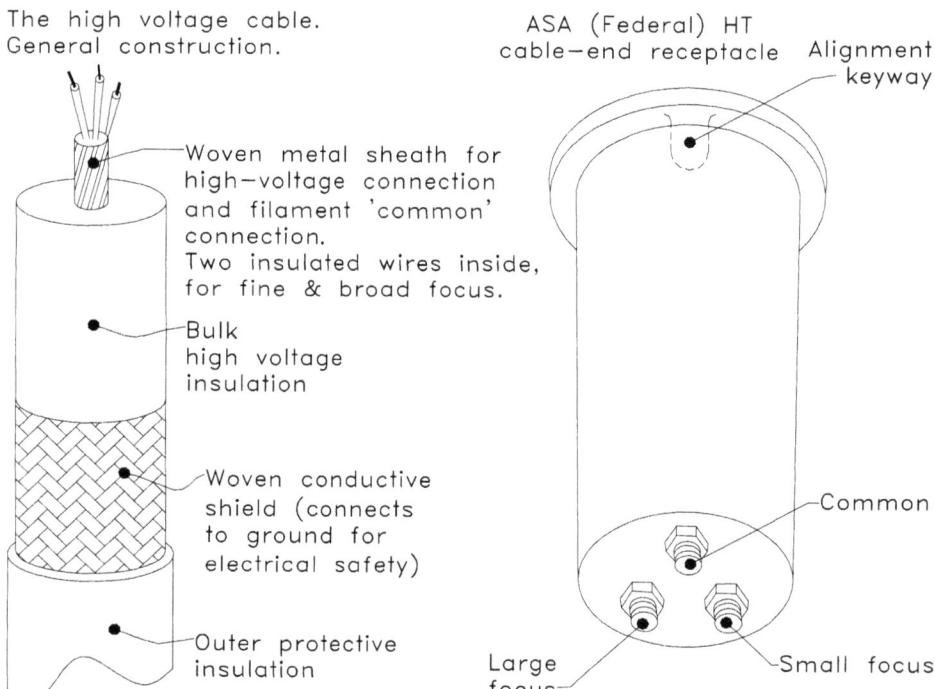

Fig E–33. The high voltage cable and standard ASA cable end receptacle

PART 6 THE X-RAY COLLIMATOR

An X-ray collimator is a device to limit radiation from the X-ray tube. A light beam is incorporated, to illuminate the patient area to be radiated.

There are several forms of collimators.

- 'Single leaf' versions, often fitted to smaller mobiles or portable generators. These are sometimes referred to as 'X-ray shutters'. The design requirement is small weight and size. The limitation is a blurred edge of the X-ray image on the film.
- 'Multiple leaf' collimators, incorporating at least two leaves or diaphragms, and in most cases, an additional small leaf positioned close to the X-ray tube focal spot. (This is to suppress off-focal radiation from the X-ray tube anode.)
- Automatic collimators. These may be either manually controlled as a standard collimator, or by remote control only, as in a fluoroscopy table. With an automatic collimator, the operator may reduce the size of the beam relative to the film area to be exposed, but cannot not exceed this area. These systems are sometimes fitted with a key switch to disable automatic collimation for special applications.
- Collimators for capacitor-discharge mobiles have an extra lead shutter to block unwanted radiation. This radiation occurs as dark-current when the capacitor is charged, or if the kV is reset to a lower value. The shutter may be motor operated and opened during preparation, or solenoid activated just prior to an exposure.
- Specialized collimators used in surgical X-ray TV systems, or angiographic equipment. These collimators often have an additional 'iris' diaphragm to limit the X-ray beam to the circular input of an image intensifier. They may also have additional rotating leaves fitted with a custom filter to reduce 'halation' due to a direct X-ray beam entering part of the image.

A sketch is provided in Fig E–34 of a standard collimator. This indicates the alignment of the light beam to the radiation field, and the lead diaphragms or leaves.

Referring to the sketch Fig E–34:

- Electrons bouncing off the focal spot, and re-landing on other areas of the anode cause off-focal radiation. This may amount to as much as 15% of the total radiation, but at reduced energy levels. The lead diaphragm 'a' reduces this off-focal radiation from the X-ray tube anode by applying a small aperture, very close to the focal spot. Some collimators

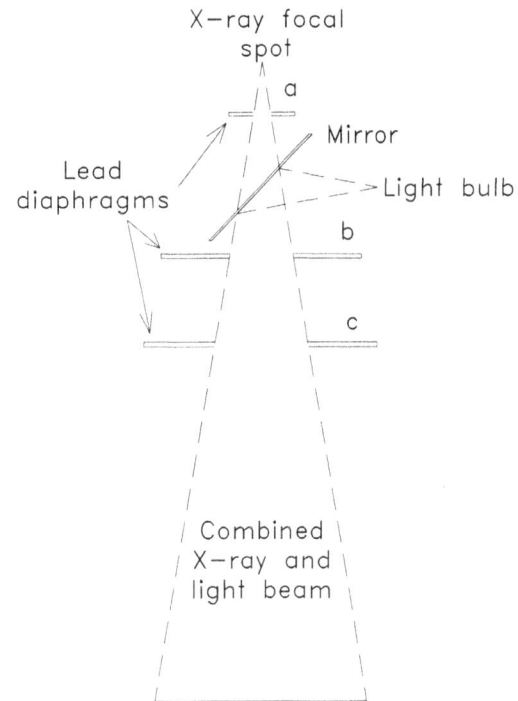

Fig E–34. Construction of the X-ray collimator

have these leaves extended above the top of the collimator body, and fit inside the collimator-mounting device.
- A mirror set at an angle of 45 degrees is mounted close to the input of the collimator. The light bulb filament is positioned the same distance from the mirror centre as the X-ray focal spot. This ensures the light beam covers the same area as the X-ray beam.
- The diaphragms 'b' and 'c' limit the actual X-ray beam. The diaphragm 'c' cleans-up the 'penumbra' that would exist at the edge of the X-ray beam. (Penumbra is caused by the focal spot size not being a true 'point' source. Eg, the larger the focal spot, the greater the penumbra.)
- Moving the mounting position of the collimator, relative to the focal spot, allows adjustment of the collimator to the X-ray source. This is often provided by four adjustment 'fingers'. With rotating collimators, the holes for the mounting screws may be enlarged, allowing several millimetres of adjustment relative to the focal spot.
- The mirror and front Perspex cover provide some of the required primary beam filtration. Additional filtration to reach the required minimum total 'half-value layer' is often placed in the X-ray tube housing throat, just before the off-focal leaves 'a'. In other collimators the added filtration may

be removable. In these cases there is often an interlock switch, to prevent operation above a specified kV.

- A scale is provided for adjusting the field size to different cassette/film sizes. This may be for the standard one-meter distance from the film, or the equivalent inch distance.

Automatic collimation

Automatic collimation is a requirement for any fluoroscopic table fitted with an image intensifier.

Automatic collimation for over-table operation is now a standard requirement for some countries with strict radiation health regulations, and is often an optional requirement in other areas.

Basic description of operation (See Fig E-35)

- The collimator is fitted with motors to operate the diaphragms. Position of these diaphragms is measured by a potentiometer, with an output voltage relative to position.
- The collimator control produces a reference voltage equivalent to the required opening. This voltage is modified depending on the film to focus distance (FFD).
- The difference between the voltage from the collimator, and the control reference voltage, is amplified and operates the motor. The motor will move the associated diaphragms in or out, until the collimator voltage matches the reference voltage.
- When automatic collimation is applied to over-table or wall-Bucky operation, the tube stand is fitted with potentiometers to measure the FFD in both the vertical direction and the horizontal direction.
- With microprocessor systems, the voltage from the collimator and height measurement potentiometers is transferred, via an analogue to digital converter, to the computer. Film and format size is entered directly to the computer. The computer then outputs a control voltage, via a digital to analogue converter, for the motor drive power circuit.
- The vertical direction only is required for over-table operation. With a wall Bucky, the horizontal measurement is used for the FFD compensation, while the vertical measurement is compared to a similar height measurement from the Bucky stand. This ensures the X-ray tube is at the correct height to match the wall Bucky. In some cases, the X-ray tube stand height is motor driven to automatically track the wall Bucky height. In other cases, the tube stand height is manually controlled until a 'ready' indicator operates.
- Elevating Bucky tables may also send a height signal to the automatic collimator control, to permit operation at different positions. Other ele-

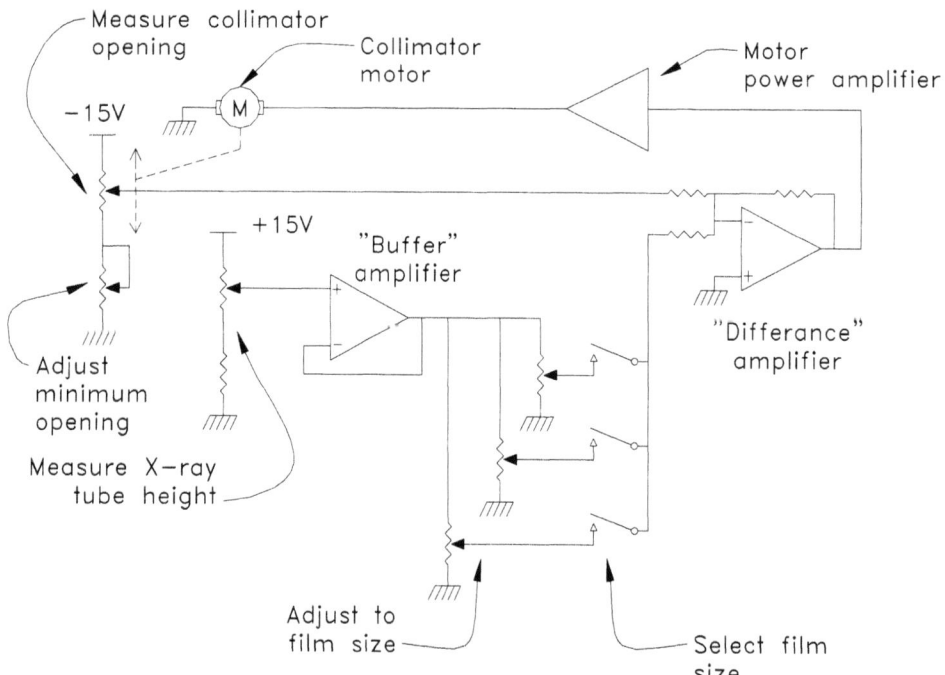

Fig E-35. Operation of an automatic collimator control system

vating tables may only permit operation when raised to a standard operating height.

- The Bucky used for these systems requires a method to determine the cassette size. This may be via potentiometers attached to the cassette tray. Some Bucky's provide a motor driven cassette tray, which passes over sensors as the cassette tray is retracted. These sensors are connected to a computer in the Bucky, which decodes the cassette size, and transmits the information to the collimator control unit.
- In the case of a fluoroscopic table, measurement of the cassette size may be via a combination of magnetically operated 'reed' switches, as the cassette is loaded into position. Many other systems also apply, including manual selection of cassette size via a selection control on the serial changer (spot filmer).
- A fluoroscopic table also selects different format sizes, to allow multiple exposures on the same film. To prevent overlap of exposures, an additional 'close to film' shutter is provided in the serial changer. These have a preset size according to the format selected, and provide a sharp delineation of the exposed film areas.

PART 7 THE X-RAY TUBE SUSPENSION

The X-ray tube stand is presented in two common forms, floor to ceiling, and ceiling mounted suspension. The ceiling mounted suspension allows maximum flexibility for a room, while the more economical floor-ceiling system is used for most general-purpose rooms. Ceiling suspended systems counterbalance the weight of the suspended X-ray tube by means of a spring and variable ratio pulley. Floor-ceiling stands may also use a similar system, or else have a counterweight installed inside the stand column. Adding or subtracting trim weights achieves adjustment of the balance point. In some cases adjustment of spring tension is also required.

To maintain the X-ray tube in the required position, manual or electrically operated locks are employed. The electrically operated locks will take the form of:

- Electromagnet. This requires power to activate the lock.
- Permanent magnet. This is an electromagnet with a permanent magnet instead of an iron core. When energized, the magnetic field generated by the electromagnet, cancels out the magnetic field of the permanent magnet.
- Solenoid operation. In this system, a spring-loaded brake pad forms the lock. On operation of the solenoid, the pad is pulled away from the brake surface.
- Rotary 'tooth' lock. The lock is fitted with two matching plates, fitted with fine 'teeth'. These are pressed together with a firm spring. On energizing an electromagnet, the spring-loaded plate is pulled away from the fixed plate, and permits rotation.

The tube-support provides indication when the tube is rotated around preset angles, commonly set at ninety-degree positions. The lateral movement will also have an indication when the X-ray tube is at the Bucky centre position. These indications may be provided by a spring-loaded ball fitting into a slot, or else by a cam operated micro-switch, or optical-sensor, operating the appropriate lock. In this case an indicator lamp is usually provided. An automatic stop at a standard height from the table Bucky, or distance from a wall Bucky, is often provided.

The height of the X-ray tube from the Bucky and the tabletop may be indicated by a set of fixed scales. In many systems height indication is provided by a digital display, operated by a potentiometer. Two potentiometers are required, one for the vertical movement, and one for the longitudinal position from the wall Bucky. Change over to the required potentiometer is performed by rotation of the X-ray tube. This system is also required if the X-ray tube is fitted with automatic collimation.

When the tube-support is used in tomographic mode, the following is required.

- Height above the table Bucky set to the required position. Some tube-supports have an interlock. This prevents operation if the height is not correct.
- The tube rotation lock released.
- The longitudinal lock or brake released.
- The height and lateral movement locks energized.

The above may be applied automatically, depending on make and model of the tube-support. In other cases, take care to ensure the locks are set correctly before operation.

PART 8 THE X-RAY GRID AND POTTER BUCKY

Contents

a. Compton, or scattered radiation
b. The Bucky grid
c. X-ray grid specifications
d. The Potter Bucky
e. Bucky systems for automatic collimation

a. Compton, or scattered radiation

When X-rays pass through atoms of any material, scattered radiation may be produced. This depends on the X-ray photon dislodging an electron. At the same time the original photon changes its direction of travel. The photon loses energy equivalent to the dislodged electron, and has a longer wavelength. This is called 'Compton' radiation.

Scattered radiation will exist even with an X-ray beam passing through air, but especially when passing objects such as a patient. For this reason radiation shields and protective clothing are required. Scattered radiation also causes fogging of the X-ray film, reducing detail and contrast. This is illustrated in Fig E–37.

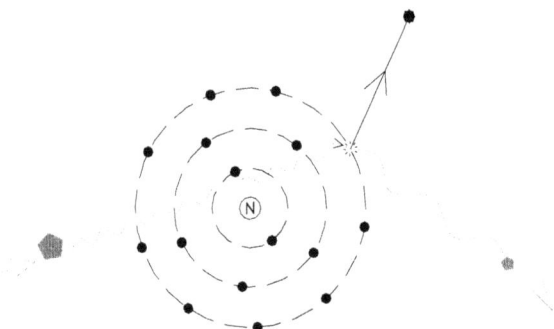

Fig E–36. Compton, or scattered, radiation

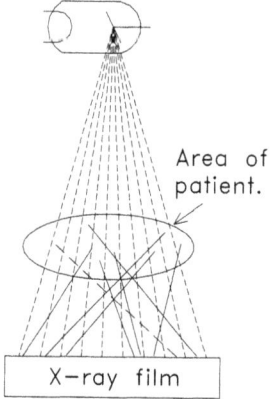

Fig E–37. Scattered radiation causes film 'fogging'

APPENDIX E. X-RAY EQUIPMENT OPERATION

b. The Bucky grid

Dr Bucky invented the X-ray grid in 1913; from this we have the name 'Bucky grid'. The grid is used to reduce the effects caused by scattered radiation. Thin strips of lead, spaced by material having a low X-ray attenuation, form the grid. This material may be wood or aluminium, and in high performance grids, carbon fibre.

While many specialized grids have been developed, the 'focussed grid' is most commonly used. This grid has the lead strips angled slightly to accept the X-ray beam from the X-ray tube, positioned at a specified distance. Although the X-ray beam is able to pass through the grid interspace material, X-rays from other directions are blocked by the lead strips, and do not enter the film.

A focussed grid is illustrated in Fig E–38, and grid action in Fig E–39.

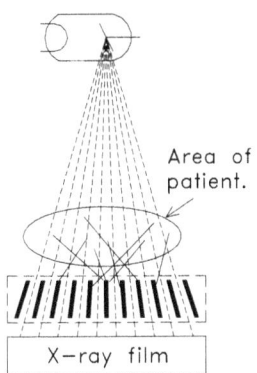

Fig E–39. The grid absorbs scattered radiation

c. X-ray grid specifications

- Grid ratio. This is the ratio of the height of the lead strips (H), and the distance between them (D). R = H/D, and is expressed as a ratio, for example, 8:1.
- Lines per inch or lines per centimetre
- Focus distance
- Interspace material
- (See Fig E–40)

As grid ratio is increased, so is a greater absorption of the useful X-ray beam. This absorption is called 'Bucky factor', and depends also on the interspace material and the number of grid lines. An increase in Bucky-factor requires an increase in X-ray output together with increased radiation to the patient.

An increase in grid ratio results in improved rejection of scatter, together with an improvement in contrast. However, the centring of the grid to the X-ray beam becomes more critical, and focal range becomes smaller.

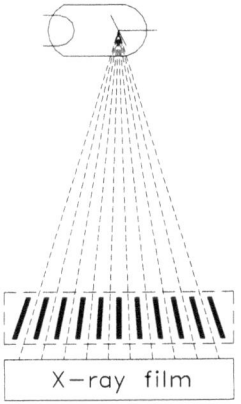

Fig E–38. The focussed grid

Table E–8. Typical specifications for X-ray grids

Lines per cm	Ratios	Focal distances
60 Lines/cm (24 Lines/inch)	8:1, 10:1, 12:1, 14:1, 16:1.	65, 70, 80, 90, 100, 120, 150, 180, 200 cm
40 Lines/cm (16 Lines/inch)	6:1, 8:1, 10:1, 12:1, 14:1, 16:1.	As above.
34 Lines/cm (14 Lines/inch)	5:1, 6:1, 8:1, 10:1, 12:1.	As above.

Table E–9. Common grid applications

Wall Bucky. Chest	12:1 ratio, 40–60 lines/cm, 150 cm focus.
Wall Bucky. Spine, Abdominal etc	10:1 ratio, 40–60 lines/cm, 100–120 cm focus.
Table Bucky	10:1 ratio. (8:1, 10:1, 12:1) 100 cm focus.

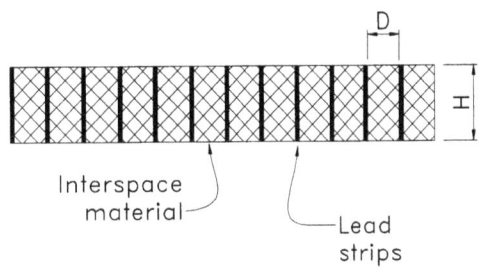

Fig E–40. Formation of an X-ray grid

d. The Potter Bucky

When a fixed grid is used, the lead strips in the grid cause a series of white lines on the film. This is where the lead strips have blocked the radiation from the X-ray tube. By moving the grid during the exposure, these lines are blurred out, and fine detail that may have been covered becomes visible. In action, the grid only travels a few centimetres in or out around the centre position.

Dr Potter invented the moving grid in 1920. This was originally called a Potter-Bucky Grid. Nowadays the device for moving the grid is commonly just called a Bucky.

There are two modes of operation for the Bucky.

- Random oscillation. The grid stars reciprocating (moving in and out) immediately the X-ray control prepares for an exposure. This method can have random appearance of grid lines, especially with short exposures. This is caused if the exposure occurs just as the grid reverses its direction of travel.
- Synchronized oscillation. The grid starts to move only after preparation and when the X-ray exposure button is pressed. After the grid has moved a short distance, a relay or switch closes, and returns the exposure request back to the X-ray control. This method ensures an exposure does not commence, while the grid is just starting to reverse its direction of travel.

The method of moving the grid varies greatly depending on design requirements.

- Linear movement provided by a reversing motor drive. Some manufactures may provide a selection of speeds. Eg, 'Par speed' or 'Super speed' as an example.
- Variable speed provided by a motor driven cam. This allows the grid to move quickly during the early part of an exposure, slowing down as the exposure progresses.
- 'Sine-wave oscillation' by solenoid activation. The grid is supported on four steel strips acting as springs. On preparation the grid is pulled to one side by the solenoid. On exposure the grid is released, and oscillates in and out until it comes to rest.

The speed at which the grid travels can sometimes cause grid lines to appear. With a single-phase generator, if the grid speed is a division of the mains frequency, the grid lines will overlap on succeeding pulses of X-ray output. Grid lines may also appear if the grid is slow, and a fast exposure is used. Some Bucky's provide rapid oscillation to overcome these problems. Unfortunately, in some cases this may cause the whole assembly to shake, producing a blurred film.

A mammography Potter-Bucky is often provided with a speed control. The speed is adjusted so the grid, during an average exposure, reaches about three quarters of the maximum travel distance. Eg, if the grid were to oscillate, grid lines are produced at the point when the grid reverses travel direction.

e. Bucky systems for automatic collimation

In some systems, automatic adjustment of the X-ray beam is provided by measurement of the cassette size. This prevents unnecessary radiation to the patient caused by incorrect adjustment of the X-ray collimator field size.

- One method measures the cassette size by two potentiometers, X and Y. These are provided in the Bucky cassette tray. On inserting the tray a plug and socket connects the potentiometers to the auto collimator system.
- Another method is provided by a microprocessor controlled Bucky. This uses a motor to drive a cassette tray in and out of the system. On inserting a cassette, the tray is pulled back inside, passing a sensor. By counting the number of motor 'steps' for the front of the cassette to reach the input sensor, the computer calculates the cassette length. The side arms gripping the cassette operate other sensors to provide the width. With the cassette in position, the microprocessor then sends the required information to the auto collimation control. These systems are required to be preset for either inch or metric cassettes.

PART 9 TOMOGRAPHY

Tomography is a technique that allows specific areas of the body to be visualized. By 'blurring out' unwanted organ outlines, the outline of organs at a specified depth are made visible. This technique may also be known as planigraphy, stratigraphy, or laminography.

When several areas of the body are superimposed on top of each other, it is difficult to visualize the organ under examination. To help with diagnosis, angled views are commonly used, as illustrated in Fig E–41.

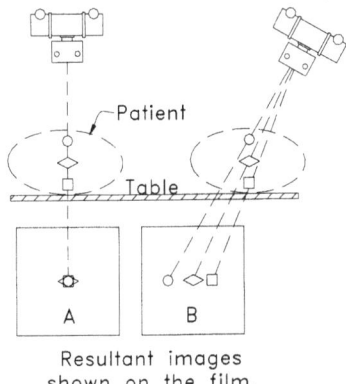

Fig E–41. Angulation provides more information

When there are a large number of similar anatomical areas, the merged outlines make diagnosis difficult. If however, the X-ray tube and film are moved simultaneously around a common axis during the exposure, the outlines of unwanted areas are blurred out. This is illustrated in Fig E–42.

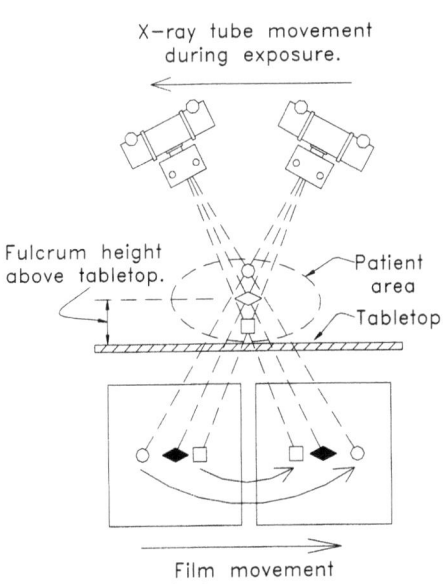

Fig E–42. Tomographic movement blurs out unwanted outlines

The common axis of rotation, or fulcrum, determines the height at which the required section will remain in 'focus'. This is called the 'focal plane'. As the angle of rotation is increased, the area or 'body section' that is in 'focus' becomes thinner. For this reason, tomographic systems allow the selection of several angles to suit the required examination.

There have been many complex systems developed to improve blurring of unwanted areas. Of these, the linear movement is commonly employed, especially as the equipment involved can be added to a standard Bucky table.

Linear tomography requires:

- A motor to move the tube stand. The motor may have several selectable speeds.
- A coupling bar between the X-ray tube and the Bucky. This bar will pass through a fulcrum, so the tube stand movement will produce an opposite movement to the Bucky. The coupling bar also aligns the X-ray beam to the Bucky.
- A system of sequence switches. These determine the start point of tube stand movement, the start and finish of X-ray exposure, and the stop position of the tube stand. These switches are often located in the fulcrum mechanism, and may be operated by a cam. In other systems, they may be included in the motor drive, and directly measure the distance of tube-stand travel.
- A method of changing the fulcrum height, relative to the tabletop. In most systems this is by directly changing the fulcrum height. Some tables (such as a remote controlled fluoroscopy table) may instead use a fixed fulcrum height, and raise or lower the tabletop.

A tomographic system may also be provided with several operation and safety interlocks:

- Tube-stand height. If not correct, no operation. (Some systems)
- Tube-stand locks. Rotation and longitudinal locks should be 'off'. Vertical and lateral locks should be 'on'. Most systems have a relay to provide this function on selection of tomography.
- Coupling bar inserted. Safety interlock is required if the fulcrum has switches controlling the tube stand motor. Earlier designs may not have this feature.
- Bucky lock 'off'. Again this depends on system design. If the Bucky has a mechanical lock, it is important the operator ensures this **is** off before operation.
- Operation of the Bucky grid movement. This may occur as soon as the tube stand starts to move, and finish either at the end of the tube-stand movement, or when the exposure handswitch is released.
- A typical timing chart is provided in Fig E–43.

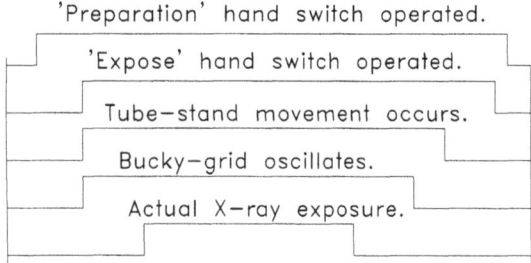

Fig E-43. Timing chart for a tomographic exposure

The exposure time is controlled by the tomography system. In use, the X-ray control timer is set to a longer time than required. This ensures the tomography system will terminate the exposure, and not the X-ray control timer.

'Electronic' tomography
- This is a system in which the mechanical fulcrum or pivot point is dispensed with. Instead three motors are required, one to move the tube stand, one to align the X-ray tube, and one to move the Bucky the required distance in the opposite direction.
- Some remote controlled tables using this system may still have a coupling bar permanently positioned. This however is required solely to keep the X-ray tube aligned to the Bucky (or serial changer).
- In operation, a potentiometer measures the relative distance of the tube stand, and another potentiometer the distance of the Bucky. (Or serial changer in the case of a fluoroscopic table). The speeds of the two motors are tightly controlled by means of a tachometer (or 'convolver') attached to the motor shaft. The information from the potentiometers and tachometers are fed via a computer to the motor drive systems.
- The advantage of the above system is convenience, and in many cases, a faster tomographic scan time. Mechanical problems due to coupling and uncoupling mechanical sections are eliminated. The disadvantage is cost and electronic complexity.

PART 10 THE FLUOROSCOPY TABLE

Contents
a. Fluoroscopy table description
b. Serial changer features
c. Table operation
d. Safety interlocks

a. Fluoroscopy table description

The fluoroscopy table is designed to allow direct viewing of the patient using continuous X-ray radiation.

Earlier tables used a fluorescent screen for this purpose. The operator is protected from direct radiation by a layer of lead glass on top of the fluorescent screen. As the image produced has a very low light level, this requires special viewing conditions. Present systems use an Image Intensifier or 'II'. This is coupled to a TV camera and monitor, and permits viewing under normal room illumination.

A serial 'changer' (or spot filmer) is used to obtain multiple radiographic exposures on a single film. The serial changer contains most of the operating controls for the table, as well as the fluorescent screen, or image intensifier.

There are two main forms of fluoroscopy tables
- Undertable X-ray tube. This is most common. The serial changer, sometimes known as a spot filmer, is mounted on top of the table. The table controls require manual operation, with the radiologist beside the patient. Due to the tube being mounted under the table, space is limited, and the distance of the focal spot to the patient is relatively short. The serial changer on these tables can be retracted, to allow full use of an undertable Bucky.
- Over-table X-ray tube. This format is normally fitted to a remote controlled table. Some versions allow for an additional control desk mounted on a trolley, for tableside operation. With the X-ray tube positioned above the tabletop, a much larger FFD is obtained. Tomographic operation is usually included, with the serial changer taking the place of a Bucky.

The table is able to tilt from horizontal to ninety degrees vertical, with the foot end towards the floor. Another tilt direction, called 'Trendelenburg' allows the head end to tilt towards the floor. Some tables may only allow about fifteen degrees of tilt in this direction. Other versions, depending on design, may allow thirty degrees, or else full ninety degrees. These are termed 90/15, 90/30, and 90/90 respectively.

Tabletop movements will normally allow for longitudinal movement beyond the head or foot end of the table body. This movement will either retract, as the table tilts to avoid collision with the floor, or else the table tilt will halt until the operator repositions the tabletop.

Note. Some over table remote control tables have no longitudinal tabletop travel. Instead the mechanism allows the undertable serial changer to travel for the full length of the tabletop. Lateral tabletop movement is available on all except the more basic tables. This movement usually has an auto stop when centred.

b. Serial changer features

- Direct fluoroscopy viewing via an image intensifier and TV combination. In this mode the film cassette is positioned to one side, together with lead shielding to protect it from scatter radiation.
- The cassette carriage is able to move the cassette in both the vertical and horizontal plane. This allows for a number of exposures to be taken on the same film. (Hence the alternate name of spot-filmer).
- Two motors usually perform movement of the carriage, with a variety of different methods to determine the stopping position. On simpler tables, movement is performed manually, with the position controlled by an electromagnet, or solenoid operated, mechanical 'stop'.
- A serial changer may accept a number of different film sizes and formats. On most serial changers, the cassette size is measured automatically as it moves into the serial changer. On older systems, the size may have to be entered manually. The formats available depend on the serial changer design and film size.
- Automatic collimation is required to limit the beam to the image intensifier field size, or to the film format size, whichever is the smallest area.
- As the collimator does not have precise registration of the X-ray beam to the film, this would result in overlap of exposures when a multiple format is selected. To prevent this, an adjustable lead 'mask' is positioned in the X-ray entrance area. The mask has a preset size to suit the format required. This allows the film to be divided into two or three vertical strips, or 'splits'.
- A second mask, sometimes attached to a compression cone, is placed in position when a 'four spot' film format is required.

- When making multiple exposures on one film, an exposure counter is required. On completion of the exposures for that film, no more exposures are permitted. Depending on the table design, the cassette may also be automatically ejected, ready for the next cassette to be inserted.
- The serial changer will have an X-ray grid fitted. In simple tables, this may be a fixed grid. When an oscillating grid movement is fitted, the X-ray exposure is normally synchronized to the grid movement. Again, a variety of methods are used for the grid movement, and in some tables the grid can be retracted to allow for non-grid exposures.

c. Table operation

On a preparation and exposure request from the table, the following takes place:

- The fluoroscopy signal from the table is locked out.
- The locks for serial changer movement, over the patient, are energized.
- Tabletop movement controls are disabled.
- The auto-collimator changes its format from the image intensifier field size, to that of the film in the cassette.
- The film cassette moves forward to the expose position.
- At the same time the X-ray control enters preparation mode.
- Once the cassette is in position, most times a short time delay is operated before permitting the actual exposure request. This is to allow any vibration or shaking to subside.
- The table waits for the 'ready' or 'preparation complete' signal from the X-ray control.
- When 'ready' is obtained from the X-ray control, and the cassette is in position, the grid will commence oscillation on pushing the 'expose' button.
- On operation of the grid-controlled exposure switch, the table sends the 'expose' request to the X-ray control.
- Once the exposure is completed, the X-ray control sends a 'time up' signal to the table. The cassette then returns to its initial position in the radiation shield area.
- Not all tables have an automatic return on end of exposure. The radiologist instead releases the exposure button after observing the exposure is completed. This especially applies in non-motorized movements, where the radiologist moves the cassette manually.

- In case the exposure button is released during exposure, this will terminate the exposure, returning the cassette to the safety area.
- At the end of a radiographic exposure, the X-ray control has a delay to allow the X-ray-tube filament temperature to drop from radiographic level to fluoroscopic heating.
- Once the X-ray control filament reset time delay has finished, and the cassette is in the radiation shield area, fluoroscopy may be resumed.

d. Safety interlocks

To provide safe operation, and reduce operator error, the following interlocks may be provided.

Radiation protection
- A switch is fitted to prevent radiation unless the image intensifier is correctly mounted. This is on tables where the image intensifier may be dismounted. This is required to enable some serial changers to be 'parked' when using the system as a Bucky-table.
- A safety switch to disable operation, when the serial changer is moved out of alignment with the X-ray beam. This is required for systems that allow the serial changer to be 'parked' for Bucky-table operation.
- On some smaller tables, a provision is made to allow the undertable tube to rotate for service, or to enable use with a wall Bucky. A safety switch is fitted to ensure in correct position for fluoroscopy.
- In many tables, a protective cover with a switch is fitted over the slot for the undertable Bucky. In these tables, the Bucky must be parked at the foot end of the table, and the Bucky-slot cover closed, to permit fluoroscopy.
- In case the cassette is in the 'loading' position, no exposures are permitted.
- Where an exposure count has been completed for a cassette, further exposures are prohibited until a fresh cassette is inserted.
- Unless the cassette is correctly positioned in front of the X-ray beam, exposure is not permitted. Note, in some cases, this is especially important in case the cassette motor drive system is not correctly adjusted.
- On some tables, a 'preparation for fluoroscopy' switch must be operated before proceeding. This switch is cancelled automatically on selection of a non-fluoroscopy technique, so the table must be deliberately re-selected every time before use.
- The X-ray automatic collimation has a facility to allow for manual control of the exposure field. This allows the field to be reduced, during fluoroscopy, and held in that position when exposing on film. Manual control is not permitted outside the area observed during fluoroscopy.
- On some remote operated tables, a key switch is fitted to allow over-ride of automatic collimation for special examinations. This switch must be in 'automatic' mode to allow fluoroscopy.
- On remote controlled tables, exposure is prohibited while the X-ray tube height is being adjusted.
- Many remote controlled tables permit the X-ray tube to rotate for other requirements. (For example, to aim at a wall Bucky). Fluoroscopy operation is disabled in this mode.
- While fluoroscopy is permitted during movement of the table or table top, this is not permitted for film operation, and all possible movements are normally locked out or disabled during radiography.
- The X-ray control is fitted with a fluoroscopy timer, usually for five minutes maximum. Some tables may have a duplicate timer for this operation.

Mechanical protection, for patient and table
- With conventional tables, a compression cone is attached under the serial changer. When in use, all compression and other movements are applied manually by the radiologist. In case the vertical movement of the serial changer is locked, tabletop movements are disabled. In another version, in case tabletop movement is energized, the vertical lock for the serial changer is immediately released.
- For remote controlled tables, the compression device is motor driven. The motor power is limited to avoid excessive compression. On movement of the serial changer or table top, the motor immediately retracts the compression cone. In another version, the motor is controlled by electronic measurement of the compression force. In case the patient is moved under compression, the motor will operate to ensure compression remains constant. These tables allow the radiologist to preset the maximum compression that may be used for a particular patient.
- A patient may place the hand outside the tabletop area. Particularly with a remote controlled table, this can lead to serious injury. In some tables, a light beam is used to detect a hand gripping the underside of a tabletop. In other systems a gel filled protective buffer is employed. This buffer has a sensor to register any change of pressure, and immediately disable the relevant tabletop movement.

- When the table is rotated, eg, from horizontal to vertical, there is the possibility of collision with an object, such as a stool etc. To reduce this possibility, a number of anti-collision systems have been used. This may be via a plate or bar on the table-base, which operates a microswitch. In other systems, a pressure pad may be installed on the floor under the table. None of these systems are perfect, and care is required to keep the floor area clear when operating the table.
- The tabletop longitudinal travel may extend a considerable distance, both towards the head or foot end. When the table is tilted, a series of cam-operated switches determine if a collision with the floor or ceiling is possible. In this case, either the table rotation is disabled, or else the tabletop is automatically retracted to a safe position. In many remote controlled tables, where a computer is used, the relevant positions of tabletop and table rotation are continuously measured. Depending on parameters entered into the computer during installation, the computer decides when operation is unsafe.
- 'Belt and braces'. A number of remote controlled tables have an emergency limit switch installed. In case the table movement does not stop after reaching the correct limit of operation, further movement trips the emergency stop, disabling the table. An operator controlled emergency operation switch is also installed at the table, as well as the control panel. This allows the operator to immediately disable all table movements in case of any unusual operation.

PART 11 THE AUTOMATIC FILM PROCESSOR

Contents

a. X-ray film properties
b. The automatic film processor
c. Processor developer section
d. Processor fixer section
e. Processor wash section
f. Processor film dryer
g. Other processor modes
h. Processor chemistry

a. X-ray film properties

- Standard X-ray film is photographic film, coated on both sides of a polyester film base. This is called 'double emulsion film'.
- A variation, 'single emulsion film', is coated on one side only. A typical application is mammography, where maximum possible image sharpness is required.
- The photosensitive material is 'silver halide', suspended in the form of small crystals in a gelatin solution. The silver halide consists of approximately 90 to 99% of silver bromide, and about 1% to 10% of silver iodide. The exact composition depending on the manufacturer and the desired characteristics.
- X-ray film has relatively poor sensitivity to X-rays. As a result, 'intensifier screens' are used to convert the X-ray energy to light energy. In a typical X-ray cassette, the film is placed between two intensifier screens. By using double emulsion film, light from the intensifier screen on each side of the film sensitises that particular layer. This process effectively doubles the film / intensifier screen sensitivity to X-rays, and permits greater film contrast.
- When an exposure is made, light from the intensifier screens cause the grains of silver halide to form a 'latent image'. Development of the film greatly magnifies this latent image, to show the visible image in the form of black metallic silver.
- After the image has been developed, the resultant image is then 'fixed'. Fixing removes the unused silver halide, which would make the film appear milky or cloudy, leaving behind the metallic silver. The fixing solution also contains a substance to harden the gelatin and make it tougher. An acid component stops any further development of the film.

- A major component of fixer is 'thiosulphate'. This is commonly called 'hypo' after an earlier chemical name of 'hyposulphite of soda'.
- After fixing the film, the film is washed to remove residual fixer, and then dried prior to viewing.
- 'Regular' X-ray film is biased to the blue region of visible light, with very poor sensitivity towards the red region. This allows the use of a red safelight in the dark room. 'orthochromatic' film is extended to the green region, and 'panchromatic' has its sensitivity extended to the red region.
- Orthochromatic film is used with current intensifier screens that have a green spectral response.

b. The automatic film processor

The following description is for a basic film processor. Larger units will have extra functions to maximize film processing time, or energy saving functions. The principle of operation, however, remains the same.

- The processor consists of three separate tanks. These contain in turn the developer, fixer, and wash water.
- The developer and fixer solutions are kept heated to a precise temperature to suit the film and chemistry used.
- A series of rollers and crossover plates transport the film through these three sections, then finally through a heated air dryer before ejecting the processed film.
- As the film passes through the chemical solutions, the developer and fixer becomes less concentrated, and requires automatic top-up or 'replenishment' to retain the correct concentration. This is done by precision metering pumps. The time these pumps operate depends on the size of the film entering the processor.
- The wash water is continuously replenished, to insure minimum residual fixer content.

c. Processor developer section

Refer to Fig E–44.

- The dotted line indicates an area outside the developer section.
- The developer supply is a pre-mixed solution of developer concentrate and water.
- When a film is inserted, sensors at the insertion point determine the film width. As the film travels into the processor, this determines the length. The electronics then calculates the developer supply pump operation time.
- The developer solution is kept under circulation by another pump. This operates continuously when the processor is ready to accept films.

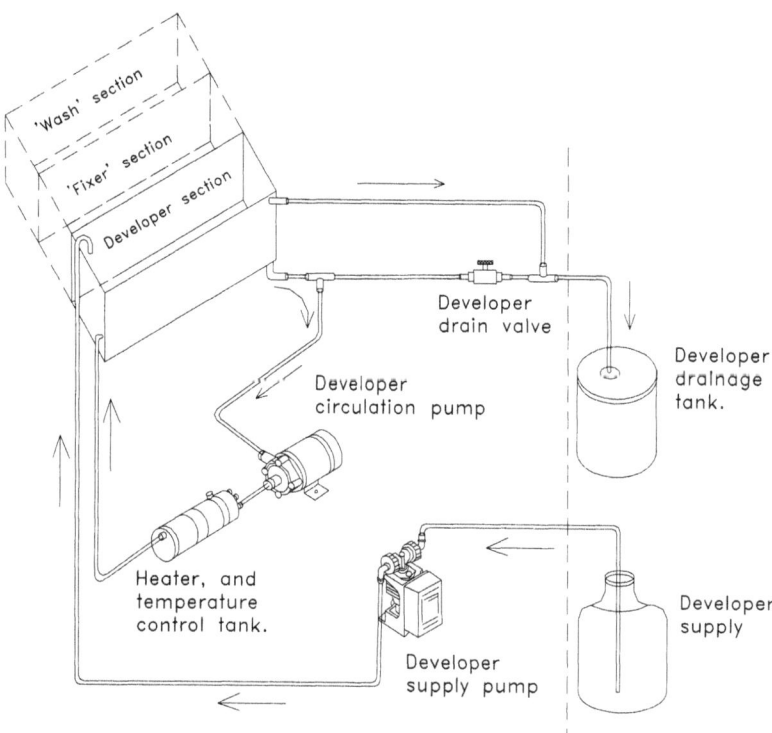

Fig E–44. Processor developer section

APPENDIX E. X-RAY EQUIPMENT OPERATION

- Developer solution passes from the circulation pump back into the tank via a temperature controlled heater. (Or heat exchanger.) The developer temperature is normally set between 34 to 36 degrees Celsius depending on developer and film combination.
- As fresh developer is pumped into the tank, excess used developer will flow from the top of the tank to a holding tank for used developer. (In some countries, though not recommended, this instead may go direct to the drain.)
- For service to the processor, a drain valve or tap is provided to empty the developer tank.

d. Processor fixer section

Referring to the diagram of the fixer section, Fig E–45

- The fixer supply is premixed fixer concentrate and water.
- The fixer supply pump is also controlled by the measurement of the film as it enters the processor.
- The rate of fixer replenishment to developer is about two to one. To obtain this the fixer pump will operate for twice the time as the developer pump. In some systems, two fixer pumps operate in parallel for a similar time as the developer pump. Other systems may instead have a larger capacity pump for the fixer.

- The fixer solution is passed through the heater and temperature control tank by the fixer circulation pump. The heater tank in this case is a dual chamber system. The major chamber of the tank is devoted to heating and controlling the developer solution. A smaller chamber of the tank allows the fixer solution to be heated, but at a slower rate than the developer. (In some larger systems, separate heating tanks are provided for fixer and developer).
- As fresh fixer is pumped into the tank, depleted fixer is passed into a storage tank or else a silver recovery unit. **Fixer contains components that are harmful to the environment, and health regulations forbid allowing this chemical to be dumped into the drainage system.**
- For service to the processor, a fixer drain valve or tap, is supplied to empty the fixer tank.

e. Processor wash section

Referring to the diagram in Fig E–46

- The water flow control valve or tap regulates the replacement rate of the wash water.
- A water filter is highly recommended to prevent sediment entering the processor.
- When a film is inserted, the solenoid operated 'water on' control valve operates to allow the wash water to be refreshed.

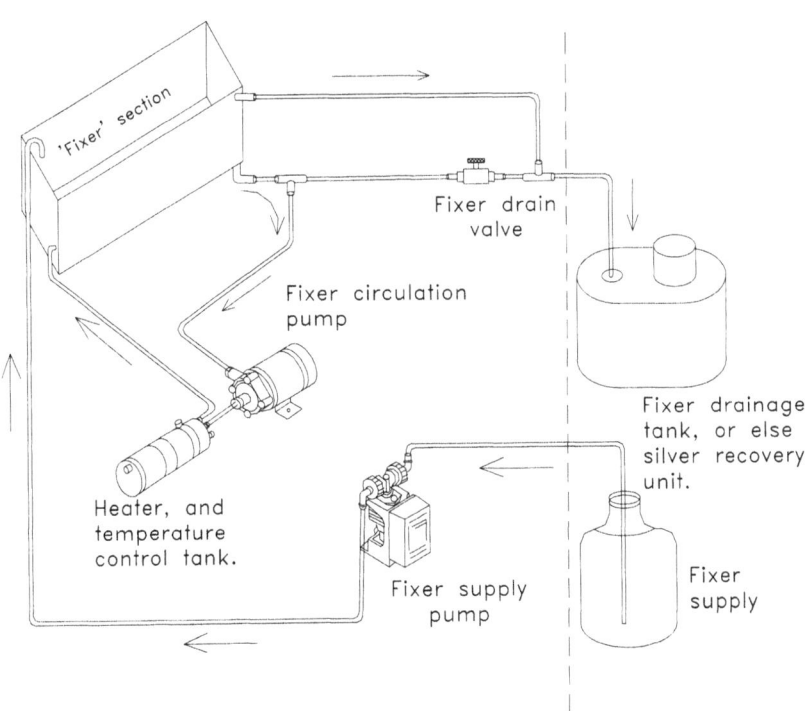

Fig E–45. Processor fixer section

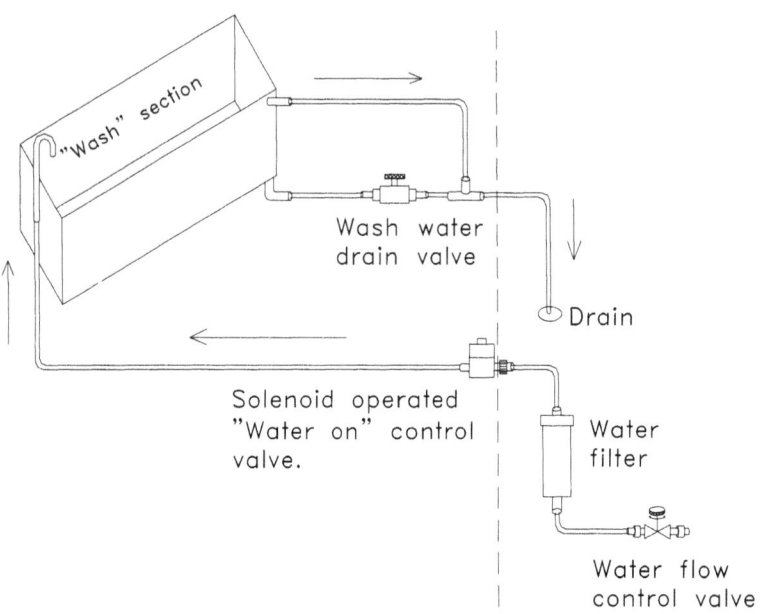

Fig E–46. Processor 'wash' section

- A timer extends the time that water flows for several minutes, and then closes to save excessive water consumption.
- Water enters the wash tank from the top via an air gap. This is a safety requirement, and prevents possible transfer of the wash water back into the water reticulation system.
- As the film passes through the transport rollers into the wash tank, the rollers remove almost all of the fixer from the film. The wash water removes the small remainder. As the concentration of fixer in the wash water is very small, the wash water is permitted to exit to the sewage drain.
- **Note.** Care should still be taken that this wash water does not enter drinking water, by draining towards where well water is obtained.
- A wash water drain valve or tap is provided for service to the processor.

f. Processor film dryer

- The transport rollers pass the film from the wash tank through the dryer section and into the film receiver.
- The dryer has a temperature-controlled heater. Air is blown past this heater and onto the film, removing the residual moisture.
- The actual temperature setting may be adjusted to suit local conditions. Eg, areas of high humidity can require increased drying temperature.
- As a safety precaution, the heating element has an 'over heat' sensing switch, as a backup to the temperature control for the heater.

g. Other processor modes

- On power up, a time delay is activated to ensure an adequate preparation time. During this period developer temperature is stabilized, and the dryer temperature is raised to the required level.
- During start-up preparation, an added amount of developer and fixer may be supplied to each tank. After about an hour of operation, this may be repeated, although a film is not inserted.
- The film transport motor operation time may be determined by the film size and processor mode. In some systems, although not processing a film, the motor will start up and run for a short period. This is to optimize condition of the rollers.
- Depending on make and model, an economy or standby mode will be entered after a preset time.
- Again, depending on make and model, after a long preset time (usually eight hours), a complete automatic shutdown may occur if the processor has not been used.
- Temperature setting may be reset to suit different film specifications, eg, in case single emulsion films are being processed. To accomplish rapid reset to a lower temperature, cold water circulates through a heat exchanger section of the heater and temper-

APPENDIX E. X-RAY EQUIPMENT OPERATION

ature control tank, or tanks. The temperature setting in a modern processor may be adjustable from 25 to 40 degrees Celsius, but is normally around 34 degrees Celsius.
- Some processors measure ambient temperature and humidity. This information is used to help optimize the dryer temperature.
- While some processors may have a separate (or combined) heater for the fixer, fixer temperature is less critical than developer. In some cases, the fixer may be kept warm simply by heat conduction, from the developer tank to the fixer tank.
- Some processors have a rinse water replenishment pump. This allows for a change of rinse water in a small trough under the film crossover rollers. In effect, this provides a small washing action when film is fed from the developer tank to the fixer tank. In some units, this is repeated for transfer from the fixer tank to the wash tank.
- 'Dwell time' is the amount of time the film is in the chemistry. Modern film processors can run as low as 45 seconds from time of the film leading edge 'in' to leading edge 'out'. 45, 60, 90, and 120 second cycles are most common. Single emulsion films may require a longer dwell time due to the thicker emulsion layer used. Typical of this is mammography film, using 120 second dwell time.
- Height of chemicals in the developer and fixer troughs may also be monitored. This can be via a float with a magnet attached, passing over a sensor.
- Modern processors now make considerable use of microprocessors. This allows for direct input by the operator for the required mode of operation, as well as displaying status and error messages.

h. Processor chemistry

- Processor chemicals are provided in concentrated form. For use, they are mixed with water, preferably filtered water, to a recommended 'specific gravity'. A floating specific-gravity gauge or 'hydrometer' is used to ensure the correct ratio of water to solution is obtained.
- When films pass through the processor, chemicals are depleted. Also a small amount is carried over to the next tank. In the developer section, oxidation occurs, and the bromide level increases. To compensate, a metered amount of developer is added for each film processed. This is called 'replenishment'.
- Developer solution is initially highly reactive until the bromide level stabilizes. When starting up a new processor, or replacing suspect chemicals with a fresh solution, 'starter' solution is added. This brings the bromide level to the correct operating level.
- In case the developer replenishment rate is excessive, then the bromide level falls, and results in overactive chemistry.
- Low developer replenishment results in low activity chemistry, and poorly developed films.
- Fixer also loses its ability to harden films, and requires replenishment.
- Typical replenishment rates for a 35 by 43 cm film are: Developer 45 to 65 cc, fixer 80 to 110 cc. The exact rate depends on the film, chemicals used, and processor make and model.

APPENDIX F
Teaching techniques

How you teach a subject is most important. The most highly skilled and knowledgeable person can fail to pass on the necessary information to students by not using the correct teaching skills or methods.

Following is a broad coverage of teaching methods, some of which will be appropriate to your situation, and some not.

This material has been reproduced from the WHO Quality assurance workbook together with additional suggestions suitable for this workbook; appended to the section 'Suggested method of teaching with this workbook'.

You should consider how you may perform and what method you will use. Choose a method suited to your subject, and what you want to achieve. Planning is all-important. Research your topic well.

Overview of teaching methods in common use

Lecture
- Stand or sit in front of a class and verbally give the relevant information.
- Suitable for large and small classes.
- Rather inflexible.
- Can be boring.
- Audio visual aids can be used.
- Printed notes can be given out in support of the spoken word.

Tutorial
- More informal than the lecture.
- Suitable for smaller classes.
- Students are encouraged to present material and enter into group discussion.
- The teacher acts more as a facilitator.
- Participants can sit around a table or in a circle.
- Feedback is important.

Practical
- Students carry out a practical exercise under teacher supervision.
- Verbal or written instructions may be made available to the student.
- The practical should be relevant to recently acquired information.

Demonstration
- Given by the teacher to illustrate a particular point.
- May be carried out as a supplement to a lecture or tutorial or an introduction to a practical.

Role Play
- Students act out specific sets of circumstances.

Reading
- Students are given topics or specific references to read up on.
- Often used as a preliminary before a tutorial or instead of a lecture.

Self directed learning
- The student is given the topic and the expected outcome.
- The student does their own research and problem solving.
- Exchange of information and group problem solving is encouraged.
- Usually followed up by a tutorial and a written confirmation of the student's knowledge.

Context based learning
- Problem solving, in groups.

Presentation
- The student researches the topic and gives a talk to other students.
- The teacher acts as facilitator and assessor.

Assignment
- The student researches a set topic and hands up a written presentation to the teacher.

APPENDIX F. TEACHING TECHNIQUES

Workbook
- A specialized book that poses questions and/or sets practical tasks. Answers are generally recorded in the book.

Book review
- To encourage students to read certain books.
- The student is asked to present a written summary and comment on the book.

Posters
- Can be done in work groups.
- May be used as teaching material at a later stage.
- Could be displayed in department to convey information.

Presentation technique

In any form of presentation by a teacher to a class, the following points should be considered:

- Stand or sit where students can see you.
- Address them clearly.
- Use language they can understand.
- Present your facts logically.
- Speak directly to the class.
- Cast your eyes around the class as you speak.
- Use visual aids where necessary.
- May use flip charts.
- In a tutorial or practical situation the teacher acts largely as a facilitator.

Teaching aids
- 35 mm slides.
- Overhead projection (OHP).
- Video.
- White board/chalk board.
- Butchers paper.
- Models.
- Charts.
- Radiographs.
- Pieces of relevant equipment.
- Printed notes.

Preparation
- Select the topic.
- Research the topic.
- Assess the educational and technical level of the students.
- Decide on the breadth and depth of the material to be covered.
- Select the method of teaching to be used.
- Select the teaching environment.
- Prepare relevant teaching notes.
- Prepare relevant teaching aids.
- The length of the session must match the time slot available.
- The amount of material to be presented must match the length of the time slot.
- In the case of a practical, ensure before hand that it will work.

Running a practical
- Identify each piece of equipment.
- Explain the procedure.
- Outline the aim.
- Demonstrate if necessary.
- Identify likely problems.
- Observe student carrying out the task.
- Comment as necessary.
- Be available to assist or answer questions.

Feedback to the student
(Following any learning activity performed by the student, the teacher should give feedback to the student regarding their performance).

- Mark/grade achieved.
- Method.
- Content.
- Technique.
- Performance.
- Presentation.
- Or whatever is appropriate.

Assessment
- Written examination.
  - Multiple choice.
  - Short answer.
  - Essay.
  - Numerical answer.
  - Problem solving.
  - Open book.
  - Fill in a missing word.
  - Tick a box.
- Practical task (written question and answer).
- Practical task (teacher observed-must have a pre-determined mark sheet).
- Assignment.
- Book review.
- Oral examination.

Grading student work
- Written.
- Verbal.
- Practical.

Forms of grading
- Pass/Fail/Referred.
- ABCDEF
  —Where A is the highest and F is the lowest.
  —A to E are graded passes.
  —E to F are graded fails.
- Distinction/Credit/Pass/Fail.
- Marks out of 100.
- Marks out of 10.
- Combined project(s)/exam(s) e.g. 50%/50% or 25%/25%/50%.
- Satisfactory/Unsatisfactory.
- No grading, the student simply attends classes and completes all work set.
- Outcome based.
- Can the student perform a task satisfactorily?
- The student must be made aware of the grading system.
- The student must be made aware of the results of any assessment.
- The teacher must allocate marks to the various parts of the material before marking takes place.

Teacher performance
- Start and finish on time.
- Plan content to fit the time frame.
- Encourage questions.
- Speak clearly.
- Maintain a friendly discipline.
- Be well prepared:
  —Knowledge.
  —Notes.
  —Teaching aids.
- Mark all work fairly and accurately.
- Return all marked work as soon as possible.
- Give feedback.
- Regularly look around your audience.
- Humour is a useful tool if used properly.
- Don't be sexist, superior, aggressive or condescending.

How good are you at teaching?
Some ways of understanding how you perform as a teacher:
- Evaluation questionnaire filled in by the student at the end of a class or series of classes.
- Ask someone to watch you teach and give you feedback.
- Watch student reaction during a class.
- Videotape a lecture, view it and then evaluate yourself.

Sample evaluation of teacher performance
- Were you able to hear? Yes/No
- Did the teacher start and finish on time? Yes/No
- Was the material presented in logical way? Yes/No
- Was the material covered adequately? Yes/No
- Was the presentation carried out satisfactorily? Yes/No
- Was the attitude of the teacher satisfactory? Yes/No
- Were the aim and objectives of the course achieved? Yes/No
- Further comment:

  _____

  _____

  _____

- Instead of Yes/No the student could be asked to choose from the following:
  —Excellent.
  —Good.
  —Acceptable.
  —Needs improvement.

Suggested method of teaching with this workbook

This book sets out to give the relevant information on the topics listed in the contents, regarding equipment maintenance, fault diagnosis or repair, then sets simple tasks that students are expected to carry out, responding to any questions asked. The teaching method used should therefore be practically oriented.

The tutor
A tutor using this book must:

- Become thoroughly familiar with the book.
- Understand its contents and know how to perform all tasks set.
- Know the answers to all questions asked.
- Understand and be able to run practical exercises and carry out any other form of teaching considered appropriate.
- Be able to select and use the most appropriate method of teaching.
- Make sure that all equipment needed is available.
- Make sure the student understands what is required of them.
- Be available to advice while the student is carrying out the exercises.
- Assess student's work fairly and accurately. Give useful feedback.
- Be sympathetic to student needs.

Method

It is suggested that a tutor use the following basic teaching format:

- Understand the **student's needs**.
- Devise an appropriate **teaching programme** (lecture, tutorial, practical etc.).
- Ensure that you have all the **equipment and teaching aids** that you require.
- You should carry out a practical exercise yourself first, to ensure it works.
- Ensure that you have the correct **answers/results**.
- Issue the workbook to the student, at least two weeks **before** they start the course, for pre reading discussion with colleagues and completion of the information about their department.
- Read the completed questionnaire, '**Student's own department**', and discuss with the student in order to determine their needs.
- Before starting the course the student must complete the **Pre test**.
- Outline the **teaching format** to the student.
- Identify the **topics** to be covered.
- Give a copy of the **teaching programme** to the student.
- Cover one topic at a time.
- With each topic, first give formal instructions covering all the **relevant information**.
- Answer any **questions**.
- Allow the students to carry out **the exercises**.
- The tutor should be available in an **advisory capacity**.
- Give the student time to **complete the exercise** and any necessary written work.
- **Assess** the practical component.
- **Assess** the answers on the task sheet and make written comment.
- **Grade** the answers and practical performance **Satisfactory/Unsatisfactory**.
- Give **feedback** to the student.
- If the student's performance is **unsatisfactory** the task must be repeated.
- Continue with the next topic when a **satisfactory** grade has been achieved.

Role Play

- In this instance of role-play, the tutor will select a small group of students, preferably no more than two or three, and present them with an equipment failure to solve.
- The equipment failure may consist of a removed fuse, disconnected plug or connection, to achieve the desired symptom. The technique used to set up the problem should not be immediately visible.
- The tutor's role is to take the place of a radiographer reporting a problem.
- The tutor will remain in the room, to answer further questions, as in a typical situation.
- Also, while directly observing progress, the tutor can quietly intervene to ensure safety requirements are met, or, after a period of time, provide hints to enable solution of the problem.
- The students should be encouraged to make as many observations as possible, reach a possible conclusion to the possible area of the problem, and devise tests to further define the possible cause.

Practice equipment

- Routine maintenance on new, 'state of the art' equipment is difficult to practice, as there will be very few problems to locate.
- In addition, modern microprocessor systems, with computer based diagnostics, present a totally different situation compared to older units in need of attention.
- An ideal situation is to have a room of older equipment. This may come from a hospital where the equipment has been upgraded.
- This equipment would be especially suitable for practice of routine maintenance, as it may be in a similar condition to older equipment in other hospitals.
- It is easier to simulate fault conditions on old equipment, and there is much reduced possibility of creating an actual problem.

Guest tutor

- In some situations, a number of students may come from hospitals that have been supplied by the same organization or manufacturer.
- Such organizations may be persuaded to supply a 'guest tutor'
- The guest tutor should be able to instruct in procedures for **specific equipment**, such as:
  —Maintenance.
  —Types of error messages.
  —Simple repairs relevant to the specific equipment.

Identification of components

To assist with recognition of components, suitable samples should be available for evaluation. These could include:
- Types of fuses
  —'Slow blow'
  —'Standard'
  —Special types used for high power.

- Relays and contactors.
- Solenoids and lock coils.
- Microswitches and optical couplers.
- Ht cable, to show the outer braid connected to the cable end.
- Old X-ray tubes with worn or cracked anodes.

Simple test tools
- A number of simple test tools have been described in the previous 'WHO Quality assurance workbook' and also in this workbook.
- The student should be encouraged to construct these tools during training. This will enable them to have useful tools when they return to their hospitals. (Once they have returned, there is less incentive to construct these items).
- The students can practise using these tools during training.
- To facilitate construction, the training centre should have suitable materials available, which the student may purchase at cost price.

Conclusion

- This section has set out to give you an overall insight into teaching.
- Much of the material given will not directly apply to your situation.
- You must choose what you feel is relevant to you and carry out your teaching to suit your own needs and those of the student.

APPENDIX G
Health and safety

This chapter, 'Health and safety' has been reproduced from the 'WHO Quality assurance workbook', by Peter J Lloyd. The contents have been edited to suit this workbook.

Health and safety issues in any work environment are very important. It is the responsibility of all heads of department to ensure that injury or sickness, due to working conditions, is kept to a minimum. Injury or sickness may increase absenteeism of staff members and reduce efficiency. Staff must not put patients, colleagues or self at risk.

X-ray departments should be prepared for emergencies such as fire, major disaster or any life-threatening situation. Radiography involves working with:

- Machinery and tools.
- Electricity.
- Hazardous chemicals.
- Radiation.

The first thing to consider is making the work environment as safe as possible, by minimizing the risk of problems arising. To achieve this, ensure that:

- Regular maintenance inspections are carried out.
- Safety procedures are followed.
- Adequate staff instruction is given.
- Safety equipment is readily available.

Machinery and tools

Regularly inspect all machinery.

- Do not attempt to repair anything you do not understand.
- If uncertain of any procedure or action, phone the service department for advice.
- Call an X-ray engineer if you are unable to fix the problem.

Take care with all moving parts, to minimize the risk of:

- Trapping fingers.
- Loose parts falling off onto staff or patient.
- Equipment moving unexpectedly and striking staff or patient.
- Staff or patient striking head on overhead equipment.

When carrying out maintenance:

- Do not stand on a tabletop to gain access to a tube stand.
- If removing items from suspended equipment such as a tube stand or vertical Bucky, this may suddenly move upwards out of control. Attach a safety restraint to prevent this possibility.
- Do not lean a ladder against a tube stand, the stand may move unexpectedly.
- When using a ladder, an assistant should hold the ladder in place.
- Whenever dismantling any section, place any small parts, such as screws, in a container to avoid loss.
- **If removing panels, ensure all power is turned off**.
- Screwdrivers should be in good condition; a blunt or wrong size screwdriver could damage a screw head, or else slip and cause injury.
- If removing a 'pin', use a 'pin punch'. Substitute tools can cause damage, or slip and cause injury.
- Using an incorrect spanner to tighten a nut may cause the spanner to slip, damaging the nut or your hands.

Electricity

> *Before investigating a possible fuse or wiring problem, always ensure power is turned off and unplugged from the power point. If the equipment is part of a fixed installation, besides switching the generator power off, ensure the isolation power switch for the room is also switched off.*
>
> *With battery operated mobiles, ensure the battery isolation switch is in the off position. If this cannot be located, contact the service department for advice before proceeding.*

Consult a qualified electrician or X-ray engineer. Regularly inspect all electrical equipment, cables and connections. **Do not attempt to repair anything you do not understand.**

When carrying out any simple maintenance, repair, or cleaning of electrical equipment:

- **Note.** Never work on the inside of the X-ray control by yourself.
- Electrical power cords or plugs should only be repaired by an electrician.
- If equipment is being worked on, place tape and a label on the power switch to avoid it being switched back on.
- With an X-ray room, locate and turn off the power isolation switch. Then place tape or a label to indicate it must not be turned back on.
- Beware of extension power cables that are locally made. There have been many cases of incorrect or loose connections creating a dangerous situation.
- Do not tamper with anything you do not fully understand.
- Unless you are qualified, restrict your actions to replacing light bulbs, simple electrical parts, tightening connections, replacing fuses and inspecting cables.
- When replacing a fuse, ensure it is the correct type and size. If in doubt, consult the service department.
- Ensure that all parts are correctly and safely installed or adjusted.
- Ensure that all protective panels are replaced.
- Report all faults to your immediate senior or through the recognised channel.
- Ensure that other members of staff are aware of any problem.

Fire

- Adequate fire fighting equipment, instructions, and evacuation procedures must be in place at all times.
- Emergency exit doors not locked or blocked.
- Illuminated EXIT signs in all public area.
- Fire alarms easily accessible.

Hazardous chemicals

(Laws and regulations to be followed)

Developer and fixer are hazardous chemicals and should be handled with care. Display manufacturer's instructions for mixing, care and first aid treatment, in a prominent place in the area in which the chemicals are to be used.

The risks involved are:

- Inhaling fumes or powders.
- Swallowing.
- Contact with the skin or eyes.

When mixing solutions:

- Work in a well-ventilated room.
- Avoid skin or eye contact with chemicals.
- Wear a mask, goggles, rubber gloves and a plastic apron.
- Avoid splashes.
- Wash all equipment used after mixing.
- Clean up any spills or splashes.

When processing films:

- Avoid skin or eye contact with chemicals.
- Ensure that the darkroom is adequately ventilated.
- Minimize splashes.
- Clean up any splashes as soon as possible.
- Replace any tank lids when finished.

Disposal of empty chemical bottles

- Should not be used as drinking water containers.
- Puncture and place in a sealed plastic bag before disposal.

Disposal of exhausted chemistry: Things NOT to do

- Do not flush into common drains or simply throw away. The chemicals may get into the local water supply or contaminate crops.
- Do not flush into a septic tank system. The chemicals will kill the 'good' bacteria and stop the breakdown of solid matter.

Disposal of exhausted chemistry: Helpful suggestions

- Ideally use a silver recovery unit and dispose of the chemistry through a recognized hazardous chemicals agency.
- Select a suitable site where the chemicals can be buried and are not likely to get into the local water supply or in any way affect humans, animals or crops.
- Further refinements of the 'bury method' is to use a sand trap first, then bury the residual sand or use an evaporative trench lined with sand and bury the sand when the water has evaporated.
- Local soil, terrain and weather conditions should be considered.

First aid treatment

- Follow manufacture's recommendations.
- Skin contact.

APPENDIX G. HEALTH AND SAFETY

—Wash thoroughly in water immediately.
- Eye contact.
  —Wash eye thoroughly, immediately.
  —Darkrooms should be equipped with emergency eye wash kits.
- Inhaled.
  —Move out into fresh air immediately.
  —Seek medical advice.
- Swallowed
  —Wash mouth and lips in clean water.
  —Seek medical advice immediately.

Radiation

Follow national laws and regulations!

- Use an ongoing personal monitoring system.
- Do not produce X-radiation unnecessarily.
- Keep clear of the primary radiation beam.
- Keep clear of any scattered radiation.
- Collimate the beam as much as practicable.
- Make sure that all items of lead rubber are in good condition and effective.
- Make sure that shielding to the control panel is effective.
- Make sure that X-ray room walls effectively protect people in adjacent areas.
- Close door to X-ray room when exposing.
- Standard radiation warning symbols must be placed on the doors of all X-ray rooms.
- Illuminated signs should be placed at the entrance to all X-ray rooms where prolonged X-ray exposures are made, warning when X-rays are being used, e.g. screening rooms.
- Make sure that all unnecessary personnel are clear of the radiation area when exposing.
- Make sure that X-ray equipment is working properly and is safe, by carrying out regular quality control checks.
- X-ray equipment should be switched off when not in use and any safety lock keys removed.

PART VI
Post test and glossary

Post test

Now that you have completed the course, your knowledge of the subjects should be much greater. You should now complete this post course test, and compare the results with those of the pre-course test.

This will allow you to assess how much knowledge you have obtained from the course, or perhaps the need for careful revision.

Name and address _____

_____

Hospital name and address _____

_____

Instructions

This is a multiple-choice test. In each question you are given three possible answers.

- Read each question carefully.
- Indicate the answer that you feel is the most accurate by placing an 'X' in front of the letter preceding it.

All questions must be answered

1. What is the **main** purpose of the metal braid around the outside of the high-tension cables?
   a) Allow the high-tension cable to become a capacitor.
   b) Provide protection against electrical shock.
   c) Reduce interference to other equipment when used with a high frequency generator.

2. The high-tension cables in a capacitor discharge mobile will:
   a) Have high voltage applied only during an exposure.
   b) Carry the maximum value of the indicated capacitor charge voltage.
   c) Each cable will carry half the indicated capacitor charge voltage.

3. The X-ray tube housing has a safety switch. This switch operates in case:
   a) The tube housing has an oil leak.
   b) The anode is overheated.
   c) The oil is too hot.

4. The effect of the added aluminium filter between the X-ray tube and the collimator is to:
   a) Improve image contrast at high kV exposures.
   b) Remove off-focal radiation from the X-ray tube.
   c) Reduce unwanted low energy X-ray photons.

5. The spinning top can be used to test the exposure times of:
   a) A single-phase full wave generator.
   b) A capacitor discharge mobile.
   c) A three-phase twelve-pulse generator.

6. The X-ray tube filament requires compensation for the space-charge effect. If not compensated, the mA output will:
   a) Fall as kV is increased.
   b) kV does not effect the emission.
   c) Rise as kV is increased.

7. You have a complaint of grid lines suddenly appearing on the film while using the table Bucky. You decide to first:
   a) Remove the grid from the table Bucky, and check the type of grid fitted.
   b) Make a test exposure, and check if the table Bucky grid is in fact oscillating.
   c) Check the control panel, to make sure the table Bucky is selected, and not the upright Bucky.

8. When using a stepwedge to check calibration reproducibility, you should:
   a) Keep the same kV, but change the mA selection, adjusting the time to obtain the same mAs.
   b) Using the same kV and mA station, make a series of five exposures, exposing a fresh area of the film each time.

c) Using the same kV setting, make two exposures, first on the fine focus, then on the large focus. Use the same mAs setting, exposing a fresh area of film each time.

9. The Bucky table lateral movement lock suddenly has reduced grip, and does not hold the table firmly in position. You decide to:
   a) Look for a faulty fuse.
   b) Request an electrician to boost the voltage to the lock coils.
   c) Check the lateral lock coils, to see if they are both warm.

10. You have been transferred to another hospital. The first thing you notice is that the collimator lamp is not very bright. You decide to:
    a) Replace the globe with a more powerful one.
    b) Order a replacement mirror.
    c) Request an electrician to check the lamp voltage, while the lamp is turned on.

11. On checking the alignment of the light beam to the X-ray field, you find it is displaced to one side. The collimator is designed to rotate. Your first action should be:
    a) Adjust the position of the globe.
    b) Adjust the position of the collimator relative to the focal spot.
    c) Rotate the collimator 908 to the left, and then 1808 to the right. Check that the light beam remains in the same position.

12. The collimator blades tend to slip to the closed position, often at an awkward time. To prevent this you should:
    a) Place sticky tape over the adjustment knobs, to prevent them moving.
    b) Remove the collimator cover, and adjust the clutch or brake pressure pad.
    c) Remove the cover, and adjust the spring tension pulling the blades together.

13. You have set up a portable X-ray unit in a village to carry out a survey. However the unit does not power up. The first check should be:
    a) Check the connections to the power plug.
    b) Look for a blown fuse in the control unit.
    c) Test the power point by plugging in a lamp.

14. A common fault with a capacitor discharge mobile is a short circuit in the cathode cable, between the X-ray tube grid and cathode connections. This results in:

    a) Although preparation is obtained, an exposure cannot be made.
    b) The capacitor will not charge to the required kV.
    c) As the generator is placed into preparation, an immediate exposure takes place, discharging the capacitor.

15. The films from the processor are discoloured, and feel sticky.
    a) The wash water temperature is too high.
    b) Poor quality films. Try a new box.
    c) Fixer is depleted or contaminated.

16. Developer temperature should be checked.
    a) Weekly.
    b) Twice a week.
    c) Daily.

17. A densitometer:
    a) Is used to print a test strip onto the film.
    b) Measures patient size to help estimate exposures.
    c) Measures the degree of film density.

18. The X-ray room has an older installation with a floor ceiling tube stand. The X-ray tube appears centred to the table Bucky, but not to the upright Bucky.
    a) Reposition the wall Bucky.
    b) Test the tube stand cross-arm in case it is not truly horizontal.
    c) Adjust the X-ray tube housing in the trunnion rings.

19. The effective calibration of the tomography fulcrum height will be changed if:
    a) The exposure time is too short.
    b) The X-ray tube is rotated from its central position.
    c) The X-ray tube is set to the incorrect height.

20. A 'slow blow' or 'delay' fuse is described as:
    a) A small glass tube, with the fuse element attached to a spring.
    b) A similar glass tube, with a rectangular fuse element.
    c) Any large ceramic style fuse, with a body diameter over 25 mm.

21. You have carried out an exposure linearity test on a generator. One of the test results shows a darker stepwedge. This can indicate:
    a) The kV at that point was out of calibration.
    b) The value of mAs had increased.
    c) Either of the above.

22. On making a high kV exposure, a fault signal occurs, together with a light film. You suspect the tube may be unstable. To improve the X-ray tube stability you should.
    a) Perform three 70 kV 200 mAs exposures to warm up the anode each morning.
    b) Place the tube in extended preparation for a period of ten seconds before exposing.
    c) Make a series of 20 mAs exposures, starting at 70 kV and increasing at 5 kV per step, up to the required kV.

23. A line-pair gauge is used to:
    a) Test the resolution of a TV imaging system.
    b) Measure the number of grid lines for an X-ray grid.
    c) Measure the conversion efficiency of an image intensifier.

24. Your fluoroscopy room has two TV monitors. At the rear of each monitor is a 75-ohm switch. You find that if this is turned on, the picture contrast and brightness is reduced.
    a) With a two monitor system, both switches should be turned 'on'.
    b) The monitor connected between the camera and the end monitor should be turned 'off', and the end monitor turned 'on'.
    c) With a two monitor system, both switches should be turned 'off'.

25. Exposures using an automatic exposure control (AEC) have increased in density. As a first response, you should:
    a) Try exposing at lower kV settings.
    b) Have the AEC recalibrated.
    c) Check the processor performance.

26. The coincidence of the X-ray and light fields of a collimator, set at 100 cm FFD, are said to be acceptable when:
    a) The X-ray field is 15 mm outside the light field.
    b) The X-ray field is 12 mm inside the light field.
    c) The X-ray field is 3 mm outside the light field.

27. After investigating a problem, you find a fuse is open circuit. This is rated at 7.5 amps. You do not have this size fuse available.
    a) Replace it with a temporary 10 amp fuse.
    b) Test first with a 5 amp fuse.
    c) Replace the wire inside the fuse with household fuse wire.

28. If about to remove an X-ray tube from the stand, you should first.
    a) Remove the collimator.
    b) Remove and label all connections.
    c) Fasten the tube stand cross-arm with a rope to prevent vertical movement.

29. On testing the generator reproducibility, you find that if preparation time is extended, the radiation increases. This is possibly due to:
    a) The X-ray tube space-charge effect.
    b) Incorrect calibration of the mA station under test.
    c) Pre-heating of the X-ray tube filament is too low.

30. A sensitometry test of the processor should be carried out:
    a) If the generator's automatic exposure control (AEC) produces light films.
    b) The processor should be checked once a month.
    c) After the processor temperature is stable.

Glossary

AEC Automatic exposure control. Often called a phototimer, or by a manufacturers name. (Eg, Iontomat, Amplimat, etc).

Acid A solution with a pH less than 7. It reacts with blue litmus paper and turns it red.

Acrylic Or 'plexiglass'. A clear plastic, used for a large variety of purposes. This material makes an excellent 'phantom' for adjusting X-ray equipment, particularly for setting up or calibrating an AEC. Filtration is approximately similar to water.

Active wire Electricity supply to a power outlet normally has an 'active' and 'neutral' connection. The active wire has the full supply voltage in reference to ground potential, while the neutral wire is the same as ground potential.

Ageing See seasoning.

Alkali A solution with a pH greater than 7. It reacts with red litmus paper and turns it blue.

Anamorphic lens A special lens that changes the shape of an image. Used to change the round image from an image intensifier to an oval image. This allows the image to fully cover all pixels of a CCD sensor in an X-ray TV camera.

Artefact Marks on a radiograph that are foreign to the image, such as scratches, fingerprints or static.

Back flush To clean a processor filter. Connect the filter 'back to front' to a water supply. This will 'flush out' the dirt held in the filter.

Base + fog The density normally found in unexposed film, caused by manufacture and storage.

Bench top processor A small, automatic film processor, best sited on a bench top. Suited to low throughput.

Beam The beam of radiation produced by the X-ray tube.

Bucky A commonly used abbreviation of the Potter-Bucky moving grid system.

Callipers A device used to measure the thickness of body parts.

CCD Charge coupled device. Used in current TV cameras. The CCD consists of a large number of light sensitive cells, which produce the TV video signal.

Central ray The centre of the X-ray beam. Often used to define the direction of the beam, or, its position, related to a body part.

Cassette A light tight holder that contains a pair of intensifying screens, between which, is placed the film.

Characteristic curve Also known as H & D curve or sensitometric curve. It is a plotted graph of the various densities of a step wedge image. Any variation in type of film/screen, exposure or processing will vary the shape, or position, of the curve.

Circuit breaker An electromechanical safety switch. On excessive current the device will switch off, performing the function of a fuse. Unlike a fuse, the switch on the circuit breaker can be manually reset.

Clearing time The time it takes for a film to lose the cloudy appearance when placed in the fixer during film processing. In other words the time it takes for the unwanted film emulsion to be dissolved off by the fixer.

Code display A numeric display on a control panel is changed to an alphanumeric display. For example, the indication of set kV on a generator indicates E2 instead of the required kV. This may be due to a fault condition, or incorrect adjustment of the control. Refer to the operation manual for a description of the displayed code, or contact the service department.

Collimator A device used to control the coverage of the X-ray beam. Also known as a light beam diaphragm (LBD).

Contactor A heavy duty relay for switching large currents. Used as the 'exposure contactor' in older X-ray controls.

Contrast The difference between the light and dark areas of a radiograph. High contrast is when there are few shades of grey between the lightest and darkest areas of the image. Low contrast is when there are many more shades of grey in the image.

Compression band A strip of material, usually linen or plastic, approximately 20 cm wide, attached at one end to a ratchet device, and at the other to a hook. It is used for compressing or immobilizing patients.

Dark current The emission, or conduction, of electrons in the absence of light. For example, electron emission from an X-ray tube cathode, without heating the filament. A capacitor discharge mobile will generate X-rays due to this dark current when the capacitors are charged. X-ray emission is prevented by a dark-current shutter in the collimator.

Densitometer A device for measuring the density of any specific spot on a radiograph, by measuring the light that is allowed to pass through it.

Density, radiographic Radiographic density is the degree of blackening of a radiograph caused by the deposit of metallic silver.

Density, tissue Tissue density is the mass of body tissue in a given volume, or the concentration of atoms. The greater the tissue density the more X-ray absorption takes place and the lighter the image on the radiograph. (Do not confuse radiographic density with tissue density.)

Detail, radiographic image The amount and quality of information contained in a radiographic image. The amount of detail seen in a radiograph is determined by image sharpness, contrast and density.

Detail intensifying-screens The name applied to a type of intensifying screen that gives better image detail, but is less responsive to radiation and therefore requires a higher exposure. Commonly used for extremities.

Development The chemical process of converting the latent film image into a visible one.

Distortion Misrepresentation of a body part outline, in the image, due to changes in X-ray beam/body part alignment or unacceptable object film distance.

Edge connector Part of a printed circuit board is shaped, so the edge of the board fits directly into a socket. The printed copper strips on the side of the board slide into contacts fitted in the socket. The copper strips are usually gold plated for reliability.

Emulsion The active layer of chemical crystals suspended in a gelatine layer, of film, which is sensitive to light and radiation. The word emulsion can also be used to describe the radiation sensitive layer of intensifying screens.

Exposure The amount of radiation produced from the X-ray tube, by a pre-determined set of exposure factors, kV, mA, seconds. In practice the term tends to be used loosely. The term 'exposure' being used to mean exposure factors.

FFD Focus/film distance. More accurately, SID (source image distance).

Focal range The range of focal film distances at which, a grid is designed to be used.

Focal spot The area on the X-ray tube anode where the X-rays are produced.

Focussed grid Grid lines are inclined toward the centre of the grid, to better accommodate the spreading X-ray beam.

Fog Radiation fogging of a film is commonly caused by scattered radiation reaching the unprocessed film. Light fog is caused by unwanted white light reaching the unprocessed film. Base fog is inherent fog, caused in film manufacture.

Filter, safelight A specialized, coloured glass window, fitted to a safelight to enable the safe handling of X-ray film.

Filter, X-ray A sheet of metal, usually aluminium, fitted to the port of an X-ray tube to filter out the long wavelength X-ray photons.

Grid A device consisting of alternate radiopaque and radiolucent strips. Designed to allow the primary X-rays to pass through, but absorb scattered radiation.

Grid control X-ray tube The X-ray tube cathode cup has a negative voltage applied to control electron emission from the cathode. This allows short controlled exposures, under continuous application of high voltage. Grid control tubes are often used in specialised X-ray theatres, while much smaller versions are used in capacitor discharge mobiles.

Grid cut off A reduction in grid efficiency due to misalignment of the X-ray beam to the grid.

Grid line The number of lead strips to the cm/inch.

Grid ratio The ratio of the height of the radiopaque (lead) strips to the distance between them (radiolucent strips).

Hand-switch In this case, the dual pressure preparation, and exposure switch, for an X-ray control. This term can still apply if there are two separate push buttons mounted on the control.

Hanger Film hanger is a stainless steel frame with clips at each corner for holding the film when manual processing takes place.

Hazardous chemicals Any chemical, which may have an injurious affect. Developer and fixer both fall into this category.

Heel effect The decrease of radiation toward the anode side of the X-ray tube. As radiation emitted from the target approaches the physical angle of the anode, eg, becomes parallel with the side of the anode, radiation is absorbed by the anode.

High-Frequency generator AC input power is rectified to become DC. After passing through an inverter, the DC voltage becomes a high frequency AC voltage. This allows a more efficient HT transformer, plus tight regulation of the kV output.

High-speed starter An electronic device, which generates a multiple of the mains power frequency for rotating the X-ray tube anode. This is normally three times the mains frequency. Eg, 150 Hz for 50 Hz input, or 180 Hz for an input of 60 Hz.

Hydrometer A sealed, weighted glass tube, with a visible scale marked on it, which will float in a liquid. Used for assessing the specific gravity (S.G.) of developer and fixer, the level of which is an indication of the concentration.

IGBT Insulated gate bipolar transistor. This device can handle very large currents, and is used to form the inverter in current high frequency generators.

Image intensifier Or II. A device to convert X-rays to visible light. Due to high operating voltages, and a small output screen compared to the input area, light output may be 9000 times brighter than a fluorescent screen. This permits use of TV cameras and many other image-recording devices.

Inhibit Stop, or prevent operation.

Inverter generator See high-frequency generator.

Intensifying screens Radiation sensitive screens, placed inside a cassette, on either side of the film. When struck by radiation the screens give off a blue or green light that has a blackening effect on the radiographic image produced. The colour of light emitted depends on the type of fluorescent materials used. Remember that the film colour sensitivity must match the colour of light given off by the screens.

Interlock A switch or safety circuit to allow operation only after a set of conditions has been obtained. For example, an interlock prevents an exposure until all preparation requirements have been completed. Eg, anode rotation & X-ray tube filament heating.

kV Kilo-voltage (1000 volts). Controls the quality (penetrating power) of the X-ray beam. Affects contrast of resultant image. (High kV–low contrast, low kV–high contrast). Affects intensity of radiation and therefore patient dose, to a lesser extent. (mAs has a greater effect on intensity and patient dose).

kVp The peak or crest value of the high voltage applied to the X-ray tube.

Latitude, exposure Exposure latitude is the range of exposure factors that will produce an acceptable image.

Latitude, film Is a film emulsion characteristic that increases or reduces exposure latitude.

LED Light emitting diode.

mA (milliampere, 1/1000th of an ampere) A radiographic exposure factor that controls the intensity of radiation, influences image density and patient dose. The current flowing in the X-ray tube during an exposure.

mAs (milliampere-seconds) A radiographic exposure factor. mA x seconds.

mAs meter A meter for measuring the product of mA over a period of time. A useful item of test equipment, used for calibrating an X-ray control. Some X-ray generators have an mAs meter on the control panel.

Medium frequency generator See high frequency generator.

Neutral wire Electricity supply to a power outlet normally has an active and a neutral connection. The active wire has the full supply voltage in reference to ground potential, while the neutral wire is close to, or the same as, ground potential.

Noise As applied to an X-ray image. This is otherwise called 'quanta mottle' or 'grain'. Electronic noise is a similar effect, called 'snow' with domestic TV reception.

Non-focused A grid that does not have focused grid lines.

NTSC National Television Systems Committee (NTSC). This is the American colour TV specification.

Off focal radiation In the X-ray tube, about 10% of the electrons striking the anode tend to bounce away from the focal spot. They are then re-attracted to the anode, but over a much larger area. Off focal radiation requires care in the collimator design to reduce this unwanted radiation.

Oxidation A weakening of developer strength caused by prolonged exposure to air.

PAL Phase Alternate Line (PAL) is a further development of the NTSC system. Commonly used in Europe. Each alternate scanning line has reversed colour phase. This cancels colour error.

Penetration The ability of the X-ray beam to penetrate structures. Determined by the energy of the beam (controlled by kV).

PH Indicates the degree of acidity/alkalinity of a solution. Water is neutral and has a pH of 7. Solutions with pH of less than 7 are acid. Solutions with pH more than 7 are alkaline. Developer is strongly alkaline with a pH of about 14. Fixer is strongly acidic with a pH of about 3.

Phantom An object that substitutes as a patient, when performing X-ray tests. The material used can be anything that is similar in density and thickness to the appropriate body part. Common materials are water or acrylic.

Plumb bob A steel or brass weight, suspended by a string. The weight has a point in its centre at one end, and a central attachment for the string at the other end.

Port This is a plastic cone inserted in the X-ray tube housing, close to the X-ray tube focal spot. This cone is sometimes called the X-ray tube 'throat'.

Potter-Bucky A moving grid system designed to reduce the amount of scattered radiation reaching the film. Often abbreviated to Bucky.

Power up Switch on the equipment power.

Profile rails The metal rails attached along the sides of a tabletop, or a Bucky. The shape, or profile, of the rails allows attachment of accessories, such as a compression band. Removal of the profile rails is usually needed to remove a tabletop or Bucky cover.

Proximal diaphragm A lead disc, with a small square hole in the centre. The disc is fitted to the exit port of the X-ray tube housing, or it may take the shape of a cone, extending close to the X-ray tube focal spot. The purpose of the proximal diaphragm is to reduce scattered or off-focal radiation, emitted by the X-ray tube.

Relay An electromagnetic switch, normally with a number of switch contacts that change their state when the relay is energised. A heavy-duty relay is often called a contactor.

SCR Silicon controlled rectifier. This device is used to replace the mechanical exposure contactor in older generators, and to form the inverter in medium frequency generators.

Seasoning A procedure to improve the stability of an X-ray tube, when operating at high kV values. The procedure involves a series of short exposures, starting at a low to medium kV, and finishing at or a little above the desired operating kV. This procedure causes the residual gas molecules to be re-absorbed in the anode.

Serial changer The device on the fluoroscopic table for holding and advancing a cassette while making a radiographic exposure. Also referred to as a spot filmer.

Solenoid An electromagnet with a hollow core. When energized an iron piston is attracted into this core. A solenoid is often used as part of a brake system, with a tube stand or other moving object.

Specific gravity The weight of a substance compared to an equal volume of water. Specific gravity measurement, using a hydrometer, can be used to measure the concentration of developer and fixer.

Spot filmer Another name for a serial changer. This derives its name from its ability to take a number of divided exposures on the one film.

Stand pipe A pipe which fits into the inside of a manual processing units wash and rinse tanks drain holes. Its height is just below the top edge of the tank, allowing the tank to fill up and drain through the top of the pipe. They are also found in the developer and fixer tanks of most automatic processing units.

Step chart A system of exposure calculation using a series of exposure factor steps. Step charts are available for kV, mA, mAs, time and FFD (SID).

Step wedge Usually made of aluminium, is a block cut to form a standard number and sized steps. Used as a test tool for various quality control tests.

Test tool Specialized items of equipment that can be used to evaluate X-ray or accessory equipment.

Thermostat A device for controlling heat output from a heating unit. Used in X-ray film processing to control the temperature of developer.

Thiosulphate Usually sodium thiosulphate. This is the fixing agent in X-ray fixing solutions.

Timer, darkroom An accurate time clock for timing X-ray film development during manual processing.

Timer, X-ray A device for determining the length of radiographic exposure in an X-ray unit.

Tomography A method of moving both the film and the X-ray tube so that a sharp image is obtained only at a particular height above the tabletop. This is obtained at the fulcrum height. Areas above or below this height are blurred out. The CT scanner was developed from the tomograph.

Trendelenburg position For some examinations, the patient is positioned head down towards the floor. This is normally applied during a fluoroscopic table examination. Typical angles used are 7.58 up to 308.

Trouble-shooting Is an expression to describe the system, or method, used to first locate the cause of a problem; and then repair or eliminate the problem.

Trunnion rings Circular rings, used to mount the X-ray tube housing onto the tube-stand. The rings are in two sections, fastened together when the housing is in place. In most cases the trunnion rings have a clamp screw, or locking knob, which when released allows the X-ray tube housing to rotate inside the rings. This allows adjustment of the beam to be perpendicular to the table.

Vane A flat piece of metal, or other material. When the vane passes through the centre of a sensor, it interrupts a light beam into a photocell. (This example is called an optical sensor.)

Washing All film must be adequately washed to remove the acid fixer and avoid future film deterioration.

Wisconsin cassette A specialised X-ray cassette, used to determine X-ray kVp. Used to check the accuracy of kVp in an X-ray unit.

www.ingramcontent.com/pod-product-compliance
Ingram Content Group UK Ltd.
Pitfield, Milton Keynes, MK11 3LW, UK
UKHW051525180426
11947UKWH00018B/1576